Chromosome Structure and Aberrations

Tariq Ahmad Bhat • Aijaz Ahmad Wani
Editors

Chromosome Structure and Aberrations

Springer

Editors
Tariq Ahmad Bhat
Department of Education
Government of Jammu and Kashmir
Srinagar, India

Aijaz Ahmad Wani
Department of Botany
University of Kashmir
Srinagar, Jammu and Kashmir, India

ISBN 978-81-322-3671-9 ISBN 978-81-322-3673-3 (eBook)
DOI 10.1007/978-81-322-3673-3

Library of Congress Control Number: 2017930685

Printed on acid-free paper

This Springer imprint is published by Springer Nature
The registered company is Springer (India) Pvt. Ltd.
The registered company address is: 7th Floor, Vijaya Building, 17 Barakhamba Road, New Delhi 110 001, India

This Book is Dedicated to

Sir Syed Ahmad Khan

(1817-1898)

A great visionary, statesman, Muslim reformer of the 19th century and founder of Aligarh Muslim University, India

Foreword

Chromosomes have always fascinated me as remarkably condensed DNA thread-like structures regulating all the phenotypic features and physiological processes of an organism. Equally fascinating is the history of its discovery. Starting right from 1869, when the genetic material was first reported as "nuclein" in pus cells by Friedrich Miescher, followed by unequivocal experimental evidence provided by Oswald Avery, Colin MacLeod and Maclyn McCarty in 1944 for DNA as the genetic material in the cell and then the revolutionary DNA double helical structure of James Watson and Francis Crick in 1953 based upon X-ray diffraction analysis done by Rosalind Franklin and Maurice Wilkins – all have proven to be turning points in our understanding of life processes. The knowledge gained as a result of global interest in the DNA molecule led us into the era of recombinant DNA technology, transgenesis, animal cloning and ultimately embarking on Human Genome Project. The DNA molecule with its fundamental nucleotide base composition has been the driving force behind all the biological phenomena. The simple organisms had few base pairs (bp), which continued to grow in number depending upon the complexity of the organisms along the evolutionary scale. The smallest number of base pairs (bp) has been reported from a virus – porcine circovirus type with 1.8×10^3 bp. The number of bp continued to increase with the increase in complexity during the course of evolution spanning billions to millions of years. The bacterium *Escherichia coli* has 4.6×10^6 bp (4288 genes), *Caenorhabditis elegans* 10×10^7 bp (19,000 genes), *Arabidopsis thaliana* 15.7×10^7 bp (25,498 genes), *Drosophila melanogaster*17.6×10^7 bp (13,600 genes), *Mus musculus* 6.4×10^9 bp (20,000 genes) and *Homo sapiens* 6.9×10^9 bp (30,000 genes). In prokaryotes, the organisms could manage to accommodate this DNA (chromosome) in their protoplasms, but with the evolution of eukaryotes, the increasing size of DNA/chromosome ultimately delimited itself from the cytoplasm by nuclear envelope. In order to accommodate large size of DNA within a limited space, the DNA molecule underwent process of packaging, for which histone protein and variety of looping, coiling and coiled coiling helped in condensation of chromosomes. During the course of evolution, the long DNA broke into smaller pieces, which later became the characteristic karyotype of the organisms. These chromosomes are in their relaxed state during interphase when the cell is not dividing but are condensed into specific number and shape of typical metaphase structure during dividing phase. This cycle of condensation and decondensation is such an organized and orchestrated sequence of events that organisms remain healthy and normal, both phenotypically and genotypically. Any mishap during interphase and more so during mitotic or meiotic dividing phases results in genetic abnormalities. There are a variety of genetic syndromes which are the result of accidental changes in the number of chromosomes or loss or rearrangement of segments of chromosomes particularly during mitotic metaphase and anaphase and meiotic prophase, which are passed on to the next generation and become the basis of mild to severe abnormalities and in several cases become fatal. These biological phenomena prompted the preparation of this book.

The history of development of concept of chromosomal basis of inheritance and the molecular structure of chromosomes is given in Chap. 3. Chapter 12 encompasses the phenomena of number of chromosomes, which is normally fixed for a particular species, and the consequences of accidental increase or decrease in the chromosomal number. Chapter 13 describes the abnormalities arising due to structural rearrangements of chromosomes or chromosomal segments such as inversions, deletions, translocations, etc., which have far-reaching genetic implications with physical, physiological and mental debilities. Chapter 6 deals with two more categories of chromosomal abnormalities, i.e. asynapsis and desynapsis. Conjugation of the chromosomes during the first meiotic prophase is referred to as synapsis, and the lack of this pairing is referred to as asynapsis. Desynapsis is the condition when chromosomes initially pair at pachytene but fail to remain paired at subsequent meiotic stages. Such conditions usually result in chromosomal imbalance. Usually, eukaryotic cells are diploid, but there are some eukaryotes that contain more than two sets of chromosomes. Such cells are called as polyploid and are common in plants. Rice, for example, are diploid plants, banana is triploid, cotton is tetraploid, wheat is hexaploid, and sugarcane is octaploid. On the basis of number of chromosomes, polyploidy can be divided into euploidy and aneuploidy. Chapter 10 deals with polyploidy and its role in crop improvement. In addition to normal karyotype, wild populations of animals, plants and fungi contain B chromosomes, also known as supernumerary or accessory chromosomes. These chromosomes are not essential for life. Most of them are entirely or mainly heterochromatic, though some contain euchromatic segments. These chromosomes have been dealt with in Chap. 2.

Chromosomes can be studied under the light microscope using different types of chromogenic dyes. Chapter 14 gives a broad outline of all the cytogenetic techniques prevalent for understanding structure of chromosome. Over the past few decades, versatile methods based on fluorescence in situ hybridization (FISH) have transformed cytogenetics into a molecular science and provided cytogeneticists with powerful new tools. FISH procedures are now routinely employed in clinical cytogenetics. Although chromosomes may appear to be static structures when viewed under a microscope, cytogeneticists know that chromosomes are actually dynamic assemblies made up of a DNA-protein complex called chromatin. This chapter takes stock of all the prevalent techniques, highlighting the principals involved in each method. Karyotyping, genetic mapping, fluorescence in situ hybridization (FISH), multiplex FISH, spectral karyotyping, flow cytometry and microarray have been described. New trends in cytogenetics to understand the molecular mechanism have been discussed under new generation sequencing.

Chapter 16 describes the FISH techniques and its applications, while Chap. 4 provides guidelines to cytological techniques written specifically for the plant chromosomes. Chapter 8 describes banding techniques to study different parts of the chromosomes. The chapter describes G, Q, R banding which requires heating in a saline buffer prior to staining with Giemsa or quinacrine and is often called T banding, since it also highlights telomeres. C stains heterochromatin in the centromeres, and NOR silver nitrate stains selectively the satellite stalk of the acrocentric chromosome. Flow cytometry is another cytogenetic technique, which measures optical and fluorescence characteristics of single chromosome preparation. The information obtained is both qualitative and quantitative. The principle involved and its applications have been discussed in Chap. 5. The 2000s witnessed an explosion of genome sequencing and mapping in evolutionarily diverse species. While full genome sequencing of mammals is rapidly progressing, the alignment of whole chromosomal regions from more than a few species is not possible. Building of comparative maps for domestic, laboratory and agricultural animals has traditionally been used to understand the underlying basis of disease-related and healthy phenotypes. These maps also provide an opportunity to use multispecies analysis as a tool to infer karyotype evolution. A record of the history of karyotype changes that have occurred during evolution has been attained comparative chromosome maps. Chapter 9 deals with karyotype evolution and its applications.

Cytomixis is an interesting phenomenon which involves migration of chromosomes from one plant cell to another through intercellular cytomictic channels. It is most frequently detectable in microsporogenesis in plants and spermatogenesis in animals. Chapter 10 deals with this phenomenon and its biological and genetic implications.

Two special topics have been discussed in Chaps. 7 and 11. Chapter 7 deals with cytogenetic studies on an Indian hyacinth *Drimia* Jacq. This herb has medicinal importance. Chapter 11 is about mutagenesis in one of the major pulses of India, *Vicia faba.* Mutagenesis is the most important method of inducing alteration by mutagens in a genotype to enlarge the variability of quantitative characters such as days to flowering, number of flowers per plant days to maturity, plant height, number of fertile branches per plant, number of pods per plant, number of seeds per pod, mean pod length, mean pod girth, seed weight, etc. in the shortest possible time. The authors found that induction of meiotic aberrations was higher in combined mutagen treatments, which is more effective in inducing additional variability than individual doses in *Vicia faba* L. var. minor and major.

I only hope that readers will enjoy going through different aspects of chromosomal structure and their aberrations, with consequential phenotypic and genotypic deviations.

Lahore, Pakistan
March 7, 2016

Abdul Rauf Shakoori

Preface

The authors take great pleasure in presenting the readers this enlarged and extensive book on chromosome with special emphasis on its structure, techniques and aberrations. The book is a valuable asset to all the stakeholders at undergraduation and postgraduation level, particularly students, teachers and researchers.

During the last few decades, there has been remarkable progress in research on various aspects of chromosome including plant and animal systems. Geneticists, evolutionary biologists, ecologists, etc. have been exploiting the various aspects of chromosomal research to understand the genetic architecture of organisms, prepare chromosome linkage maps, understand evolutionary relationship among organisms and groups of organisms and understand speciation, adaptation and modes of invasion of plant and animal species. The structural organization of chromosome has been greatly understood enabling the identification of gene sequences and mapping of gene loci and their regulatory aspects. The manipulation and engineering of chromosomes have facilitated their transfer across kingdoms for genetic improvement of crop and animal species. The present book is intended to fulfil the needs of students, teachers, researchers and all stakeholders who are engaged in the study of evolution, biodiversity, mutation breeding, plant breeding, chromosome manipulation and genetic and physical mapping. The book comprises of 16 chapters. Beginning with the introduction of chromosome and organization of genetic material, these first two chapters provide a deep insight into the different aspects of chromosome with detailed information and some more emphasis on organization of eukaryotic chromosomes. This is followed by chapters on B chromosome and high-throughput analytical techniques like chromosome banding, FISH, GISH, flow cytometry, etc. to give the reader an easy understanding of these techniques and their applications in chromosomal studies. This is followed by a series of chapters on induced chromosomal aberrations, asynapsis and desynapsis, cytomixis and structural and numerical aberrations with special reference to plant and animal chromosomes. The chapters on cytogenetic studies of Indian *Drimia* Jacq. (Urgineoideae: Hyacinthaceae), karyotype evolution and its utility and polyploidy provide valuable information. The possible mechanisms and their consequences and possible role in genetic analysis are emphasized in these chapters. The chapters of the book are from eminent authors and researchers of the world scientific community who are working on different aspects of chromosome research.

In presenting this book, we wish to express our gratitude to Prof. A.H. Khan and Prof. M. Anis of the Department of Botany, Aligarh Muslim University, Aligarh. Without whose help and encouragement, we would have never become the students of cytogenetics. We are thankful to the authors who contributed chapters for this book which is the outcome of their decades of research work. Our thanks especially go to Prof. Abdul Rauf Shakoori, distinguished national professor and director and professor emeritus, for his significant contribution in the form of well-oriented and innovative chapters for this book. Many of the ideas in the book are the outcome of teacher training programmes, conferences, workshops and seminars. We wish to express our gratefulness to all teachers and researchers associated with these programmes for their suggestions and advice without holding them responsible for any shortcoming in the book.

We are also grateful to our students whom we taught all these years, because it is through teaching them in the classroom that we learnt much of cytogenetics we know.

We are also thankful to Manjusha Nalamolu, PI of the book, for her wholehearted cooperation and sympathetic assistance whenever we required. We are also thankful to Springer for publishing our book.

In the end, we wish to acknowledge our debt to our family whom we left waiting on several evenings, while we were busy in finalizing the manuscript or the illustrations for this book.

Srinagar, India Tariq Ahmad Bhat
Srinagar, Jammu and Kashmir, India Aijaz Ahmad Wani

Contents

Contributors

Saira Aftab School of Biological Sciences, University of the Punjab, Lahore, Pakistan

Khalid Al-Ghanim Department of Zoology, College of Science, King Saud University, Riyadh, Saudi Arabia

Zahoor A. Beigh Cytogenetics and Reproductive Biology Laboratory, Department of Botany, University of Kashmir, Srinagar, Jammu and Kashmir, India

Tariq Ahmad Bhat Department of Education, Government of Jammu and Kashmir, Srinagar, India

James A. Birchler Division of Biological Sciences, University of Missouri, Columbia, MO, USA

Jahangir A. Dar Cytogenetics and Reproductive Biology Laboratory, Department of Botany, University of Kashmir, Srinagar, Jammu and Kashmir, India

Kátia Ferreira Marques de Resende Lab of Plant Cytogenetics, Department of Biology/DBI, Federal University of Lavras/UFLA, Lavras, BG, Brazil

Ryan N. Douglas Division of Biological Sciences, University of Missouri, Columbia, MO, USA

Mohd Gulfishan School of Life and Allied Health Sciences, Glocal University, Saharanpur, India

Mihir Halder Center of Advanced Study, Department of Botany, University of Calcutta, Kolkata, West Bengal, India

Sumita Jha Center of Advanced Study, Department of Botany, University of Calcutta, Kolkata, West Bengal, India

Mahpara Kashtwari Cytogenetics and Reproductive Biology Laboratory, Department of Botany, University of Kashmir, Srinagar, Jammu and Kashmir, India

M.M. Lekhak Angiosperm Taxonomy Laboratory, Department of Botany, Shivaji University, Kolhapur, Maharashtra, India

Sayantani Nath Center of Advanced Study, Department of Botany, University of Calcutta, Kolkata, West Bengal, India

Irshad A. Nawchoo Economic Botany and Reproductive Biology Laboratory, Department of Botany, University of Kashmir, Srinagar, Jammu and Kashmir, India

Uzma Qaisar School of Biological Sciences, University of the Punjab, Lahore, Pakistan

Peerzada Arshid Shabir Economic Botany and Reproductive Biology Laboratory, Department of Botany, University of Kashmir, Srinagar, Jammu and Kashmir, India

Abdul Rauf Shakoori School of Biological Sciences, University of the Punjab, Lahore, Pakistan

Farah Rauf Shakoori Department of Zoology, University of the Punjab, Lahore, Pakistan

Asima Tayyeb School of Biological Sciences, University of the Punjab, Lahore, Pakistan

Aijaz Ahmad Wani Department of Botany, University of Kashmir, Srinagar, Jammu and Kashmir, India

P.B. Yadav Angiosperm Taxonomy Laboratory, Department of Botany, Shivaji University, Kolhapur, Maharashtra, India

S.R. Yadav Angiosperm Taxonomy Laboratory, Department of Botany, Shivaji University, Kolhapur, Maharashtra, India

Showkat A. Zargar Cytogenetics and Reproductive Biology Laboratory, Department of Botany, University of Kashmir, Srinagar, Jammu and Kashmir, India

About the Editors

Dr. Tariq Ahmad Bhat has received his M.Sc and Ph.D from Aligarh Muslim University, Aligarh India. He is working as lecturer Botany in Department of Education, Govt. of Jammu and Kashmir, India. He is engaged in active research of chromosome analysis and mutation breeding of legumes. He is actively involved in genetic improvement of legumes and medicinal plants. He has published about 60 research papers, review articles and book chapters in reputed national and international journals and books. He has teaching experience of 14 years. Dr. Tariq Ahmad Bhat has been conferred with the Best Innovative Science Teacher Award 2014 by the Department of Science and Technology, Government of Jammu and Kashmir, India.

Dr. Aijaz Ahmad Wani has done his Ph.D. in the field of Mutation Breeding from Aligarh Muslim University. He is working as Senior Assistant Professor in the Department of Botany, University of Kashmir India. He is actively engaged in teaching of Cytogenetics, Crop Genetics and Molecular Breeding, and Genetic Engineering of Plants. He has more than 45 publications in National and International Journals. His current research interests are Assessment and Conservation of Germplasm of Rosaceous Fruits with special reference to apple (*Malus* x *domestica* Borkh.) and Apricot (*Prunus armeniaca* L.) and Genetic Improvement of some important medicinal plants with special reference to saffron (*Crocus sativus* L.). His significant contributions are Development of Germplasm Repository and first Mapping Population of apple in Jammu and Kashmir State.

Dr. Aijaz Ahmad Wani can be mailed at aijazbotku@gmail.com/aawani@kashmiruniversity.ac.in

Introduction to Chromosome

1

Abdul Rauf Shakoori

Abstract

Chromosomes are vehicles of inheritance and reside inside the nucleus. First named so by Waldeyer-Hartz in 1988, the chromosomes were associated with physical basis of heredity by Weissmann in 1892 and then vehicles of genetic information or genes by Boveri and Sutton in 1902. Avery, MacLeod, and McCarty provided convincing evidence in 1944 that the genetic material in chromosomes is made up of deoxyribonucleic acid (DNA). Watson and Crick, based on the X-ray diffraction data of Wilkins and Franklin, proposed double helical structure of DNA, in which the genetic information resides in the form of triplet codons. The prokaryotic chromosomes are in the form of naked circular DNA molecules. In eukaryotes, however, the DNA in association with histone proteins forms 10 nm fiber which is variously coiled to form a thick and highly condensed chromosome. The number and structure of chromosomal set is unique for each species of animals and plants. Any deviation in the number or rearrangement of segments of chromosomes causes aberrations, which are manifested in a variety of abnormal phenotypes and/or physiological disorders. The development of understanding of structure, chemical composition, and functions of different segments of chromosomes is intimately related with the refinement of techniques used for studying chromosomes. The techniques like banding pattern, in situ hybridization, and its ever-increasing variety of modifications are broadening our understanding of chromosomes and opening new avenues for increasing productivity by genetic manipulations.

A.R. Shakoori (✉)
School of Biological Sciences, University of the Punjab,
Quaid-i-Azam Campus, Lahore 54590, Pakistan
e-mail: arshaksbs@yahoo.com; arshakoori.sbs@pu.edu.pk

T.A. Bhat, A.A. Wani (eds.), *Chromosome Structure and Aberrations*,
DOI 10.1007/978-81-322-3673-3_1

Keywords
History of discovery of chromosomes • Chromosomal basis of inheritance • Packaging of chromosomes • Chromosomal aberrations • In situ hybridization

1.1 Introductory Remarks

Chromosomes, which reside in the cell nucleus, are the vehicles of inheritance. They are darkly stained basophilic threadlike structures visible during dividing phase of cell cycle. During interphase the chromosomes are not visible as discrete structures but instead a darkly stained material within the confines of double porous membrane becomes obvious as nucleus. The nucleus contains chromatin, which duplicates (replicates) during S phase of interphase and then gradually and imperceptibly condenses to form variable number of chromosomes, the number being specific for each species of plant and animal.

1.2 Early History of Chromosome

The early development of understanding of chromosomal structure was intimately linked with the improvement of resolution of a compound microscope, use of new chemicals as fixatives for preserving cellular structures, and variety of new dyes that would preferentially stain nuclei and chromosomes.

In 1865, Johann Gregor Mendel published findings of his experiments on inheritance of seven pairs of contrasting characters in the garden pea. He attributed this inheritance to certain particles in the cells (now genes), which could exist at least in two different forms or alleles, one being dominant over the other. Hence, the heterozygotes with two different alleles of a gene will show characteristics of the dominant allele. The recessive allele, though not lost, will express only under homozygous condition.

The term chromatin was first coined by Walther Flemming (1843–1905) in 1989, while the term chromosomes was coined by Heinrich Wilhelm Gottfried von Waldeyer-Hartz (1836–1921) in 1888. Walther Flemming was the first to relate chromatin with chromosomes, but since he was unaware of the work of Johann Gregor Mendel (1822–1884) on heredity, he was unable to relate the chromosomes with heredity. His discovery of mitosis and chromosome is considered one of the 100 most important scientific discoveries of all times. August Weismann (1834–1914) described structures called "idi" or biophores (Lebenstrager in German) on chromosomes which are much more closer to the genes in modern genetics than to Mendelian alleles. However, the work of Waldeyer-Hartz and others demonstrated that the chromosomes of animals and plants behaved virtually identical in mitosis and meiosis. Based on this evidence, Weissmann claimed in 1892 that chromosomes were the physical basis of heredity. Dutch geneticist Hugo de Vries (1848–1935), Carl Correns (1864–1933), and Erich von Tschermak (1871–1962) arrived at the same conclusion after rediscovery of Mendelian's principles of inheritance in 1901. Within a year, Theodor Heinrich Boveri (1862–1915) and Walter Stanborough Sutton (1876–1916) independently reported that the chromosomes carry genes and hence were the first to develop chromosomal theory of inheritance also known as Boveri-Sutton chromosome theory in 1902. During the following 90 years, each organism was shown to have its genes arranged linearly along the length of each chromosome. Moreover, each gene occupied a unique location on the chromosome. Thomas Harrison Montgomery (1873–1912) while describing gametogenesis of hemipterans showed the presence of homologous pairs of chromosomes and their segregation at meiotic anaphase. Emil Heitz in 1928 observed two types of chromatins in plants – euchromatin and heterochromatin, and later he observed the same differentiation in the polytene chromosomes of dipteran insects. Thomas Hunt Morgan (1866–1945) elucidated the role of chromosome in heredity for which he got Nobel Prize in Physiology or Medicine in 1933. He demonstrated the presence of genes on the chromosomes in 1910. He provided definitive evidence of the chromosomal nature of inheritance, while working on *Drosophila*. The chromosomal theory of inheritance precisely holds that genes are linearly arranged on the chromosomes.

1.3 Chemical Nature of Chromosome

Friedrich Miescher in 1869 had already identified an acidic material in the pus cells of wounds of soldiers, which he called as "nuclein." On reflecting back, it was already known during the late nineteenth century that DNA was the major component of cell nucleus. In 1924, Robert Feulgen developed a method for staining nuclei and chromosomes with acid fuchsin after acid hydrolysis of DNA. This Feulgen reaction allowed quantitative estimation of the DNA content of tissues and even of individual nuclei by cytophotometric methods. It was revealed that the cells of each organism had a characteristic amount of DNA, which coincided with the number of chromosomes. The amount of DNA in the nondividing diploid somatic cells was found to be twofold to that of the haploid germ cells.

Since at that time protein was the focus of researchers as vehicle of hereditary characters which are passed on from one generation to another, little attention was paid to Miescher's observation, until 1928 when Frederick Griffith reported a "transforming principle" as possible carrier of genetic information. Oswald Avery, Colin MacLeod, and Maclyn McCarty in 1940–1944 provided very convincing evidence for DNA as genetic material. Further evidence was provided by Alfred Hershey and Martha Chase who followed the radioactively labeled DNA and proteins of bacteriophage after exposure to *Escherichia coli*, in 1952. During infection, the DNA labeled with radioactive ^{32}P was detected inside the bacterium, whereas ^{35}S-labeled protein coat was left outside during infection. From this experiment, they proved that the new viruses with specific viral coat proteins were synthesis from genetic information received from the viral DNA.

Torbjorn Caspersson (1910–1997) studied the chromosome's nucleic acid composition. He concluded that genes are made of molecules of nucleic acid. Later Cyril Dean Darlington (1903–1981) published an article on "Chromosome Chemistry and Gene Action" in Nature in January 1942. David Chargaff, James Norman Davidson, and their colleagues had in the meantime worked out the chemical structure of DNA. Four types of nucleotides containing four different nitrogenous bases (abbreviated as A, T, C, and G) were found to be linked to each other by phosphodiester linkage to form a polynucleotide chain. In 1950–1953, James Watson along with Francis Crick proposed double helical structure, based on the X-ray diffraction data of DNA generated by Maurice Wilkins and Rosalind Franklin. It comprises two antiparallel helical chains that are wound round each other and are held together by two hydrogen bonds between A and T and three hydrogen bonds between C and G. These complementary nitrogenous bases which are located at the opposite strands form the basis of the proposed double helical structure of DNA. The sequence of bases in one strand, therefore, allows prediction of sequence of bases in the complementary strand. Furthermore, the genetic information in each chromosome could be attributed to the specific sequence of bases in its long DNA molecule. Francis Crick showed that this genetic information is, indeed, encoded in the successive triplets of bases (later designated as triplet codons) in DNA and in the messenger RNA (mRNA) that is transcribed from one strand of the DNA. Marshall Warren Nirenberg, Heinrich J. Matthaei, Severo Ochoa, Har Gobind Khorana, and Philip Leder deciphered the complete genetic code and demonstrated that triplet codons specify each amino acid in a growing polypeptide chain. In 1968, Robert W. Holly, Har Gobind Khorana, and Marshall W. Nirenberg received the Nobel Prize in Physiology or Medicine for their interpretation of genetic code and its functions in protein synthesis. During the subsequent years, Archibald Garrod, George W. Beadle, and Edward L. Tatum provided experimental evidence that each gene acts through synthesis of single enzyme, which then became the basis of the "one gene-one enzyme" hypothesis. Later on "one gene-one enzyme hypothesis" was replaced with the "one gene-one polypeptide hypothesis" when it was realized that genes not only synthesize enzymes but also synthesize nonenzymatic proteins and even single polypeptide chain.

Although the number of human chromosomes was first of all recorded by Theophilus Painter in 1923 as 48, accurate studies of human chromosomes became possible only after improvement of several techniques to study chromosomes such as improved cell culture methods, improved chromosome spreading techniques, and use of colchicine to block cells in metaphase. These improvements enabled Joe Hin Tjio and Albert Levan to establish in 1956 that the correct human diploid chromosome number is 46. The same year, Charles E. Ford and J.L. Hamerton showed that spermatogonia during meiotic division have 23 pairs of chromosomes. The methods for studying chromosomes continued to improve. Blood samples and amniotic fluid are now found to be convenient sources for studying chromosomes.

1.4 Chromosome Packaging

Prokaryotes typically have a single circular chromosome. In most bacteria the chromosome size varies from 130,000 bp to over 14,000,000 bp. The genes are often organized in operons. Prokaryotes generally have one copy of each chromosome, but there are examples which have multiple copies. *Buchnera aphidicola*, a member of Proteobacteria, which is the primary endosymbiont of aphids, for example, has 10–400 copies of chromosome per cell, and a Gram-positive bacterium, *Epulopiscium fishelsoni* which has symbiotic relationship with surgeonfish, has up to 100,000 copies of chromosome.

The chromosomes in eukaryotes have very condensed structure. The condensation and compactness is achieved through packaging in which long DNA molecules are coiled to fit into a small space of chromosome. This packaging is 750–1000 fold. Nucleosomes, formed by wrapping of DNA molecule around an octamer of histones, are linearly arranged to form a 10 nm-thick fiber. The 10 nm fiber is further variously coiled at different level to form 30 nm-thick fiber, 300 nm-thick fiber, and 700 nm-thick chromatid. During earlier stages of cell division, the chromatids gradually undergo a series of condensation steps till they cease to function as accessible genetic material. The compact form, during metaphase, gives the individual chromosome a four-arm structure, in which a pair of sister chromatids are attached to each other at the centromere. Later during anaphase, the two chromatids are separated from each other when spindle fibers pull the two chromatids apart. Each daughter cell inherits one set of chromatids.

So apparently structureless chromatin is converted into very well-defined condensed chromosome, the number of which is specific for a species. For example, human cells are diploid and have 23 pairs of chromosomes, of which 22 pairs are autosomes and one pair of sex chromosomes. Wheat is hexaploid. The wild wheat has haploid number of 7 chromosomes. The bread wheat has 28 and 42 chromosomes. Some animals and plant species are polyploid. They have more than two sets of homologous chromosomes.

1.5 Chromosomal Aberrations

The chromosomes contain genetic information which is essential for a normal life. Any change in the number or structure of chromosomes may lead to death, mental retardation, a variety of physical and genetic abnormalities, infertility, or spontaneous abortions. The chromosomal aberrations in the form of deviation from the normal state of diploidy (monosomy, trisomy) or rearrangements in the form of deletions, insertions, duplications, and inversions lead to genetic imbalance, which end up in a variety of syndromes. Some of the human genetic disorders due to gain or less of segments of DNA are cri du chat syndrome (deletion of part of the short arm of chromosome 5), Down syndrome (trisomy 21), Edward syndrome (trisomy 18), Jacobson syndrome (terminal 11q deletion), Klinefelter syndrome (XXY), Patau syndrome (trisomy 13), triple XXX syndrome, Turner syndrome (X instead of

XX and XY), YYY syndrome, and Wolf-Hirschhorn syndrome (partial deletion of short arm of chromosome 4).

The introduction of chromosome banding techniques revolutionized human cytogenetics. In 1970, Caspersson et al. discovered that quinacrine mustard produces consistent fluorescent banding patterns that are so distinctive that every chromosome can be individually identified. This discovery which was designated as Q banding was followed by several improved banding techniques such as G (Giemsa) banding, R (reverse G) banding, C (centromeric heterochromatin) banding, etc., which facilitated the identification of a wide range of previously undetectable structural chromosome abnormalities, such as translocations, inversions, deletions, and duplications. One very fundamental discovery about chromosome structure was that the DNA in each chromosome band replicated during S phase of the cell cycle (Latt 1973).

Development of methods for manipulating DNA ushered in new molecular era. J. Marmur, P. Doty, S. Spiegelman, and D. Gillespie showed that the two strands of DNA molecule could be easily separated and then can be made to reanneal with their complementary strands, even in the presence of noncomplementary DNA. Such molecular hybridization during the later years provided foundation for developing a powerful tool in human cytogenetics in which the cytological preparation of chromosomes and nuclei were used for in situ hybridization with labeled DNA. Several thousand loci have been mapped by fluorescence in situ hybridization (FISH) which has proven to be a more reliable, less time consuming, and more precise method for mapping genes, compared to the autoradiographic detection of radioactively labeled DNA fragments. Autoradiographic in situ hybridization led to the fundamental discovery of introns and exons in the genes. The introns or intervening sequences interrupted the continuity of protein-coding portion of most of the eukaryotic genes. The development of nucleotide sequencing methods by Walter Gilbert and Fred Sanger made it possible to characterize precisely the genes and other parts of the genome. Polymerase chain reaction (PCR) was another powerful tool invented by Kary Banks Mullis which has revolutionized many aspects of chromosome studies. This technique permitted rapid amplification of any short fragment of DNA, yielding up to a million-fold increase in the number of copies.

Molecular hybridization has been extensively used (i) to construct genetic linkage maps, (ii) to determine the origin of a deleted chromosome or of the extra chromosome in a trisomic individual, and (iii) to show that some individuals with a normal chromosome number received both their copies of a particular chromosome from the same parent. This has also led to the discovery of a novel mechanism of gene regulation, called genomic imprinting in which either of the two copies (maternal or the paternal) of a gene is inactivated. For a vast majority of autosomal genes, both alleles express simultaneously. In mammals, however, one of the two alleles of some genes is imprinted or silenced, which means that gene expression occurs only from one allele. As of 2014, about 150 imprinted genes are known in mouse and about 75 are known in human.

1.6 Artificial Chromosomes

Artificial chromosomes are synthetic chromosomes which have fragments of DNA or genomic DNA integrated into a host chromosome. This is usually achieved by integrating a nonviral/nonbacterial DNA into a bacterial chromosome and then expressing it within the host. Once expressed, the host chromosome containing the integrated DNA fragment will be replicated when the host cell undergoes replication. These artificial chromosomes can be used for transfection of other cells, thus introducing new DNA in the transfected cells. The bacterial plasmids can accommodate only up to 10,000 base pairs, whereas artificial chromosomes can contain anywhere from 300,000 bp for bacterial artificial chromosome (BAC) to 1,000,000 bp for yeast artificial chromosome (YAC). Although YAC can accommodate more base pairs than BAC, BACs are more common than YACs because they are more stable. YACs may also produce chimeric effects, while BACs will not. The artificial chromosomes are used mainly for studying DNA fragments. In fact the use of artificial chromosomes has revolutionized every aspect of biological studies.

YACs replicate in yeast cells. They consist of telomeres which are involved in the replication and stability of linear DNA, origin of replication sequences (autonomously replicating sequences, ARS) necessary for the replication in yeast cells, yeast centromere for the attachment of chromosomes with the spindle fibers during mitosis, a selectable marker for identification in the yeast cell such as uracil (URA3) and tryptophan biosynthesis (TRP1) gene, ampicillin resistance gene for selective amplification, and recognition sites for restriction enzymes such as *Eco*R1 and *Bam*H1. To enable YAC's expression in *Escherichia coli*, bacterial origin of replication (*E. coli* ori) and a selectable marker should also be part of YAC structure. Yeast cells transformed with YAC containing foreign DNA will produce YAC clones for further analysis.

Bacterial artificial chromosomes (BACs) have the advantage of studying larger genes, several genes at once, or entire viral genomes. Foreign DNA/gene of interest is ligated into LacZ gene of bacterial F plasmid (that confers ability to conjugate) cut with appropriate restriction enzyme. This BAC is electroporated into *E. coli* cells, which are then grown, and the positive clones isolated as white colonies from X-gal/IPTG medium. Currently BACs are the most common vector for large inserts such as eukaryotic genome projects.

Mammalian artificial chromosomes (MACs) are conceptually similar to YACs, but instead of yeast sequences, they may contain minimum mammalian or human DNA elements required for the maintenance of chromosome function. The human artificial chromosomes (HACs) which are constructed artificially in cultured human cells comprise multiple copies of telomere sequences TTAGGG and the commonly used repeated centromere sequences of human chromosomes in addition to the usual origin of replication, selectable markers, etc. These MAC or HAC DNAs are grown as YACs or, more recently, as BACs. On transfection of suitable cell line, the MAC/HAC DNAs behave like normal chromosomes just like those in the cell. They undergo accurate segregation and interact with the normal complement of proteins at telomeres and centromeres. The primary aim of the artificial chromosomes is to use them as vectors for delivery of large fragments of DNA to mammalian cells and/or to the whole animals for expression of large genes or sets of genes. Their involvement in genome mapping is of secondary importance. The engineered mammalian chromosomes can be used for preparation of stable cell lines with high expression for manufacture of recombinant proteins at industrial level. To date the most successful technique has been based on artificial chromosome expression system.

The area of artificial chromosomes is still passing through its developmental phase, and although the introduced genes have been shown to express, their practical application still needs to be effectively demonstrated.

1.7 Chromosome Painting

Chromosome painting enables visualization of specific chromosome regions or the entire chromosome through hybridization with fluorescently labeled chromosome-specific DNA probes. These probes with multiple fluorochromes on hybridization produce multicolored or painted effect with a unique color at each site of hybridization. This technique gives remarkable insight into the numerical and chromosomal aberrations associated with human diseases. For many years, this technique was used for study of animal chromosomes only for clinical and cancer cytogenetics, radiation biology, and nuclear topography, but now this technique is being successfully used in plants too.

Glossary

A

Artificial chromosomes These are synthetic chromosomes which have fragments of DNA integrated into a host chromosome. These chromosomes are useful in cloning large fragments of DNA.

C

Chromosome painting It is a term used to describe the direct visualization of specific chromosomes or parts of chromosomes in metaphase spreads and in interphase nuclei, through hybridization with fluorescently labeled chromosome-specific DNA probes.

E

Engineered chromosomes These are the chromosomes which have a large transgene carrying capacity, are non-integrating, and stably express in the eukaryotic cells

H

Human artificial chromosome Human artificial chromosome (HAC) is a microchromosome that can act as a new chromosome in a population of human cells. That is, instead of 46 chromosomes, the cell could have 47 with the 47th being very small, roughly 6–10 Mb in size instead of 50–250 Mb for natural chromosomes, and able to carry new genes introduced by researchers. HAC contains minimum human DNA elements such as telomere and centromere which are required for maintenance of chromosomal function.

References and Additional Reading

Arnold J (1879) Uber feinere Struktur der Zellen unter normalen und pathologischen verhaeltnissen. Virchows Arch Pathol Anat 77:181–206

Bauman JG, Wiegant J, Borst P, van Dujin P (1980) A new method for fluorescence microscopical localization of specific DNA sequences by in situ hybridization of fluorochrome labelled RNA. Exp Cell Res 128:485–490

Brown CJ, Hendrick BD, Rupert JL, Lafreniere RG, Xing Y, Lawrence J, Willard HF (1992) The human *XIST* gene: Analysis of a 17 kb inactive X-specific RNA that contains conserved repeats and is highly localized within the nucleus. Cell 71:527–542

Capanna E (2000) Chromosome yesterday: a century of chromosome studies. In: Almo E, Redi CA (eds) Chromosome today, vol 13. Birkhaeuser Verlag, Basel, pp 3–22

Caspersson T (1934) Durchfiltrierung von Thymonucleinsaeure. Biochem Z 270:161–163

Caspersson T (1937) Aufbau der Strukturen des Zellkernes. Acta Med Skand 73(Suppl 8):1–151

Caspersson T, Zech L, Johansson C (1970) Differential banding of alkylating fluorochromes in human chromosomes. Exp Cell Res 60:315–319

Darlington CD (1942) Chromosome chemistry and gene action. Nature 147:66–69

Federoff N, Botstein D (1992) The dynamic genome. Barbara McClinktock's ideas in the century of genetics. Cold spring Harbor Laboratory, Plainview

Flemming W (1879) Beitraege zur Kenntniss die Zelle und Lebenserscheinungenhrer. Theil II. Arch Mikrosk Anat 18:151–258

Flemming W (1898) Ueber die Chromosomenzahl beim Menschen. Anat Anz 14:171–174

Graves JAM, Disteche CM, Toder R (1998) Gene dosage in the evolution and function of mammalian sex chromosomes. Cytogenet Cell Genet 80:94–103

Hamerton JL (1971) Human cytogenetics, vol II. Academic, New York, p 65

Harris H (1995) The cells of the body. A history of somatic cell genetics. Cold Spring Harbor Laboratory, Plainview

Heitz E (1928) Das Heterochromatin der Moose. Jahrb Wiss Bot 69:762–818

Heitz E (1935) Chromosomenstruktur und Gne. Z inductive Abstammungs- und Verebungslehre (now Mol Gen Genet) 70:402–447

Heng HHQ, Squire J, Tsui LC (1992) High resolution mapping of mammalian genes by in situ hybridization to free chromatin. Proc Natl Acad Sci U S A 89:9509–9513

Idziak D, Betekhtin A, Wolny E, Lesniewska K, Wright J, Febrer M, Bevan MW, Jenkins G, Hasterok R (2011) Painting the chromosomes of Brachpodium – current status and future prospects. Chromosoma 120:469–479

ISCN (1995) An international system for human cytogenetic nomenclature. In: Mitelman F (ed) Karger, Basel

Jacobs PA, Baikie AG, Court Brown WM (1959) Evidence for existence of a human "superfemale". Lancet ii:423–425

Kallioniemi A, Kallioniemi OP, Sudar D, Rutovitz D, Gray JW, Waldman F, Pinkle D (1992) Comparative genomic hybridization for molecular cytogenetic analysis of solid tumors. Science 258:818–821

Kennard ML (2011) Engineered mammalian chromosomes in cellular protein production. Future prospects. Methods Mol Biol 738:217–238

Langer PR, Waldrop AA, Ward DC (1981) Enzymatic synthesis of biotin-labeled polynucleotides: novel nucleic acid affinity probes. Proc Natl Acad Sci U S A 78:6633–6637

Latt SA (1973) Microfluorometric detection of deoxyribonucleic acid replication in human metaphase chromosomes. Proc Natl Acad Sci U S A 70:3395–3399

Lejeune J, Gautier M, Turpin R (1959) Etude des chromosomes somatiques de neuf enfants mongoliens. C R Acad Sci 248:1721–1722

Lewin B (1997) Genes VI. Oxford University Press, Oxford

Lichter P, Cremer T, Borden J, Manuelidis L, Ward DC (1988) Delineation of individual human chromosomes in metaphase and interphase cells by in situ suppression hybridization using recombinant DNA libraries. Hum Genet 80:224–234

Lyon MF (1961) Gene action in the X-chromosome of the mouse (Mus *musculus* L). Nature 190:372–373

Makino S (1975) Human chromosomes. Lgaku Shoin, Tokyo

Miller OJ (1995) The fifties and the renaissance in human and mammalian cytogenetics. Genetics 139:489–494

Miller OJ, Therman E (2001) Human chromosomes. Springer Verlag, New York

Montgomery TH (1901) A study of the chromosome of the germ cells of metazoan. Trans Am Philos Soc 20:154–236

Ohno S (1967) Sex chromosomes and sex-linked genes. Springer, Heidelberg

Pardue ML, Gall JG (1969) Molecular hybridization of radioactive DNA to the DNA of cytological preparations. Proc Natl Acad Sci USA 64:600–604

Parra I, Windle B (1993) High resolution visual mapping of stretched DNA by fluorescent hybridization. Nat Genet 5:17–21

Pinkel D, Straume T, Gray J (1986) Cytogenetic analysis using quantitative, high-sensitivity, fluorescence hybridization. Proc Natl Acad Sci U S A 83:2934–2938

Reid T, Baldini A, Rand TC, Ward DC (1992) Simultaneous visualization of seven different DNA probes by in situ hybridization using combinatorial fluorescence and digital imaging microscopy. Proc Natl Acad Sci U S A 89:1388–1392

Rudkin GT, Stollar BD (1977) High resolution detection of DNA-RNA hybrids in situ by indirect immunofluorescence. Nature 265:472–473

Schrock E, du Manoir S, Veldman T, Schoell B, Weinberg J, Ferguson-Smith MA, Ning Y, Ledbetter DH, Bar-Am I, Soenksen D, Garini Y, Ried T (1996) Multicolor spectral karyotyping of human chromosomes. Science 273:494–497

Solinas-Toldo S, Lampel S, Stilgenbauer S, Nickolenko J, Benner A, Dohner CT, Lichter P (1997) Matrix-based comparative genomic hybridization: biochips to screen for genomic imbalances. Genes Chromosome Cancer 20:399–407

Speicher MR, Gwyn Ballard S, Ward DC (1996) Karyotyping human chromosomes by combinatorial multi-fluor FISH. Nat Genet 12:368–375

Strachan T, Read AP (1996) Human molecular genetics. Wiley-Liss, New York

Sutton W (1902) The chromosomes in heredity. Biol Bull 4:231–248

Tijo JH, Levan A (1956) The chromosome number of man. Hereditas 42:1–6

Von Hansemann D (1891) Ueber pathologische Mitosen. Virchow's Arch Path Anat Etc 123:356–370

Watson JD, Crick FHC (1953a) Molecular structure of nucleic acid. Nature 171(4356):737–738

Watson JD, Crick FHC (1953b) General implication of the structure of the deoxyribonucleic acid. Nature 171(4361):964–967

B Chromosomes

2

Ryan N. Douglas and James A. Birchler

Abstract

Supernumerary B chromosomes (Bs) have been observed in over 2000 plant, animal, and fungal species. Bs are nonessential and may be deleterious to the host genome at high copy numbers. In order to maintain their presence in a population, they frequently display non-Mendelian inheritance via mitotic or meiotic drive mechanisms. Advances in sequencing technology have recently enabled researchers to confirm earlier assumptions that Bs often arise as amalgamations of normal chromosomes. Because Bs are nonessential, they tend to accumulate transposons, repetitive DNA, and organellar DNA. Some Bs have also been shown to harbor transcribed genes and noncoding loci. Not surprisingly, the presence and/or transcription of Bs may affect transcription of the host genome, and much effort has been expended investigating this possibility. The dispensable nature of Bs makes them excellent tools for studying centromere biology, as well as making them amenable to serving as platforms for genetic engineering. Here, we review the origin, composition, inheritance mechanisms, transcription, and potential uses of Bs from multiple kingdoms.

Keywords

B chromosome • Supernumerary chromosome • Nondisjunction • Selfish DNA • Drive mechanism • Genome evolution

---

R.N. Douglas • J.A. Birchler (✉)
Division of Biological Sciences, University of Missouri, 65211 Columbia, MO, USA
e-mail: DouglasRN@missouri.edu; BirchlerJ@missouri.edu

T.A. Bhat, A.A. Wani (eds.), *Chromosome Structure and Aberrations*,
DOI 10.1007/978-81-322-3673-3_2

2.1 Introduction

E. B. Wilson was the first to observe supernumerary chromosomes when he discovered them in the insect *Metapodius* (Wilson 1907), and they were first identified in plants in the mid-1920s when Gotoh (Gotoh 1924) and Kuwada (Kuwada 1925) observed them in rye (*Secale cereale*) and maize (*Zea mays*), respectively. However, these chromosomes were not called supernumerary chromosomes until 1927 (Longley 1927), and they were not referred to as B chromosomes (Bs) until L. F. Randolph coined the term to distinguish them from normal A chromosomes (As) (Randolph 1928). Bs are remarkable in that they are nonessential to the host genome and may even be deleterious, yet they survive in numerous populations across multiple eukaryotic kingdoms. Indeed, since their discovery over one hundred years ago, Bs have been identified in over 2000 species of plants, animals, and fungi (Jones and Rees 1982).

There are a few important characteristics a chromosome must possess to be identified as a B chromosome and not another class of special chromosome such as sex chromosomes; L, E, or S chromosomes; or chromosome segments (Jones and Rees 1982; Jones 1991).

1. Bs are nonessential chromosomes. They are not found in all members of a population, and copy number can vary among individuals possessing Bs. In fact, B copy number can even vary between cells of an individual. The grasses *Aegilops speltoides* and *Ae. mutica* lack Bs in their roots but possess them in their aerial tissues, including gametes (Mendelson and Zohary 1972; Ohta 1996). Many mammals have also been found to have varying numbers of Bs depending on tissue type (Vujošević and Blagojević 2004).
2. Bs have a morphology that is distinct from the As, and they are often heterochromatinized (Fig. 2.1; Jones and Rees 1982). Bs in plants are often smaller than the smallest A, but mammalian Bs are often within the size range of the As (Jones and Rees 1982; Vujošević and Blagojević 2004). Although Bs often arise from As (discussed below), Bs are not homologous to As, and Bs do not pair or recombine with As during meiosis.
3. Bs often, but not always, display non-Mendelian inheritance. In plants, this typically occurs via postmeiotic nondisjunction in gametophytes (Jones 1991), but it can also occur premeiosis as in the migratory locust *Locusta migratoria* (Nur 1969), during meiosis as in the lily *Lilium callosum* (Kayano 1957), or after fertilization of the oocyte in the parasitic wasp *Nasonia vitripennis* (Nur et al. 1988). Few mammals have Bs that exhibit drive mechanisms, but most that do have meiotic drive in females (Vujošević and Blagojević 2004).

Due largely to varying efforts of cytological investigation across taxa, there are likely many species possessing Bs that remain unknown at the present time. It has been estimated that Bs occur in 15 % of eukaryotic species (Beukeboom 1994), and at present, the vast majority of species known to have Bs are in the plant kingdom (Jones et al. 2008). Among animals, Bs are rare in mammals (1.2 % of species,

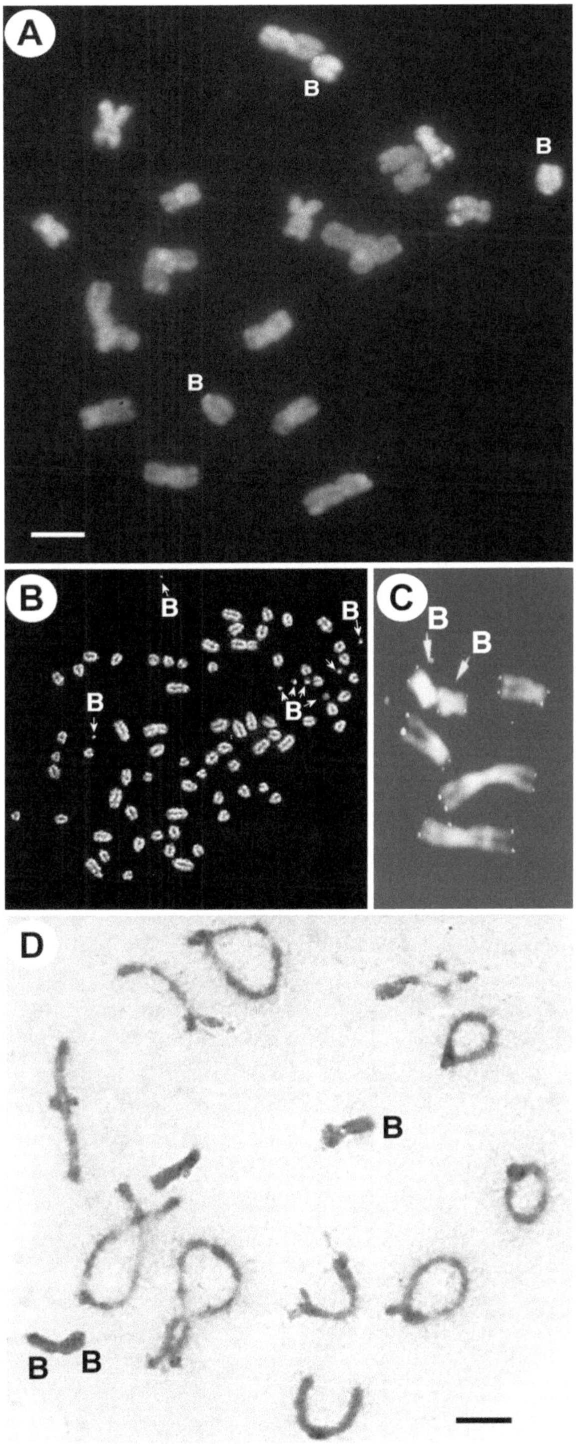

Fig. 2.1 Photomicrographs showing the distinctive morphology of B chromosomes in several species. (**a**) Maize (*Zea mays*) with three B chromosomes in somatic metaphase (Photo by Ryan N. Douglas). (**b**). Siberian roe deer (*Capreolus pygargus*) with seven B chromosomes in somatic metaphase (Modified with permission from Trifonov et al. (2013). Transcription of a protein-coding gene on B chromosomes of the Siberian roe deer (*Capreolus pygargus*) BMC Biology 11:90). (**c**) *Brachyscome dichromosomatica* with two B chromosomes in somatic metaphase (With kind permission from Springer Science + Business Media: Plant Systematics and Evolution, A monophyletic origin of the B chromosomes of *Brachyscome dichromosomatica* (Asteraceae), 219, 1999, pp. 127–135, Houben et al., Fig. 2.2). (**d**). *Eyprepocnemis plorans* with three B chromosomes in diplotene (With kind permission from Springer Science + Business Media: Molecular Genetics and Genomics, B chromosomes showing active ribosomal RNA genes contribute insignificant amounts of rRNA in the grasshopper *Eyprepocnemis plorans*, 289, 2014, pp. 1209–1216, Ruiz-Estévez et al., Fig. 2.2). *Bars* = 5 μm

mostly in order Rodentia) and completely absent in birds (Vujošević and Blagojević 2004). The distribution of Bs across flowering plant (angiosperm) taxa is nonrandom, but widespread (Burt and Trivers 1998; Palestis et al. 2004b; Trivers et al. 2004; Levin et al. 2005). Among angiosperms, 8 % of studied monocots have Bs, while only 3 % of eudicots have Bs (Levin et al. 2005). Within the monocots, there are noticeable differences of B frequency between different families. Some monocot families lack Bs entirely, while others, such as the Melanthiaceae, have Bs in over 40 % of the studied species (Levin et al. 2005). Frequency of Bs is independent of genome ploidy, Bs are more common in families with larger genomes, and Bs are found more frequently in mammals with acrocentric chromosomes (Burt and Trivers 1998; Palestis et al. 2004a; Trivers et al. 2004; Levin et al. 2005). Interestingly, there have been several reports in vertebrates suggesting that presence of Bs is positively correlated with pollution and/or stressful climactic environments (Pauls and Bertollo 1983; Giagia et al. 1985; Belyanin et al. 1994; Vujošević and Blagojevic 2000; Feldberg et al. 2004; Oliveira et al. 2008).

Because Bs are dispensable components of the genome, the role they play in the life of the host organism has long been of interest to researchers. Bs may have negative, neutral, or positive effects on their host. Bs are often viewed as selfish or parasitic elements that hijack the cellular machinery required to replicate and transmit As to daughter cells (Östergren 1945; Jones and Rees 1982; Camacho et al. 2000). Generally, as the number of Bs increases, cell size and mass increase (Müntzing and Akdik 1948), the length of time required to complete mitosis increases (Evans et al. 1972), and nuclear protein and RNA levels decrease (Kirk and Jones 1970). Perhaps not surprisingly, the idea that Bs are parasitic elements has been supported by numerous instances where the presence of Bs, especially in high numbers, is detrimental to the host and causes reduced fitness (Jones and Rees 1982). In these cases, equilibrium is reached whereby the selection against Bs is balanced by the drive mechanisms that allow Bs to persist despite their negative effects (Camacho et al. 2000). Incidentally, outcrossing species are more likely to have Bs than inbreeding species, as inbreeding lines that lack Bs will outcompete lines that contain detrimental Bs (Puertas et al. 1987; Burt and Trivers 1998; Palestis et al. 2004b). In contrast to the apparently harmful nature of most Bs studied so far, some Bs are considered to be beneficial, or heterotic (White 1973). The heterotic theory suggests that the presence of small numbers of Bs provides an adaptive advantage to the host and thus allows Bs to persist, even if a drive mechanism is absent. However, a drive mechanism could evolve on the beneficial B, thus allowing numbers to rapidly increase and transforming it into a parasitic chromosome (Camacho et al. 2000).

In this chapter, we will examine the different types of non-Mendelian inheritance that allow selfish Bs to persist, explore the origins of Bs and their sequence compositions in multiple species, take a detailed look at how Bs affect their host genomes, and review applications for using Bs to study aspects of chromosome biology.

2.2 B Chromosome Inheritance Mechanisms

The detrimental nature of most Bs at high copy numbers would lead to their extinction in the absence of a drive mechanism. Many Bs have evolved a mitotic or meiotic drive mechanism that allows them to accumulate and be inherited at greater than expected ratios based on Mendelian genetics. Depending on the species, the accumulation of Bs may occur only in the female, only in the male, or in both sexes, and there are some instances where drive through one sex is balanced by drag through the other. Most plant Bs undergo nondisjunction during gametophytic mitoses, while most animal Bs have accumulation mechanisms in germ tissue before or during meiosis (Jones and Rees 1982). However, as will be discussed later, not all Bs possess a drive mechanism.

Unlike plants, animal germ lines do not replicate their meiotic end products, so animal drive mechanisms occur either before or during meiosis. Premeiotic drive is probably best described in males of the order Orthoptera, which includes crickets, grasshoppers, and locusts. Male-specific premeiotic drive has been described in the grasshoppers *Calliptamus palaestinensis* (Nur 1963), *Camnula pellucida* (Nur 1969), *Neopodismopsis abdominalis* (Rothfels 1950), and the migratory locust *L. migratoria*, where the accumulation mechanisms have been particularly well studied. The Bs of *L. migratoria* are present in the early embryo, but they nondisjoin during development and eventually accumulate in the germ lines of males (Nur 1969; Kayano 1971; Viseras et al. 1990; Pardo et al. 1995a, b). The nondisjunction of the Bs occurs as a result of mitotic instability and is detectable in 3-day old embryos of both sexes. This instability peaks in 5-day old embryos, somehow driving an accumulation of Bs specifically to the testes of males (Nur 1969; Kayano 1971; Viseras et al. 1990; Pardo et al. 1995b). Although Bs do not accumulate premeiotically in the female germ line, they do preferentially segregate to the oocyte, providing meiotic drive on the female side in *L. migratoria* (Pardo et al. 1995b).

Meiotic drive is common in animals with Bs that possess drive mechanisms but is relatively rare in plants compared to postmeiotic drive. However, the first case of meiotic drive of a B was observed in the lily *L. callosum* (Kayano 1957). When a male containing one B was crossed to a female containing zero Bs, 50 % of the progeny had one B and 50 % had zero Bs as expected by Mendelian genetics. However, when a female containing one B was crossed by a male containing zero Bs, 80 % of the progeny contained one B, showing a female-based meiotic drive. In angiosperms, the ovule contains an embryo sac that is set up in an asymmetric manner. There is a chalazal end, which serves as an attachment point for the embryo sac, opposite a micropylar end, which provides a small opening to allow a pollen tube to reach and fertilize the egg cell. In looking at 284 embryo-sac mother cells in metaphase I of meiosis, Kayano (1957) observed that 63.7 % of univalent Bs were outside the metaphase plate on the micropylar side, which is the side that will eventually give rise to the egg cell. After also analyzing embryo-sac mother cells in metaphase II, anaphase I, and anaphase II, Kayano found that 80 % of the Bs were in micropylar nuclei, thus corroborating the results of his crossing study. Although the presence of more than one B in *L. callosum* causes reductions in pollen viability and seed fertility (Kimura and Kayano 1961), this meiotic drive system ensures the survival of the Bs. It has also been suggested that an asymmetric alignment of Bs on the meiotic spindle, where they are preferentially found near the egg pole, may be behind the meiotic drive of the grasshopper *Myrmeleotettix maculatus* (Hewitt 1976).

Less direct evidence for meiotic drive exists in mammals, and relatively few species have been thoroughly investigated. The mouse *Apodemus peninsulae* accumulates Bs in spermatocytes during zygotene and pachytene (Kolomiets et al. 1988), females of the lemming *Dicrostonyx torquatus* eliminate univalent Bs from polar bodies and incorporate them into secondary oocytes (Gileva and Chebotar 1979), more Bs are found in germinal tissue than somatic tissue in the pocket mouse *Perognathus baileyi* (Patton 1977), and Bs were found to randomly segregate at metaphase II of meiosis in germinal tissue of males in the silver fox *Vulpes vulpes* (Radzhabli et al. 1978). Laboratory studies with the black rat *Rattus rattus* have shown that Bs are mitotically stable and have Mendelian transmission rates through males, but the Bs have enhanced, non-Mendelian transmission in females (Yosida 1980; Stitou et al. 2004). In controlled crosses between males with one B and females with zero Bs, only 43 % of the offspring had a B chromosome. In contrast, when a female with one B was crossed to a male with zero Bs, 67 % of the progeny possessed at least one B chromosome (Stitou et al. 2004). Additional evidence for meiotic drive came from a study of 1170 mammalian karyotypes that suggested nonrandom chromosome segregation during female meiosis is a major driver of mammalian karyotype evolution (Pardo-Manuel de Villena and Sapienza 2001).

The majority of angiosperm Bs utilize postmeiotic drive mechanisms to maintain themselves, and the best studied of these systems are those of the grasses rye and maize. The rye system is unusual in that nondisjunction occurs in both male and female reproductive organs; in maize, nondisjunction only occurs in pollen. Here, the systems of both species will be described in detail.

Bs in rye and maize have both evolved to make use of postmeiotic mitoses to accumulate. Meiosis during angiosperm microsporogenesis reduces the genome to a haploid state. Following meiosis, the microspores undergo two rounds of mitosis to generate the microgametes. The first mitosis produces a vegetative cell, which will ultimately produce a pollen tube, and a haploid generative cell. The generative cell undergoes a second round of mitosis to produce two haploid sperm cells. One sperm cell will fertilize the egg cell to form a diploid zygote, while the other sperm cell combines with the two polar nuclei of the megagametophyte's central cell to form a triploid endosperm in a process unique to angiosperms known as double fertilization.

In rye, the Bs progress through male meiosis normally, but during anaphase of the first postmeiotic mitosis the two B chromatids lag behind and do not separate (Hasegawa 1934). It was noted that the B chromatids seemed to stick together at two loci, one on each side of the centromere. The B chromatids preferentially nondisjoin to the generative nucleus, producing a nucleus that contains two Bs. The Bs disjoin normally during the second postmeiotic mitosis, yielding two sperm cells that each carry two Bs (Fig. 2.2a). However, Hasegawa (1934) also noted that Bs do not

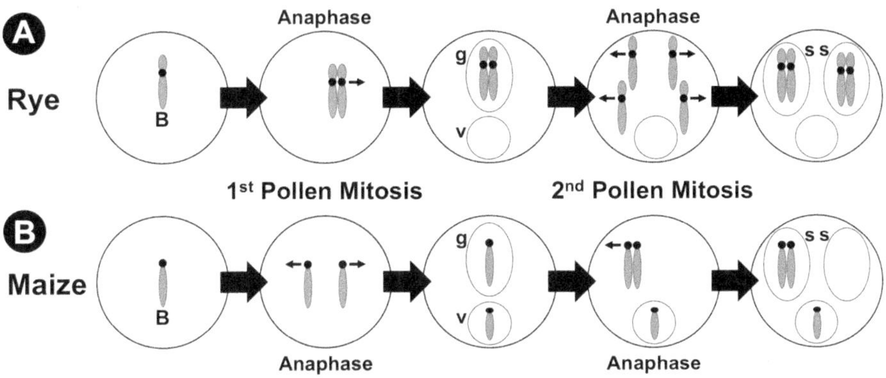

Fig. 2.2 Rye (*Secale cereale*) and maize (*Zea mays*) B chromosomes (B) undergo nondisjunction during different stages of microsporogenesis. A chromosomes are not shown, as they separate normally during all stages of postmeiotic mitoses. (**a**) Following meiosis, rye Bs undergo nondisjunction during anaphase of the first pollen mitosis, which produces a generative cell (*g*) and a vegetative cell (*v*), which will give rise to the pollen tube. The generative cell undergoes a second pollen mitosis, and the rye Bs disjoin normally, producing two sperm cells (*s*) that each have two Bs. (**b**) Maize Bs separate normally during anaphase of the first pollen mitosis, which places one B in the generative cell and one B in the vegetative cell. However, the Bs can nondisjoin during anaphase of the second pollen mitosis, which produces one sperm cell with two Bs and one with zero. The sperm cell containing the Bs preferentially fertilizes the egg cell at a frequency of 60–70 %

always undergo nondisjunction during the first postmeiotic mitosis. In his study, nondisjunction directed the Bs to the generative nucleus 81 % of the time and to the vegetative nucleus 6 % of the time, and the Bs disjoined normally in 13 % of cases. A similar mechanism in female gametogenesis allows rye Bs to accumulate in egg cells (Håkansson 1948). The Bs frequently nondisjoin during the first megaspore mitosis, and the nucleus on the micropylar end of the megaspore often receives the Bs. This nucleus eventually gives rise to the egg cell, allowing accumulation of Bs in female gametes in rye.

A potential mechanism explaining the nondisjunction of rye chromosomes remained out of reach for many years. A clue that the process was under genetic control came when a deletion derivative of a B was found that lacked the heterochromatic distal tip of the long arm (Müntzing 1948). This new B (defB) was unable to nondisjoin, indicating that the long arm of the B contained at least one factor required for nondisjunction. The factor was found to act in *trans*, as defB is able to nondisjoin when in the presence of a standard B (Lima-de-Faria 1962) or even simply in the presence of the distal tip of the standard B long arm (Endo et al. 2008). In order to determine if the rye genome was responsible for B nondisjunction, researchers introduced standard rye Bs into wheat *Triticum aestivum*, *Triticale*, and *Secale vavilovii* (Lindström 1965; Müntzing 1970; Kishikawa and Suzuki 1982; Puertas et al. 1985; Niwa et al. 1997; Endo et al. 2008). Standard rye Bs continue to nondisjoin in the various genetic backgrounds, which suggests the process of nondisjunction is entirely controlled by the B.

Two B-specific, tandemly arrayed repeats were identified within the subtelomeric heterochromatic DNA that is absent in defB: the 1.1-kb D1100 (Sandery et al. 1990) and the 3.9-kb E3900 (Blunden et al. 1993). Detailed analysis of D1100 and E3900 sequences showed that they are composed of many unrelated DNA fragments, suggesting that they had a de novo origin within the B itself (Langdon et al. 2000). This region is epigenetically marked by the presence of conflicting histone modifications: trimethylated histone H3 at lysine-4 (H3K4), which promotes transcription in euchromatin, and trimethylated H3K27, which is found predominantly in heterochromatic regions of the rye genome (Carchilan et al. 2007). Additionally, this repetitive region decondenses during interphase, unlike subtelomeric regions of the As, which remain condensed. Both repeat families lack open reading frames (Langdon et al. 2000), but they are transcriptionally active, particularly in anthers (Carchilan et al. 2007). Variously sized, polyadenylated, noncoding RNAs are produced from D1100 and E3900 loci (Carchilan et al. 2007), giving rise to the suggestion that these RNAs may have a catalytic or structural function required for nondisjunction of rye Bs. Recently, Banaei-Moghaddam et al. (2012) examined the rye B loci that stick together during anaphase of the first postmeiotic mitosis. They found that rye Bs have B-specific repeats in the pericentromeric regions that are involved in forming pericentromeric heterochromatin. Heterochromatin and pericentromeric repeats play a role in sister chromatid cohesion, so it has been postulated that the pericentromeric B-specific repeats play an essential role in the nondisjunction of rye Bs (Banaei-Moghaddam et al. 2012). After over 75 years of research, a model explaining the nondisjunction of rye Bs in pollen emerged and will be summarized here. Rye Bs have unique pericentromeric repeats that allow enhanced sister chromatid cohesion during the first postmeiotic pollen mitosis. This cohesion is under the control of a *trans*-acting factor located on the distal tip of the B's long arm. Because of unequal spindle formation in the microspore, similar to the meiotic nondisjunction of *L. callosum*, the Bs preferentially sort into the generative nucleus. After the first pollen mitosis completes, the segregation of Bs proceeds normally, and the two sperm nuclei receive an unreduced number of Bs.

In contrast to rye, maize Bs pass normally through the egg parent but accumulate through the pollen parent. Maize microsporogenesis proceeds normally through the first postmeiotic mitosis, but Bs can nondisjoin during the second pollen mitosis (Fig. 2.2b). Nondisjunction occurs in 50–100 % of pollen grains, and the sperm possessing Bs preferentially fertilizes the egg 60–70 % of the time, while the sperm lacking Bs fertilizes the polar cells (Roman 1948). A mechanism to explain preferential egg fertilization by the sperm nucleus containing Bs remains unknown. Additionally, maize Bs are able to proceed through meiosis as univalents (Carlson and Roseman 1992). Like rye, the nondisjunction of maize Bs is genetically controlled by distinct regions of the B itself. Many studies examining maize B nondisjunction have utilized B-A translocations. B-A translocations arise from reciprocal translocations between standard A chromosomes and supernumerary Bs (Roman 1948; Beckett 1978). The result is an A-B chromosome, where the centromere from the A is attached to part of the B, and a B-A chromosome, where the B centromere is attached to part of an A chromosome arm. The B-A chromosome can only

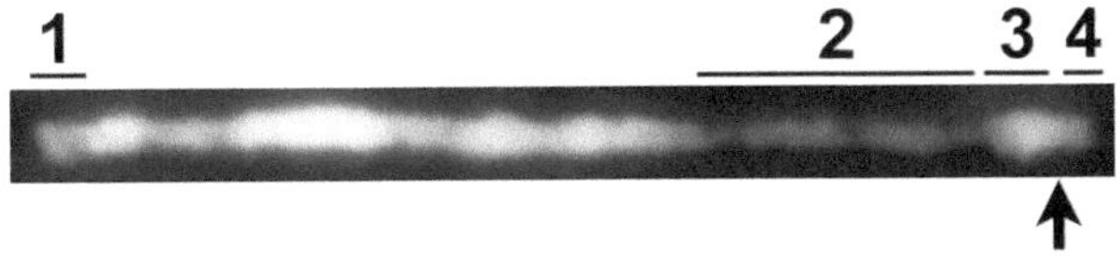

Fig. 2.3 A stretched, computationally straightened, maize B chromosome during pachytene. The four regions involved in enabling and/or enhancing B chromosome nondisjunction are labeled above, and an arrow denotes the position of the centromere. See the text for details about each region (Modified with kind permission from Springer Science + Business Media: Chromosome Research, Localization and transcription of a retrotransposon-derived element on the maize B chromosome, 15, 2007, pp. 383–398, Lamb et al., Fig. 2.1)

undergo nondisjunction when the A-B chromosome is present (Roman 1949); the B-A chromosome may also nondisjoin in the presence of a standard B chromosome (Carlson 1974). Researchers have identified three regions on the long arm of the maize B that are essential for, or enhance, nondisjunction, and a fourth region, consisting of the centromere and short arm of the B, that is required as the target of nondisjunction (Carlson and Chou 1981). Structurally, the maize B is subtelocentric in that the centromere appears near the very end of the chromosome, with a very small, heterochromatic short arm (Region 4). The long arm of the B has a small heterochromatic region adjacent to the centromere (Region 3), followed by a long euchromatic region (Region 2), then a large heterochromatic region, and, finally, a distal, euchromatic tip (Region 1; Fig. 2.3).

Region 1 was identified using the B-A translocation line TB-8La. The 8-B component of TB-8La carries only the distal euchromatic end of the B. As 8-B is required for the B-8 counterpart to nondisjoin, this euchromatic tip is required, in *trans*, for maize B nondisjunction (Ward 1973). A deletion derivative of TB-9Sb has a disjoining B-9 chromosome that lacks the euchromatic region nearest the centromere and was used to identify region 2. However, the addition of a standard B allows this B-9 to nondisjoin, indicating that the proximal euchromatic region acts in *trans* to promote nondisjunction (Carlson 1974). Region 2 was further delineated by a series of X-ray-generated translocations between the long arm of chromosome 10 and the B (Lin 1978). Region 3 contains DNA sequences that are partly homologous to the 180-bp knob repeat found in some heterochromatic regions of maize As (Peacock et al. 1981). This is intriguing, because certain As with distal knob sequences may undergo chromosomal loss in the presence of Bs (Rhoades et al. 1967; Rhoades and Dempsey 1972). During the second postmeiotic pollen mitosis, knob sequences on As may stick together if a B is present, producing chromosome bridges, although it is important to note that the knob repeat is unlikely to be the only target of the *trans* nondisjunction signal. Several maize minichromosomes containing only the B centromere, and no knob repeat, are able to nondisjoin in the presence of Bs (Han et al. 2007a). The bridges eventually break, leading to loss of chromosomal fragments. It was proposed that the *trans* signals required for nondisjunction of the B act on region 3, which functions as a receptor site. This proposed mechanism is strikingly similar to the system suggested in rye, whereby noncoding

RNA transcripts from repetitive regions of the distal tip of the rye B may be acting in *trans* on receptor sites adjacent to the centromere. Additional evidence for the importance of region 3 in nondisjunction of the maize B was provided by Carlson and Chou (1981). They identified a pseudoisochromosome from TB-9Sb that contains a complete B-9 on one side of the centromere, while a B-9 on the other side of the centromere lacks region 3. When the pseudoisochromosome undergoes misdivision, many derivatives can be recovered, including telocentric chromosomes that either possess or lack region 3. The derivatives that possess region 3 undergo nondisjunction, while the derivatives that lack region 3 separate properly. Region 4 includes the B centromere and the small short arm of the B. Independent studies by Lin (1979) and Carlson and Chou (1981) suggested that the presence of the centromere and/or the short arm can enhance the rate of B nondisjunction as nondisjunction rates dropped in the absence of the short arm or in the presence of a damaged B centromere. However, a more recent study determined that centromere function and nondisjunction are independent of each other (Han et al. 2007b). A B centromere was translocated to the short arms of normal chromosomes 7 and 9, respectively, where the centromeres were epigenetically inactivated. By themselves, the chromosomes containing inactive B centromeres were unable to undergo nondisjunction. Though in the presence of a normal B, nondisjunction properties were transferred to the B centromere-containing chromosomes. Han et al. (2007b) thus showed that nondisjunction does not require an active centromere but that the B centromere region (Region 4) is required for nondisjunction. Additionally, several B misdivision derivatives containing little more than the B centromere are able to nondisjoin in the presence of a B (Han et al. 2007a). These observations strongly suggest that the B centromere and/or surrounding region is a potential target for B nondisjunction in maize. Although regions of the B have been identified as playing a role in nondisjunction, a detailed mechanism explaining maize B nondisjunction remains unknown.

Some Bs do not rely on meiotic or mitotic drive mechanisms to accumulate in their hosts. In the case of the parasitic wasps *Nasonia vitripennis* and distantly related *Trichogramma kaykai*, Bs are transmitted in a truly unique fashion. In *N. vitripennis*, males are haploid and develop from unfertilized eggs, while females are diploid and develop from fertilized eggs. The B is only transmitted through sperm, and when present, it destroys the paternal genome (Werren et al. 1987; Nur et al. 1988). Therefore, the presence of the B turns a diploid, female embryo into a haploid, male embryo that can pass the B to future generations. The process of paternal-specific genome destruction is possible because, in insects, the maternal and paternal genomes are separated from each other during the first embryo mitosis by a nuclear envelope that surrounds each genome (Kawamura 2001). Although the paternal genome replicates during S phase and enters mitosis, it soon becomes hypercondensed, does not resolve into individual chromosomes, does not segregate, and is eventually lost during the first embryonic mitosis when the B is present (Swim et al. 2012). The B itself somehow escapes the fate of the paternal genome and ensures its inheritance at a frequency of 94–100 % (Beukeboom and Werren 1993). Recently, it was discovered that the *N. vitripennis* B always inhabits the anterior-most edge of the paternal nucleus after fertilization, far from the standard paternal genome, where it may receive an epigenetic advantage over the paternal genome (Swim et al. 2012). Incidentally, the paternal genome was found to have prolonged phosphorylation of H3 serine-10 compared to the maternal genome and greatly enriched binding by the condensin component SMC2, which affects its ability to move through mitosis (Swim et al. 2012). Swim et al. (2012) have hypothesized that the *N. vitripennis* B may encode an enzyme that affects only euchromatin, thereby not affecting the heavily heterochromatic B, or that the presence of the B mis-regulates a euchromatin-modifying enzyme present in the standard genome.

If a B is truly beneficial to its host, it may not need a drive mechanism to persist. It should also be noted that some Bs that lack drive mechanisms suffer a net loss from the parents in some species. Up to 40 % of plant species with Bs are thought to lack drive mechanisms (Jones 1995). Such is the case of the B in the chive *Allium schoenoprasum* (Bougourd and Parker 1979). Individuals containing Bs have an advantage over individuals that lack Bs during the seed and seedling stages of the life cycle (Holmes and Bougourd 1989). In particular, the presence of a B allows *A. schoenoprasum* to germinate more quickly and have higher subsequent survival in drought conditions (Holmes and Bougourd 1991; Plowman and Bougourd 1994). There are also Bs in animals that exist in a neutral state without an apparent drive mechanism. The grasshopper *Eyprepocnemis plorans* has multiple, stable, types of Bs, but most are randomly inherited, with no drive or drag mechanism observed in males or females (López-León et al. 1992).

2.3 Origins and Sequence Content of B Chromosomes

The origin of Bs has long been debated, although host genomes have generally been accepted as the donors of B sequences. Recent advances in sequencing technology have shown that, indeed, many Bs arose from host genome As, although some Bs almost certainly began as interspecific alien chromosomes. Generally, Bs are largely heterochromatic, devoid of protein-encoding loci, and enriched for repetitive DNA sequences and transposons. Here, we will review the apparent origins, and DNA content, of Bs in several plant and animal species.

The most straightforward identification of B origin comes from the de novo creation of an apparent B in the herb *Plantago lagopus*. A spontaneous trisomy arose in one line, where roughly 3 % of pollen mother cells exhibited chromosomal nondisjunction of the A genome. Over approximately 10 years, the progeny of crosses between the trisomic line and normal euploids gave rise to many more aneuploids, including a line that was trisomic for chromosome 2 (Sharma and Koul 1984; Sharma et al. 1985a, b). The extra copy of chromosome 2 rapidly changed its morphology and sequence composition, eventually acquiring many characteristics of a B (Dhar et al. 2002). Since three copies of chromosome 2 were present, the host plant would be able to survive damage to one copy, making chromosome 2 particularly susceptible to recoverable chromosome breakage. Chromosome 2 contains arrays of 5S, 18S, 5.8S, and 25S rRNA genes, all of which are present on the apparent

B. Indeed, rRNA genes are found on Bs in many species (reviewed in Jones 1995). In their hypothesis, Dhar et al. (2002) suggested that breaks in each arm of chromosome 2 could have led to the formation of a ring chromosome via fusion of the broken ends. Subsequent fragmentation of the ring chromosome could have produced a small chromosome containing little more than a functional centromere and rDNA repeats. Ring and miniature chromosomes were observed cytologically during the development of the apparent B. Next, rapid amplification of 5S rDNA occurred, telomere sequences were recruited, and a centromeric misdivision produced the isochromosome that was ultimately observed. The newly formed apparent B is mostly heterochromatic; it is mainly composed of 5S rDNA interspersed with *Ty1/Copia* and *Ty3/Gypsy* transposons; it does not pair with any As during meiosis, and was inherited at a greater frequency than expected, suggesting a functional drive mechanism is already present (Dhar et al. 2002; Kour et al. 2013).

The first B to be sequenced with next-generation sequencing technologies was that of rye, and the results unequivocally showed that the rye B originated from parts of the A genome (Martis et al. 2012). Rye Bs were sorted with flow cytometry and then sequenced at 0.9-fold coverage with Roche 454 sequencing technology. The resulting sequence was then compared to flow sorted rye As, as well as genomic DNA from plants with and without Bs. When the sequences were compared, it became clear that the origin of the rye B, after multiple rearrangements, lies principally with rye chromosomes 3RS and 7R. Additionally, small blocks of homology were observed for many other regions of the A genome, suggesting that the rye B truly is an amalgamation of the rye A chromosomes. The rye B also contains many plastid- and mitochondrion-derived sequences, mostly in pericentromeric regions. Similarly, the B of *Ae. speltoides* was recently found to also contain many organelle DNA insertions (Ruban et al. 2014). The size of the rye B was estimated to be 580 Mbp, and phylogenetic analyses suggested the B originated 1.1–1.3 million years ago, which fit with the estimated age of *Secale* at 1.7 million years ago.

One of the biggest surprises about the DNA content of the rye B was that it contained nearly 5000 nonredundant sequences with high homology to plant genes (Martis et al. 2012). Since Bs are thought to be nonfunctional, whether or not any of these potential genes are functional was an important question to address. Earlier studies used cDNA-amplified fragment length polymorphism (AFLP) experiments to conclude that the rye B was transcribed at a rate of only 5 % as compared to transcription of the As (Carchilan et al. 2009). Banaei-Moghaddam et al. (2013) examined transcription of 15 gene-like loci on the rye B and concluded that earlier work underestimated transcription of the rye B. It was discovered that approximately 15 % of the gene-like loci in rye are transcribed in a tissue-specific manner and that reduced selection has rendered many of the gene-like loci as nonfunctional pseudogenes. However, the transcribed pseudogenes are homologous to their A counterparts, suggesting that B-derived transcripts may affect the A genome. That the gene-like loci are mostly nonfunctional is no surprise; for an intact gene to be present on a B, it likely either had a recent insertion or it may be required for the maintenance and/or transmission of the B itself.

Because Bs are nonessential and are composed of various A fragments, it has been hypothesized that Bs have a distinct evolutionary path from their host genomes. Evidence for this hypothesis was uncovered by a detailed analysis of repetitive DNA sequences on the rye B (Klemme et al. 2013). Rye Bs are rich in repetitive DNA, including many transposable elements, which is not surprising given that 90 % of the A genome is repetitive. *Ty3/Gypsy* elements represent nearly half of the mobile elements on the rye B, while *T1/Copia* elements account for 9 % of the mobile elements on the B. Many repetitive elements are equally distributed on As and Bs, but one of the most interesting observations by Klemme et al. (2013) was that the retroelement *Sabrina* is nearly absent on Bs compared to As. In contrast, the *Copia* element Sc36c82 and *Gypsy*-like element *Revolver* are highly enriched on Bs compared to As. *Sabrina* is an ancient retroelement that is inactive in rye (Shirasu et al. 2000), while Sc36c82 and *Revolver* are active. A model was proposed whereby the rye B contained a comparable amount of *Sabrina* elements to the As when it first formed. Subsequently, *Sabrina* inactivated and was diluted on the B as active elements, such as *Revolver* and Sc36c82, moved in. The B was more amenable to insertion events than the As due to the lower selection pressure allowed by its dispensable nature. In addition to the repeats E3900 and D1100 (described above), Klemme et al. (2013) identified nine B-enriched repeats. Most clustered in the distal end of the long arm responsible for rye B nondisjunction near E3900 and D1100, but some localize to the pericentromere, where the chromatin appears to stick together during the nondisjunction process. Interestingly, all but one of the newly described repeats is transcribed in the anthers of plants containing a B.

Maize Bs have been well studied, but little information is known about their genetic content and evolutionary history due to the highly repetitive nature of their transposable elements. The maize B shares many retrotransposons with the As, and the B typically contains more copies per unit of sequence than the As for retrotransposon types that are shared (Theuri et al. 2005). Based on cytological analysis, it has been predicted that the maize B is roughly 0.66x the size of chromosome 10, which would equal approximately 100 Mbp. A centromere-enriched, B-specific repeat (ZmBs) was the first DNA sequence cloned from the maize B (Alfenito and Birchler 1993). Genomic DNA, with and without Bs, was hybridized to a λ phage library to identify ZmBs. The 1.4-kb tandemly arrayed ZmBs shows homology to plant and animal telomeres, as well as to 90-bp of the 180-bp maize knob repeat, and the centromere of chromosome 4 (Page et al. 2001). A second B-specific sequence, pBGBM18.2, was identified as a 637-bp random amplification of polymorphic DNA (RAPD) marker between maize lines that possessed or lacked Bs (Stark et al. 1996). pBGBM18.2 contains sequences that are unique to the B, as well as sequences found on A chromosomes, including a potential coding region for *Prem2*, a maize long terminal repeat (LTR) retroelement. Lamb et al. (2007) used pBGBM18.2 to identify a ~22-kb element on the maize B, *StarkB*, that is located within the distal heterochromatin of the long arm of the B. *StarkB* is not organized into arrays; it contains sequences unique to the B as well as repeats known from As, it is transcribed, and individual elements can be distinguished from each other by differing insertion and deletion events. Taking advantage of the fact that some LTR retrotransposons are in common between *StarkB* and the As, their divergence was calculated, and it was determined that the maize B is at least two million years old (Lamb et al. 2007). A third B-specific repeat, the ~1.5-kb CL-repeat, was identified and found to also be distributed throughout heterochromatin on the long arm of the B (Cheng and Lin 2003; Cheng and Lin 2004; Chien et al. 2015).

Some epigenetic histone modifications have been studied on the maize B. In general, methylation is associated with heterochromatin and gene silencing in eukaryotes, while acetylation is associated with euchromatin and gene transcription (Martin and Zhang 2005; Clayton et al. 2006). Perhaps not surprisingly, marks of euchromatin, such as H3K4me2, and acetylation of H3 and H4 were found to be depleted in the heterochromatic regions of the maize B (Jin et al. 2008). Epigenetic marks associated with heterochromatin, such as H3K9me2, H3K27me1, and H3K27me2, were found on the B in heterochromatic regions, although H3K27me1 is depleted as compared to As. Both A and B centromeres in maize were found to have H3K9me2, but not H3K4me2, which is in stark contrast to animals, which have H3K4me2, but not H3K9me2, in their centromeres (Jin et al. 2008).

Although centromeres are usually very difficult to sequence due to their repetitive nature, many individual sequence components are often known in well-studied species. In maize, the A centromeres are mainly composed of tandem arrays of a 156-bp satellite repeat called CentC and interspersed centromeric retrotransposon of maize (CRM) family members (Ananiev et al. 1998; Miller et al. 1998; Presting et al. 1998); B centromeres contain the same sequences, plus ZmBs (Jin et al. 2005). CentC, ZmBs, and CRM sequences are found throughout the long arm of the maize B, but the sequences do not recruit the kinetochore proteins required to form an active centromere (Lamb et al. 2005). Active centromeres are epigenetically marked by the presence of a centromere-specific histone H3 variant called CENH3 in plants, or CENP-A in animals (Henikoff et al. 2001); CENH3 is predominately associated with CRM2 in at least some maize centromeres (Wolfgruber et al. 2009). Numerous centromere misdivision derivatives of the B-A translocation stock TB-9Sb have been recovered, several of which have functional B centromeres of various sizes that were used to define the minimal centromeric components of the maize B (Carlson and Chou 1981; Kaszás and Birchler 1996; Jin et al. 2005). Fluorescence in situ hybridization of stretched chromatin fibers from canonical Bs and centromere misdivision derivatives established that the maize B centromere contains a ≥3.7 Mbp array of ZmBs, and the core of the centromere, determined by CENH3 binding, comprises ~700-kb of ZmBs, CentC, and CRM sequences (Jin et al. 2005). The analysis showed that the core of the B centromere contains five distinct regions of ZmBs interspersed by CentC and CRM elements. Using an antibody against 5-methylcytosine, the core of the B centromere was found to be hypomethylated compared to the rest of the centromere. In particular, the five ZmBs regions in the core were methylated, but the CentC and CRM elements did not have detectable cytosine methylation (Koo et al. 2011).

B chromosomes were recently discovered in a *Drosophila melanogaster* stock (Bauerly et al. 2014). The Bs seem to have arisen within the past decade in a line possessing a mutant allele of *matrimony* (*mtrm126*), although it is not believed that the mutation to *mtrm* played a role in creating the original B, and it is unclear whether there was a single origin or multiple origins for the Bs that were observed. However, Bauerly et al. (2014) argue that the missegregation of achiasmate chromosomes in *mtrm126* heterozygous females may allow for the nondisjunction of the Bs, which could explain how some individuals accumulate up to 10 Bs. Similar to other described Bs, the *D. melanogaster* Bs are morphologically distinct from As, mostly heterochromatic, and mitotically unstable. Another insect with Bs is the grasshopper *E. plorans*. However, *E. plorans* is unusual in that it is known to have over 50 cytologically distinct types of Bs (López-León et al. 1992). These Bs arise from As or sex chromosomes via centromere misdivision, pericentric inversions, centric fusions, inverse tandem fusions, or deletions (López-León et al. 1992; Camacho et al. 2000). Although *E. plorans* Bs often have differences in drive (some B varieties lack drive entirely; others are passed to 70 % of offspring) and their effect on the host (some are deleterious; others are neutral), they all contain a 180-bp satellite repeat (satDNA) and rDNA (Muñoz-Pajares et al. 2011). One of the B variants, B_2, appears to have been derived from the *E. plorans* X chromosome, as the two chromosomes contain identical arrangements of satDNA and rDNA (López-León et al. 1994). The rDNA on the Bs of *E. plorans* is transcribed in some individuals, albeit at extremely low levels compared to A-located rDNA (Ruiz-Estévez et al. 2012, 2014). Further elucidation of the origin of these Bs likely lies with next-generation sequencing.

At least 21 cichlid fish species, including several species from Lake Victoria in Africa, are known to harbor supernumerary chromosomes (Valente et al. 2014). The B from the cichlid fish *Astatotilapia latifasciata* was recently sequenced with next-generation technologies by examining individual genomes with and without Bs, along with microdissected Bs (Valente et al. 2014). It was discovered that the *A. latifasciata* B contains sequences from almost all of the normal chromosomes, and the major sequence donors were likely linkage groups 1, 3, and 9. Work comparing the *A. latifasciata* Bs to Bs from other Lake Victorian cichlids, such as *Lithochromis rubripinnis* and *Haplochromis chilotes*, suggested that a proto-B was present before the species split from each other (Yoshida et al. 2011; Valente et al. 2014). It was postulated that the *A. latifasciata* B started as a segmental duplication of a normal chromosome that included a centromere. The new chromosome may have been the target for additional A chromosome DNA fragments before it became an isochromosome that was bombarded by additional A chromosome DNA fragments and duplication events (Valente et al. 2014). The *A. latifasciata* B contains many transposable elements, but the relative amounts are similar to the A chromosomes in other cichlid species. As in rye, many gene-like loci were identified on the *A. latifasciata* B, although many are likely to be pseudogenes. Intriguingly, some of the gene-like loci that appear to be intact are involved in microtubule organization, kinetochore structure, recombination, and cell-cycle progression (Valente et al. 2014). It is tempting to postulate that these B-located genes may be essential to promoting B transmission in *A. latifasciata*, especially because B-specific RNA transcripts have been detected from similar genes in the Lake Victoria cichlid *Pundamilia nyererei*, but additional studies are needed.

Another well-studied B is that of the aster *Brachyscome dichromosomatica*. The species only has two pairs of chromosomes but can have multiple Bs of two different sizes (Smith-White 1968). The larger Bs, which can be present in up to three copies and were observed in ~12 % of plants, are mitotically stable, but the smaller Bs, termed micro Bs, are infrequently encountered (3.8 %) and unstable during mitosis and meiosis (Carter 1978). The first sequence obtained from the large *B. dichromosomatica* B, via subtractive hybridization, was a family of 176-bp repeats called Bd49 that was highly abundant near the B centromere (John et al. 1991). Bd49 is so abundant that it accounts for roughly 10 % of the ~336-Mbp large B sequence (Leach et al. 1995). The repeat family is also found on the As of *B. dichromosomatica*, as well as several sister taxa that lack Bs (John et al. 1991; Leach et al. 1995). Like many other species, rRNA gene clusters are present on the large *B. dichromosomatica* B, but they are not transcriptionally active (Donald et al. 1995, 1997). Similar to maize, the Bs of *B. dichromosomatica* are depleted in H4 acetylation as compared to the As, despite having approximately the same amount of heterochromatin as the As, suggesting they are not transcriptionally active in general (Houben et al. 1997a, b). Micro-isolation and DOP-PCR revealed the sequence of two tandem repeats, Bdm29 and Bdm54, that hybridize strongly to micro Bs during FISH. Interestingly, Bdm29 was also found on the large B and other *Brachyscome* Bs, suggesting it either evolved independently multiple times or arose early in *Brachyscome* evolution and may be important for maintenance of Bs in the genus (Houben et al. 1997a; Houben et al. 2001). Another tandemly arrayed repeat, the 92-bp Bds1, was identified on the micro B. It is not detectable on the large B, but it is found on some As (Houben et al. 2000). Due to the fact many known B sequences are present on *B. dichromosomatica* As, it might be tempting to suggest that the B arose from the A, as is strongly suspected in rye and maize. However, several lines of evidence exist that suggest the *B. dichromosomatica* B may have arisen as an alien chromosome from a sister taxon such as *B. campylocarpa* or *B. lineariloba*. The *B. dichromosomatica* large B undergoes late condensation, which is more similar to *B. campylocarpa* chromosomes than *B. dichromosomatica* As. There is extensive range overlap between *B. dichromosomatica* and *B. lineariloba*, and hybrids have been confirmed. A variant of *B. lineariloba* C possesses a chromosome that is morphologically similar to the large B of *B. dichromosomatica*. A Bd49-like repeat is found in *B. lineariloba* C, and the ITS1 and ITS2 regions of rRNA genes on the large B from *B. dichromosomatica* are nearly as similar to *B. lineariloba* C as they are to *B. dichromosomatica* A chromosome loci (Leach et al. 2004).

Although many Bs, particularly those that have been well characterized and at least partially sequenced, appear to have their origins in the As of their host genome, at least three additional cases in plants and animals provide evidence that some Bs may have alien origins. In the grasses of the genus *Coix*, Sapre and Deshpande (1987) observed one or two Bs in *C. aquatica* after crosses with the allopatric sister species *C. gigantea*. The fish *Poecilia formosa* is a female, unisexual species that requires sperm from *P. mexicana* or *P. latipinna* to initiate clonal embryogenesis after elimination of the paternal genome (Schartl et al. 1995). However, Schartl et al. (1995) discovered that some male *P. mexicana* and *P. latipinna* possess Bs that occasionally escape paternal genome elimination and pass some genetic material to the *P. formosa* offspring. Similarly, a centric fragment from the parasitic wasp *Nasonia giraulti* was identified in *N. vitripennis*, where its transmission rate increased over several generations, until it was transmitted through males at a frequency of nearly 100 % (Perfectti and Werren 2001).

2.4 What Are B Chromosomes Doing?

One of the most important questions regarding Bs is: what effects do they have on their host genomes? Historically, Bs were viewed as inert entities, but extensive study across numerous species has shown that many Bs possess transcribed gene-like sequences that could affect the host genome, and some Bs have been viewed as potential sex chromosomes (reviewed in Camacho et al. 2011). The cichlid fish *Lithochromis rubripinnis* was discovered to have Bs, but only in females (Yoshida et al. 2011). When a female with zero Bs was crossed to a male, the male/female offspring ratio was split 1:1. However, when females had one or two Bs, the offspring ratio skewed to ~75–100 % female, suggesting the presence of loci on the B that contribute to sex determination. Although the exact mechanism behind the sex-biasing properties of the *L. rubripinnis* B remains unknown, potential protein-encoding sequences were identified on the B. Yoshida et al. (2011) argued that the *L. rubripinnis* B might be a proto-sex chromosome based on similarities with the indispensible W sex chromosome of the New Zealand frog *Leiopelma hochstetteri* and the Y chromosome of some insects such as *Drosophila* and *Cacopsylla peregrine* (Green 1988; Hackstein et al. 1996; Nokkala et al. 2003).

As described in the previous section, various loci including transposons, repeats, organellar DNA, rDNA, pseudogenes, and genes that look functional are transcribed in Bs of many species. The first evidence of gene activity on a B came from transcribed rDNA loci on the B of the smooth hawk's-beard *Crepis capillaris* (Leach et al. 2005). The proto-oncogene *C-KIT* has been identified on Bs of multiple canids, including *V. vulpes*, the Japanese raccoon dog *Nyctereutes procyonoides procyonoides*, and the Chinese raccoon dog *N. p. viverrinus* (Graphodatsky et al. 2005; Yudkin et al. 2007). Early reports suggested *C-KIT* was an intact gene that might be functional due to its presence in species that diverged more than 12.5 million years ago. However, recent research has suggested that B-derived *C-KIT* is likely either not translated or not functional in the studied canids (Makunin et al. 2014). The Siberian roe deer *Capreolus pygargus* B contains at least three genes, *FPGT*, *LRRIQ3*, and *TNNI3K*, that were duplicated from the A genome (Trifonov et al. 2013).Trifonov et al. (2013) were able to confirm the existence of B-specific RNA transcripts from *FPGT*, thus demonstrating the transcription of B loci in vertebrates for the first time. However, to date, there is no definitive evidence in any species that a B-produced RNA transcript is transcribed into a functional protein.

Given that many Bs arose from the A genome, it is fair to wonder what effects partial genome duplication, and subsequent transcription, may have on the host organism when Bs are present. The presence of Bs has occasionally been shown to have an effect on transcription of loci in the A genome, especially if small RNAs generated by duplicated sequences on the B targeted similar A loci. Electrophoretic patterns of the A-located esterase E-1 changed in the presence of Bs in the lily *Scilla autumnalis* (Oliver et al. 1982). The B of the yellow-necked mouse *Apodemus flavicollis* is largely euchromatic, which has suggested that it may harbor functional genes that could affect the A genome (Tanić et al. 2005). Four A-located *A. flavicollis* genes with differential expression in the presences of Bs have been identified. Transforming growth factor β, which is involved in cell-cycle progression, shows decreased expression, while the other three genes show increased expression (Tanić et al. 2005; Adnađević et al. 2014). Of the three genes that increase their expression in the B, one was identified as encoding a hypothetical protein, one encodes a fragile histidine triad protein, and the third is a chaperonin containing TCP-1 subunit (CCT6B). Tanić et al. (2005) hypothesized that the B could be making use of CCT6B to ensure its survival because CCT6B is involved in the folding of actin and tubulin, both of which are required for chromosome movement during cell division. However, they also noted that it was unclear whether the increase in *CCTB6* transcription was due to the presence of a duplicated copy on the B or from a B-located *trans*-acting factor that changed the transcription of the A genome *CCTB6*. Much work has also been performed looking for cDNA-amplified fragment length polymorphisms in maize and rye. Several B-specific transcripts that could affect the A genome have been identified, but the studies did not closely examine that possibility (Carchilan et al. 2009; Lin et al. 2014).

In addition to simply affecting the transcription of the A genome, there appears to be a tug-of-war occurring between the A genome that seeks to remove Bs, and Bs, that attempt to persist, in some species including maize and *E. plorans*. If a host organism finds a way to entirely suppress the drive mechanism of a B that provides no benefit to the host, the B will begin a march towards extinction unless the B evolves a way to counteract host suppression. Carlson (1969) discovered an inbred maize line that lost preferential fertilization when crossed as a female by pollen containing a B-9 translocation. Later, Rosato et al. (1996) identified maize genotypes that had either high or low B transmission, and these lines were used to identify an A chromosome locus called *mBt* (male B transmission) that appears to control the preferential fertilization of the egg by B-containing pollen (González-Sánchez et al. 2003). It was found that the egg cell controls which sperm pollinates it, and sperm with Bs preferentially fertilize egg cells containing the high-transmission *mBth* allele; egg cells with the low-transmission *mBtl* allele are fertilized at random. It is believed that *mBt* plays a role in normal fertilization but that Bs hijack *mBth* to enhance their transmission; the *mBtl* allele serves to counteract B accumulation. Additionally, another A chromosome locus involved in B transmission, called *fBt* (female B transmission), was identified (González-Sánchez et al. 2003). Plants that are homozygous for *fBth* do not lose univalent Bs, thus allowing enhanced transmission. In contrast, univalent Bs tend to be lost during meiosis in the presence of the dominant *fBtl* allele, which provides a defense mechanism against B accumulation in maize.

As in maize, the A genome of *E. plorans* also appears to be in conflict with invading Bs (Herrera et al. 1996; Camacho et al. 1997; Perfectti et al. 2004). As briefly described above, *E. plorans* has multiple types of Bs, and the majority of them seem to have no net drive or drag, that is, they are randomly inherited (López-León et al. 1992). However, extensive research has suggested that the presumed ancestral B type, B_1, was neutralized by the A genome, then replaced by a new B, B_2, in a subset of the *E. plorans* population. In turn, B_2 was replaced by B_{24} after B_2 lost its drive mechanism, but the neutralization of B_{24} is ongoing, as its transmission dropped by ~33 % over 6 years in one population (Perfectti et al. 2004). Although a mechanism has not been discovered, the repeated loss and gain of drive mechanisms in *E. plorans* suggests that this type of battle between parasitic Bs and host genomes may be ongoing in many species that contain Bs that appear to lack accumulation mechanisms at present.

Most Bs are thought to be genetic parasites of their hosts, and the mechanisms of the A genome to rid themselves of Bs seems to support that notion. Not surprisingly, the presence of Bs often has a negative effect on host fitness. The presence of Bs, particularly in odd numbers, has been shown to be deleterious in many plants, including rye and maize (Jones and Rees 1982). Rye shows reduced vigor, delayed germination, reduction in mature plant size, and reduction in seed set with increasing numbers of Bs. In fact, rye with 8 Bs failed to set any seed (Müntzing 1943; Moss 1966). Bs in maize have similar phenotypic effects as rye, and some plants display longitudinal white striping on their leaves as B number increases (Randolph 1941; Kato 1970; Jones and Rees 1982; Staub 1987). Many animals also exhibit reduced vigor, delayed development, and/or reduced fertility when Bs are present (Jones and Rees 1982).

Some potentially beneficial effects have been associated with Bs in several species. One well-studied example is the B of the fungus *Nectria haematococca*. This species possesses three genes on its B that enhance its pathogenicity toward the common pea *Pisum sativum* (Miao et al. 1991; Han et al. 2001). The B genes do this by giving *N. haematococca* resistance to pisatin, an antibiotic produced by *P. sativum*. Similarly, a line of oat *Avena sativa* was discovered to have enhanced resistance to *Puccinia coronata*, a fungus that causes crown rust, and necrosis of maize caused by *Wheat streak mosaic virus* is delayed 15 % when Bs are present (Dherawattana and Sadanaga 1973; McGirr and Endrizzi 1978). As briefly described above, Bs allow the chive *A. schoenoprasum* to germinate more quickly in drought conditions, while the presence of Bs in the orchid *Listera ovata* allows it to better survive in wet habitats (Vosa 1983; Holmes and Bougourd 1989). Many have postulated that the Bs can serve as a source of additional genetic diversity, and Rhoades (1968) noted that Bs enhance recombination in maize. Recent work by Montiel et al. (2014) showed that Bs may also benefit their hosts by acting as a sponge for absorbing potentially deleterious retrotransposons. *E. plorans* R2 is a non-LTR retroelement that inserts only into 28S rDNA loci. The inactive 28S rDNA loci on *E. plorans* Bs harbor up to ten times more R2 insertions than the 28S rDNA loci in the A genome, thus removing many potential hazards to the host genome.

2.5 Applications for B Chromosomes

Because Bs are dispensable, they are amenable to being used by researchers for a variety of studies for which As are unsuitable. Maize Bs have been widely used to map loci on A chromosomes (Beckett 1994), to study centromere biology (Kaszás and Birchler 1996; Kaszás et al. 2002; Jin et al. 2005; Han et al. 2009), to study gene dosage (Birchler 1994), and to create engineered minichromosomes (Yu et al. 2007). As described above, B-A translocations arise as reciprocal translocations between Bs and As, and the B-A chromosome can undergo nondisjunction if the A-B, containing the distal tip of the B long arm, is present. This allows for the creation of sperm with varying copy numbers for specific A chromosome arms. B-A translocations exist for all maize A chromosome arms except 8S. Researchers are able to cross recessive mutants to the suite of B-A translocation stocks, and if the recessive mutant phenotype is observed in the F_1 generation, the location of the gene is identified to a specific chromosome arm. Additional B-A translocations often exist to allow the researcher to further pinpoint the location of the locus of interest (Beckett 1994).

Maize Bs have been used as platforms to create engineered minichromosomes that have the potential to allow straightforward stacking of transgenes at one locus (reviewed in Birchler 2015). Briefly, a construct containing telomere repeats can be bombarded into tissue, and a successful integration will cleave the chromosome at the insertion point. Truncated Bs are more readily recovered than truncated As because Bs are dispensable (Yu et al. 2007); cells containing truncated Bs do not suffer from aneuploidy, which can occur when As are truncated.

2.6 Conclusions

Bs were first discovered over a century ago, and in many ways, they still remain a mystery today. Next-generation sequencing technologies have begun to solve the origins of several Bs, but much work remains to be performed to establish whether or not protein-encoding genes are harbored on any Bs. As work with Bs progresses, we will undoubtedly gain new insights into genome evolution, and we will likely clarify the origins of many Bs. Continued work with B centromeres may unravel the secrets of the drive mechanisms that permit their existence in many species.

Glossary

B chromosome A special class of nonessential chromosome that is present in addition to the normal complement of chromosomes. They may vary in copy number among individuals in a population or among cells in an individual organism. They do not pair with normal chromosomes during meiosis, and they may possess drive mechanisms to enable their survival.

Drive mechanism A mechanism that allows a chromosome to be recovered at rates greater than predicted by Mendelian genetics.

Nondisjunction The failure of sister chromatids to separate properly during mitosis or meiosis. The result of nondisjunction is a pair of aneuploid daughter cells, where one cell contains two copies of a chromosome and the other cell possesses zero copies.

References

Adnađević T, Jovanović V, Blagojević J et al (2014) Possible influence of B chromosomes on genes included in immune response and parasite burden in *Apodemus flavicollis*. PLoS One 9:e112260. doi:10.1371/journal.pone.0112260

Alfenito MR, Birchler JA (1993) Molecular characterization of a maize B chromosome centric sequence. Genetics 135:589–597

Ananiev E, Phillips R, Rines H (1998) Chromosome – specific molecular organization of maize (*Zea mays* L.) centromeric regions. Proc Natl Acad Sci U S A 95:13073–13078

Banaei–Moghaddam A, Schubert V, Kumke K et al (2012) Nondisjunction in favor of a chromosome: the mechanism of rye B chromosome drive during pollen mitosis. Plant Cell 24:4124–4134

Banaei–Moghaddam A, Meier K, Karimi–Ashtiyani R, Houben A (2013) Formation and expression of pseudogenes on the B chromosome of rye. Plant Cell 25:2536–2544

Bauerly E, Hughes S, Vietti D et al (2014) Discovery of supernumerary B chromosomes in *Drosophila melanogaster*. Genetics 196:1007–1016

Beckett JB (1978) B-A translocations in maize. J Hered 69:27–36

Beckett JB (1994) Locating recessive genes to chromosome arm with B-A translocations. In: Freeling M, Walbot V (eds) The maize handbook. Springer–Verlag, New York, pp 315–327

Belyanin A, Boyeskorov G, Lyapunova E (1994) B–chromosomes in *Apodemus flavicollis* from Eastern Europe. Pol Ecol Stud 20:523–526

Beukeboom L (1994) Bewildering Bs: an impression of the 1st B–chromosome conference. Heredity 73:328–336

Beukeboom LW, Werren JH (1993) Deletion analysis of the selfish B chromosome, paternal Sex ratio (PSR), in the parasitic wasp *Nasonia vitripennis*. Genetics 133:637–648

Birchler JA (1994) Dosage analysis using B-A translocations. In: Freeling M, Walbot V (eds) The maize handbook. Springer–Verlag, New York, pp 328–329

Birchler JA (2015) Engineered minichromosomes in plants. Chromosome Res 23:77. doi:10.1007/s10577-014-9454-4

Blunden R, Wilkes TJ, Forster JW, Jimenez MM (1993) Identification of the E3900 family, a second family of rye B chromosome specific repeated sequences. Genome 36:706–711

Bougourd SM, Parker JS (1979) The B–chromosome system of *Allium schoenoprasum*. Chromosoma 75:369–383

Burt A, Trivers R (1998) Selfish DNA and breeding system in flowering plants. Proc Biol Sci 265:141–146

Camacho J, Shaw M, López-León M et al (1997) Population dynamics of a selfish B chromosome neutralized by the standard genome in the grasshopper *Eyprepocnemis plorans*. Am Nat 149:1030–1050

Camacho JPM, Sharbel TF, Beukeboom LW (2000) B–chromosome evolution. Phil Trans R Soc Lond B 355:163–178

Camacho J, Schmid M, Cabrero J (2011) B chromosomes and sex in animals. Sex Dev 5:155–166

Carchilan M, Delgado M, Ribeiro T et al (2007) Transcriptionally active heterochromatin in rye B chromosomes. Plant Cell 19:1738–1749

Carchilan M, Kumke K, Mikolajewski S, Houben A (2009) Rye B chromosomes are weakly transcribed and might alter the transcriptional activity of A chromosome sequences. Chromosoma 118:607–616

Carlson WR (1969) Factors affecting preferential fertilization in maize. Genetics 62:543–554

Carlson WR (1974) B chromosomes induce nondisjunction in 9 B^9 pollen. Maize Gen Coop Newsl 48:81

Carlson WR, Chou T–S (1981) B chromosome nondisjunction in corn: control by factors near the centromere. Genetics 97:379–389

Carlson WR, Roseman R (1992) A new property of the maize *B* chromosome. Genetics 131:211–223

Carter CR (1978) The cytology of *Brachyscome*. II. The subgenus Metabrachycome: a general survey. Aust J Bot 26:699–706

Cheng YM, Lin BY (2003) Cloning and characterization of maize B chromosome sequences derived from microdissection. Genetics 164:299–310

Cheng YMM, Lin BYY (2004) Molecular organization of large fragments in the maize B chromosome: indication of a novel repeat. Genetics 166:1947–1961

Chien YL, Lin CY, Lo KL, Cheng YM (2015) Development and mapping of CL – repeat display markers on the maize B chromosome. Cytogenet Genome Res 144:227. doi:10.1159/000370173

Clayton A, Hazzalin C, Mahadevan L (2006) Enhanced histone acetylation and transcription: a dynamic perspective. Mol Cell 23:289–296

Dhar M, Friebe B, Koul A, Gill B (2002) Origin of an apparent B chromosome by mutation, chromosome fragmentation and specific DNA sequence amplification. Chromosoma 111:332–340

Dherawattana A, Sadanaga K (1973) Cytogenetics of a crown rust – resistant hexaploid Oat with 42 + 2 fragment chromosomes. Crop Sci 13:591–594

Donald T, Leach C, Clough A, Timmis J (1995) Ribosomal RNA genes and the B chromosome of *Brachyscome dichromosomatica*. Heredity 74:556–561

Donald TM, Houben A, Leach CR, Timmis JN (1997) Ribosomal RNA genes specific to the B chromosomes in *Brachyscome dichromosomatica* are not transcribed in leaf tissue. Genome 40:674–681

Endo T, Nasuda S, Jones N et al (2008) Dissection of rye B chromosomes, and nondisjunction properties of the dissected segments in a common wheat background. Genes Genet Syst 83:23–30

Evans GM, Rees H, Snell CL, Sun S (1972) The relationship between nuclear DNA amount and the duration of the mitotic cycle. Chromosome Today 3:24–31

Feldberg E, Porto JIR, Alves–Brinn MN et al (2004) B chromosomes in Amazonian cichlid species. Cytogenet Genome Res 106:195–198

Giagia E, Soldatovic B, Savic I (1985) Karyotype study of the genus *Apodemus* (Kaup, 1829) populations from the Balkan Peninsula. Acta Vet Beograd 25:289–298

Gileva E, Chebotar N (1979) Fertile XO males and females in the varying lemming, *Dicrostonyx torquatus pall. (1779)*. Heredity 42:62–77

González–Sánchez M, González–González E, Molina F et al (2003) One gene determines maize B chromosome accumulation by preferential fertilisation; another gene(s) determines their meiotic loss. Heredity 90:122–129

Gotoh K (1924) Uber die chromosomenzahl von *Secale cereale* L. Bot Mag Tokyo 38:135–152

Graphodatsky A, Kukekova A, Yudkin D et al (2005) The proto – oncogene C – KIT maps to canid B–chromosomes. Chromosome Res 13:113–122

Green DM (1988) Cytogenetics of the endemic New Zealand frog, *Leiopelma hochstetteri*: extraordinary supernumerary chromosome variation and a unique sex–chromosome system. Chromosoma 97:55–70

Hackstein J, Hochstenbach R, Hauschteck–Jungen E, Beukeboom L (1996) Is the Y chromosome of Drosophila an evolved supernumerary chromosome? Bioessays 18:317–323

Håkansson A (1948) Behaviour of accessory rye chromosomes in the embryo–sac. Hereditas 34:35–59

Han Y, Liu X, Benny U et al (2001) Genes determining pathogenicity to pea are clustered on a supernumerary chromosome in the fungal plant pathogen *Nectria haematococca*. Plant J 25:305–314

Han F, Gao Z, Yu W, Birchler JA (2007a) Minichromosome analysis of chromosome pairing, disjunction, and sister chromatid cohesion in maize. Plant Cell 19:3853–3863

Han F, Lamb J, Yu W et al (2007b) Centromere function and nondisjunction Are independent components of the maize B chromosome accumulation mechanism. Plant Cell 19:524–533

Han F, Gao Z, Birchler JA (2009) Reactivation of an inactive centromere reveals epigenetic and structural components for centromere specification in maize. Plant Cell 21:1929–1939

Hasegawa N (1934) A cytological study on 8–chromosome rye. Cytologia 6:68–77

Henikoff S, Ahmad K, Malik H (2001) The centromere paradox: stable inheritance with rapidly evolving DNA. Science 293:1098–1102

Herrera J, López-León M, Cabrero J et al (1996) Evidence for B chromosome drive suppression in the grasshopper *Eyprepocnemis plorans*. Heredity 76:633–639

Hewitt GM (1976) Meiotic drive for B–chromosomes in the primary oocytes of *Myrmekotettix maculatus* (Orthoptera: Acrididae). Chromosoma 56:381–391

Holmes D, Bougourd S (1989) B–chromosome selection in *Allium schoenoprasum*. I. Natural populations. Heredity 63:83–87

Holmes D, Bougourd S (1991) B–chromosome selection in *Allium schoenoprasum* II. Experimental populations. Heredity 67:117–122

Houben A, Belyaev ND, Leach CR, Timmis JN (1997a) Differences of histone H4 acetylation and replication timing between A and B chromosomes of *Brachyscome dichrosomatica*. Chromosome Res 5:233–237

Houben A, Leach C, Verlin D et al (1997b) A repetitive DNA sequence common to the different B chromosomes of the genus *Brachyscome*. Chromosoma 106:513–519. doi:10.1007/PL00007689

Houben A, Wanner G, Hanson L et al (2000) Cloning and characterisation of polymorphic heterochromatic segments of *Brachyscome dichromosomatica*. Chromosoma 109:206–213

Houben A, Verlin D, Leach C, Timmis J (2001) The genomic complexity of micro B chromosomes of *Brachyscome dichromosomatica*. Chromosoma 110:451–459

Jin W, Lamb J, Vega J et al (2005) Molecular and functional dissection of the maize B chromosome centromere. Plant Cell 17:1412–1423

Jin W, Lamb J, Zhang W et al (2008) Histone modifications associated with both A and B chromosomes of maize. Chromosome Res 16:1203–1214

John UP, Leach CR, Timmis JN (1991) A sequence specific to B chromosomes of *Brachyscome dichromosomatica*. Genome 34:739–744

Jones RN (1991) B–chromosome drive. Am Nat 137:430–442

Jones RN (1995) B chromosomes in plants. New Phytol 131:411–434

Jones RN, Rees H (1982) B chromosomes. Academic, London

Jones RN, Viegas W, Houben A (2008) A century of B chromosomes in plants: so what? Ann Bot 101:767–775

Kaszás E, Birchler JA (1996) Misdivision analysis of centromere structure in maize. EMBO J 15:5246–5255

Kaszás É, Kato A, Birchler J (2002) Cytological and molecular analysis of centromere misdivision in maize. Genome 45:759–768

Kato TA (1970) Influence of B–chromosomes on 4 characters. Maize Gen Coop Newsl 44:18–21

Kawamura N (2001) Fertilization and the first cleavage mitosis in insects. Dev Growth Differ 43:343–349

Kayano H (1957) Cytogenetic studies in Lilium callosum. III. Preferential segregation of a supernumerary chromosome in EMCs. Proc Jpn Acad 33:553–558

Kayano H (1971) Accumulation of B chromosomes in the germ line of *Locusta migratoria*. Heredity 27:119–123

Kimura M, Kayano H (1961) The maintenance of supernumerary chromosomes in wild populations of *Lilium callosum* by preferential segregation. Genetics 46:1699–1712

Kirk D, Jones RN (1970) Nuclear genetic activity in B chromosome rye, in terms of quantitative interrelationships between nuclear protein, nuclear RNA and histone. Chromosoma 31:241–254

Kishikawa H, Suzuki A (1982) Cytological study on hypo – pentaploid *Triticale* with four B chromosomes of rye. Jpn J Genet 57:17–24

Klemme S, Banaei–Moghaddam A, Macas J et al (2013) High – copy sequences reveal distinct evolution of the rye B chromosome. New Phytol 199:550–558

Kolomiets OL, Borbiev TE, Safronova LD et al (1988) Synaptonemal complex analysis of B–chromosome behavior in meiotic prophase I in the East–Asiatic mouse *Apodemus peninsulae* (Muridae, Rodentia). Cytogenet Genome Res 48:183–187

Koo D–H, Han F, Birchler JA, Jiang J (2011) Distinct DNA methylation patterns associated with active and inactive centromeres of the maize B chromosome. Genome Res 21:908–914

Kour G, Kaul S, Dhar M (2013) Molecular characterization of repetitive DNA sequences from B chromosome in *Plantago lagopus* L. Cytogenet Genome Res 142:121–128

Kuwada Y (1925) On the number of chromosomes in maize. Bot Mag Tokyo 39:227–234

Lamb JC, Kato A, Birchler JA (2005) Sequences associated with A chromosome centromeres are present throughout the maize B chromosome. Chromosoma 113:337–349

Lamb J, Riddle N, Cheng Y–M et al (2007) Localization and transcription of a retrotransposon – derived element on the maize B chromosome. Chromosome Res 15:383–398

Langdon T, Seago C, Jones RN, Ougham H (2000) De novo evolution of satellite DNA on the rye B chromosome. Genetics 154:869–884

Leach CR, Donald TM, Franks TK et al (1995) Organisation and origin of a B chromosome centromeric sequence from *Brachyscome dichromosomatica*. Chromosoma 103:708–714

Leach CR, Houben A, Timmis JN (2004) The B chromosomes in brachyscome. Cytogenet Genome Res 106:199–209

Leach C, Houben A, Field B et al (2005) Molecular evidence for transcription of genes on a B chromosome in *Crepis capillaris*. Genetics 171:269–278

Levin D, Palestis B, Jones R, Trivers R (2005) Phyletic hot spots for B chromosomes in angiosperms. Evolution 59:962–969

Lima–de–Faria A (1962) Genetic interaction in rye expressed at the chromosome phenotype. Genetics 47:1455–1462

Lin B–Y (1978) Regional control of nondisjunction of the B chromosome in maize. Genetics 90:613–627

Lin B–Y (1979) Two new B–10 translocations involved in the control of nondisjunction of the B chromosome in maize. Genetics 92:931–945

Lin H–Z, Lin W–D, Lin C–Y et al (2014) Characterization of maize B–chromosome – related transcripts isolated via cDNA–AFLP. Chromosoma 123:597–607

Lindström J (1965) Transfer to wheat of accessory chromosomes from rye. Hereditas 54:149–155

Longley AE (1927) Supernumerary chromosomes in *Zea mays*. J Agric Res 35:769–784

López-León MD, Cabrero J, Camacho JM et al (1992) A widespread B chromosome polymorphism maintained without apparent drive. Evolution 46:529–539

López-León MD, Neves N, Schwarzacher T et al (1994) Possible origin of a B chromosome deduced from its DNA composition using double FISH technique. Chromosom Res 2:87–92

Makunin A, Dementyeva P, Graphodatsky A et al (2014) Genes on B chromosomes of vertebrates. Mol Cytogenet 7:99

Martin C, Zhang Y (2005) The diverse functions of histone lysine methylation. Nat Rev Mol Cell Biol 6:838–849

Martis MM, Klemme S, Banaei–Moghaddam AM et al (2012) Selfish supernumerary chromosome reveals its origin as a mosaic of host genome and organellar sequences. Proc Natl Acad Sci U S A 109:13343–13346

McGirr SC, Endrizzi JE (1978) The effects of B, K10, and AR chromosomes on the resistance of maize to viral infection. Genetics 90:331–338

Mendelson D, Zohary D (1972) Behavior and transmission of supernumerary chromosomes in *Aegilops speltoides*. Heredity 29:329–339

Miao VP, Matthews DE, VanEtten HD (1991) Identification and chromosomal locations of a family of cytochrome P–450 genes for pisatin detoxification in the fungus *Nectria haematococca*. Mol Genet Genomics 226:214–223

Miller JT, Dong F, Jackson SA et al (1998) Retrotransposon – related DNA sequences in the centromeres of grass chromosomes. Genetics 150:1615–1623

Montiel E, Cabrero J, Ruiz–Estévez M et al (2014) Preferential occupancy of R2 retroelements on the B chromosomes of the grasshopper *Eyprepocnemis plorans*. PLoS One 9:e91820. doi:10.1371/journal.pone.0091820

Moss JP (1966) The adaptive significance of B chromosomes in rye. Chromosomes Today 1:15–23

Muñoz–Pajares A, Martínez–Rodríguez L, Teruel M et al (2011) A single, recent origin of the accessory B chromosome of the grasshopper *Eyprepocnemis plorans*. Genetics 187:853–863

Müntzing A (1943) Genetical effects of duplicated fragment chromosomes in rye. Hereditas 29:91–112

Müntzing A (1948) Cytological studies of extra fragment chromosomes in rye. Hereditas 34:435–442

Müntzing A (1970) Chromosomal variation in the Lindström strain of wheat carrying accessory chromosomes of rye. Hereditas 66:279–285

Müntzing A, Akdik S (1948) The effect on cell size of accessory chromosomes in rye. Hereditas 74:41–56

Niwa K, Horiuchi G, Hirai Y (1997) Production and characterization of common wheat with B chromosomes of rye from Korea. Hereditas 126:139–146

Nokkala S, Grozeva S, Kuznetsova V, Maryanska–Nadachowska A (2003) The origin of the achiasmatic XY sex chromosome system in *Cacopsylla peregrina* (Frst.) (Psylloidea, Homoptera). Genetica 119:327–332

Nur U (1963) A mitotically unstable supernumerary chromosome with an accumulation mechanism in a grasshopper. Chromosoma 14:407–422

Nur U (1969) Mitotic instability leading to an accumulation of B chromosomes in grasshoppers. Chromosoma 27:1–19

Nur U, Werren JH, Eickbush DG, Burke WD (1988) A"selfish" B chromosome that enhances its transmission by eliminating the paternal genome. Science 240:512–514

Ohta S (1996) Mechanisms of B–chromosome accumulation in *Aegilops mutica* Boiss. Genes Genet Syst 71:23–29

Oliveira C, Foresti F, Hilsdorf A (2008) Genetics of neotropical fish: from chromosomes to populations. Fish Physiol Biochem 35:81–100

Oliver JL, Posse F, Martinez–Zapater JM et al (1982) B–chromosomes and E–1 isozyme activity in mosaic bulbs of *Scilla autumnalis* (Liliaceae). Chromosoma 85:399–403

Östergren G (1945) Parasitic nature of extra fragment chromosomes. Bot Notiser 2:157–163

Page BT, Wanous MK, Birchler JA (2001) Characterization of a maize chromosome 4 centromeric sequence: evidence for an evolutionary relationship with the B chromosome centromere. Genetics 159:291–302

Palestis B, Burt A, Jones R, Trivers R (2004a) B chromosomes are more frequent in mammals with acrocentric karyotypes: support for the theory of centromeric drive. Proc Biol Sci 271:S22–S24

Palestis BG, Trivers R, Burt A, Jones RN (2004b) The distribution of B chromosomes across species. Cytogenet Genome Res 106:151–158

Pardo MC, López-León M, Viseras E et al (1995a) Mitotic instability of B chromosomes during embryo development in *Locusta migratoria*. Heredity 74:164–169

Pardo MC, López-León MD, Cabrero J, Camacho JP (1995b) Transmission analysis of mitotically unstable B chromosomes in *Locusta migratoria*. Genome 37:1027–1034

Pardo-Manuel de Villena F, Sapienza C (2001) Female meiosis drives karyotypic evolution in mammals. Genetics 159:1179–1189

Patton JL (1977) B–chromosome systems in the pocket mouse, *Perognathus baileyi*: meiosis and C–band studies. Chromosoma 60:1–14

Pauls E, Bertollo L (1983) Evidence for a system of supernumerary chromosomes in *prochilodus scrofa* steindachner, 1881 (pisces, prochilodontidae). Caryologia 36:307–314

Peacock W, Dennis E, Rhoades M, Pryor A (1981) Highly repeated DNA sequence limited to knob heterochromatin in maize. Proc Natl Acad Sci U S A 78:4490–4494

Perfectti F, Werren J (2001) The interspecific origin of B chromosomes: experimental evidence. Evolution 55:1069–1073

Perfectti F, Corral JM, Mesa JA et al (2004) Rapid suppression of drive for a parasitic B chromosome. Cytogenet Genome Res 106:338–343

Plowman A, Bougourd S (1994) Selectively advantageous effects of B chromosomes on germination behaviour in *Allium schoenoprasum* L. Heredity 72:587–593

Presting G, Malysheva L, Fuchs J, Schubert I (1998) ATY3/GYPSY retrotransposon – like sequence localizes to the centromeric regions of cereal chromosomes. Plant J 16:721–728

Puertas M, Romera F, Peña A (1985) Comparison of B chromosome effects on *Secale cereale* and *Secale vavilovii*. Heredity 55:229–234

Puertas MJ, Ramirez A, Baeza F (1987) The transmission of B chromosomes in *Secale cereale* and *Secale vavilovii* populations. II. Dynamics of populations. Heredity 58:81–85

Radzhabli SI, Isaenko AA, Volobuev VT (1978) Investigation of the nature and role of additional chromosomes in silver fox. Genetika 24:438–443

Randolph LF (1928) Types of supernumerary chromosomes in maize. Anat Rec 41:102

Randolph L (1941) Genetic characteristics of the B chromosomes in maize. Genetics 26:608–631

Rhoades MM (1968) Studies on the cytological basis of crossing over. In: Peacock WJ, Brock RD (eds) Replication and recombination of genetic material. Australian Academy of Science, Canberra, pp 229–241

Rhoades MM, Dempsey E (1972) On the mechanism of chromatin loss induced by the B chromosome of maize. Genetics 71:73–96

Rhoades M, Dempsey E, Ghidoni A (1967) Chromosome elimination in maize induced by supernumerary B chromosomes. Proc Natl Acad Sci U S A 57:1626–1632

Roman H (1948) Directed fertilization in maize. Proc Natl Acad Sci U S A 34:36–42

Roman H (1949) Factors affecting mitotic nondisjunction in maize. Rec Genet Soc Am 18:112

Rosato M, Chiavarino AM, Naranjo CA et al (1996) Genetic control of B chromosome transmission rate in *Zea mays* ssp. *Mays* (poaceae). Am J Bot 83:1107–1112

Rothfels KH (1950) Chromosome complement, polyploidy and supernumeraries in *Neopodismopsis abdominalis* (Acrididae). J Morphol 87:287–315

Ruban A, Fuchs J, Marques A, Schubert V et al (2014) B chromosomes of *Aegilops speltoides* are enriched in organelle genome-derived sequences. PLoS One 9:e90214

Ruiz–Estévez M, López-León M, Cabrero J, Camacho J (2012) B–chromosome ribosomal DNA is functional in the grasshopper *Eyprepocnemis plorans*. PLoS One 7:e36600. doi:10.1371/journal.pone.0036600

Ruiz–Estévez M, Cabrero J, Camacho JP, López-León M (2014) B chromosomes in the grasshopper *Eyprepocnemis plorans* Are present in all body parts analyzed and show extensive variation for rDNA copy number. Cytogenet Genome Res 143:268–274

Sandery MJ, Forster JW, Blunden R, Jones RN (1990) Identification of a family of repeated sequences on the rye B chromosome. Genome 33:908–913

Sapre AB, Deshpande DS (1987) Origin of B chromosomes in *Coix* L. through spontaneous interspecific hybridization. J Hered 78:191–196

Schartl M, Nanda I, Schlupp I et al (1995) Incorporation of subgenomic amounts of DNA as compensation for mutational load in a gynogenetic fish. Nature 373:68–71

Sharma PK, Koul AK (1984) Genetic diversity among Plantagos III. Primary trisomy in Plantago lagopus L. Genetica 64:135–138

Sharma PK, Langer A, Koul AK (1985a) Genetic diversity among plantagos V. Transmission of the additional chromosome in a triplo–4 individual of *Plantago lagopus* L. Genetica 67:131–135

Sharma PK, Langer A, Koul AK (1985b) Genetic diversity among Plantagos VII. Nature of aneuploids in the progeny of aneutriploid x diploid plants of Plantago lagopus L. Genetica 67:137–144

Shirasu K, Schulman AH, Lahaye T, Schulze–Lefert P (2000) A contiguous 66–kb barley DNA sequence provides evidence for reversible genome expansion. Genome Res 10:908–915

Smith–White S (1968) *Brachyscome lineariloba*: a species for experimental cytogenetics. Chromosoma 23:359–364

Stark EA, Connerton I, Bennett ST et al (1996) Molecular analysis of the structure of the maize B–chromosome. Chromosome Res 4:15–23

Staub RW (1987) Leaf striping correlated with the presence of B chromosomes in maize. J Hered 78:71–74

Stitou S, Zurita F, Guardia R et al (2004) Transmission analysis of B chromosomes in Rattus rattus from Northern Africa. Cytogenet Genome Res 106:344–346

Swim M, Kaeding K, Ferree P (2012) Impact of a selfish B chromosome on chromatin dynamics and nuclear organization in *Nasonia*. J Cell Sci 125:5241–5249

Tanić N, Vujošević M, Dedović–Tanić N, Dimitrijević B (2005) Differential gene expression in yellow – necked mice *Apodemus flavicollis* (Rodentia, Mammalia) with and without B chromosomes. Chromosoma 113:418–427

Theuri J, Phelps–Durr T, Mathews S, Birchler JA (2005) A comparative study of retrotransposons in the centromeric regions of A and B chromosomes of maize. Cytogenet Genome Res 110:203–208

Trifonov V, Dementyeva P, Larkin D et al (2013) Transcription of a protein – coding gene on B chromosomes of the Siberian roe deer (*Capreolus pygargus*). BMC Biol 11:90

Trivers R, Burt A, Palestis BG (2004) B chromosomes and genome size in flowering plants. Genome 47:1–8

Valente G, Conte M, Fantinatti B et al (2014) Origin and evolution of B chromosomes in the cichlid fish *Astatotilapia latifasciata* based on integrated genomic analyses. Mol Biol Evol 31:2061–2072

Viseras E, Camacho J, Cano MI, Santos JL (1990) Relationship between mitotic instability and accumulation of B chromosomes in males and females of *Locusta migratoria*. Genome 33:23–29

Vosa C (1983) The ecology of B–chromosomes in *Listera Ovata* (L.) R. Br. (Orchidaceae). Caryologia 36:113–120

Vujošević M, Blagojevic J (2000) Does environment affect polymorphism of B chromosomes in the yellow – necked mouse *Apodemus flavicollis*? Mamm Biol 65:313–317

Vujošević M, Blagojević J (2004) B chromosomes in populations of mammals. Cytogenet Genome Res 106:247–256

Ward EJ (1973) Nondisjunction: localization of the controlling site in the maize B chromosome. Genetics 73:387–391

Werren J, Nur U, Eickbush D (1987) An extrachromosomal factor causing loss of paternal chromosomes. Nature 327:75–76

White MD (1973) Animal cytology and evolution, 3rd edn. Cambridge University Press, London

Wilson EB (1907) The supernumerary chromosomes of hemiptera. Science 26:870–871

Wolfgruber T, Sharma A, Schneider K et al (2009) Maize centromere structure and evolution: sequence analysis of centromeres 2 and 5 reveals dynamic loci shaped primarily by retrotransposons. PLoS Genet 5:e1000743. doi:10.1371/journal.pgen.1000743

Yoshida K, Terai Y, Mizoiri S et al (2011) B chromosomes have a functional effect on female Sex determination in lake victoria cichlid fishes. PLoS Genet 7:e1002203. doi:10.1371/journal.pgen.1002203

Yosida TH (1980) Cytogenetics of the black rat. University Park Press, Baltimore

Yu W, Han F, Gao Z et al (2007) Construction and behavior of engineered minichromosomes in maize. Proc Natl Acad Sci U S A 104:8924–8929

Yudkin DV, Trifonov VA, Kukekova AV et al (2007) Mapping of KIT adjacent sequences on canid autosomes and B chromosomes. Cytogenet Genome Res 116:100–103

Organization of Genetic Material into Chromosomes

3

Abdul Rauf Shakoori

Abstract

The genetic information of most living organisms is stored in deoxyribonucleic acid (DNA). DNA usually exists as a double helix, with two strands of opposite polarity, held together by hydrogen bonds between the complementary bases. Adenine is paired with thymine with double hydrogen bonds, whereas guanine is paired with cytosine with triple hydrogen bonds. The specific base sequence and their complementarity make it a unique feature for each organism to store and transmit genetic information. Ribonucleic acid (RNA) usually exists as a single-stranded molecule containing uracil instead of thymine. The DNA molecules present in the bacterial chromosomes are circular and contain about 50 negatively supercoiled domains. Each eukaryotic chromosome contains one giant molecule of DNA packaged into 10 nm ellipsoidal beads called nucleosomes. The 10 nm fibers are further coiled into chromatin fibers about 30 nm in diameter, which are organized into domains by scaffolds comprising nonhistone chromosomal proteins. The chromosomes are maximally condensed during metaphase of meiosis and mitosis. The centromeres, which are spindle fiber attachment regions, and telomeres, which are the ends of chromosomes, have unique structures that facilitate their respective functions.

Keywords

Chromatin • Centromere • Telomere • Nucleosome • Histones • Prokaryotic chromosome • Eukaryotic chromosome • Double helix • High-mobility-group proteins • Nonhistone proteins • Telomerase • Shelterin

A.R. Shakoori (✉)
School of Biological Sciences, University of the Punjab,
Quaid-i-Azam Campus, Lahore 54590, Pakistan
e-mail: arshaksbs@yahoo.com; arshakoori.sbs@pu.edu.pk

T.A. Bhat, A.A. Wani (eds.), *Chromosome Structure and Aberrations*,
DOI 10.1007/978-81-322-3673-3_3

3.1 Chromatin

A cell comprises a mass of protoplasm containing genetic material and is surrounded by the plasma membrane. The genetic material may be in the form of a DNA molecule floating in the protoplasm as is the case in prokaryotic cells or enclosed in a nuclear envelope formed by two membranes, as is the case in eukaryotic cells (Fig. 3.1). The nuclear envelope is punctuated with nuclear pores and is supported by network of intermediate filaments just beneath the inner nuclear membrane, known as nuclear lamina, and another less organized network of intermediate filaments on the outer side of outer nuclear membrane. The eukaryotic cells therefore have a well-defined nucleus, whereas prokaryotes do not have any well-demarcated nucleus.

In a eukaryotic cell, the genetic material acquires the shape of an intricately compact mass of chromatin fibers surrounded by nuclear envelope during interphase. During this phase of cell cycle (Fig. 3.2), the chromatin can be differentiated into Feulgen-positive condensed areas known as heterochromatin and Feulgen-negative less condensed or non-condensed areas known as euchromatin. Heterochromatin remains condensed during interphase and early prophase. Heterochromatin is usually concentrated beneath the nuclear lamina, which is adjacent to inner nuclear membrane; around the nucleolus, there it is known as nucleolus-associated chromatin. Heterochromatin is also present in the form of dark patches in the chromatin. In a chromosome, the heterochromatic region is preferentially localized in the peri-centromeric region, telomeres, or adjacent to nucleolar organizer. In some cases the entire chromosome may become heterochromatic. Heterochromatic areas are genetically inactive, because of their condensed nature. The euchromatin areas represent genetically active regions of the chromatin.

Nuclei contain a constant amount of DNA, and no genetic information is lost during the differentiation of various somatic cells. DNA in nuclei is stained using Feulgen reaction, which can be measured by using a special microscope – cytophotometer. Haploid cells have half the amount of DNA, whereas in liver some cells are polyploidy and these cells have correspondingly higher DNA content (Table 3.1).

Chromatin can be isolated biochemically by purifying nuclei and then lysing them in the hypotonic solution. Prepared this way chromatin appears viscous, gelatinous substance that contains DNA, RNA, basic proteins called histones, and non-histone proteins (Table 3.2).

When the cell at interphase stage of cell cycle enters into the mitotic phase, chromatin undergoes drastic morphological changes. The thin, elongated chromatin fibers gradually become shorter and thicker and ultimately acquire shape of a chromosome. Chromosomes are at their shortest size during metaphase stage (Fig. 3.3a, metaphase stage; Fig. 3.3b, M). Prior to metaphase, the thin, elongated, and lightly stained chromatin fiber undergoes various degrees of coiling leading to the typical metaphase chromosomes. The division of the chromosome is also very obvious, though the two daughter chromosomes are still bound to each other in the region of centromere. In fact the duplication of chromosome had already taken place during S phase of interphase, but since the chromosomes were too thin, the split was invisible. The split became visible during mitosis because of shortening and thickening of daughter chromosome (Fig. 3.3b).

A

B

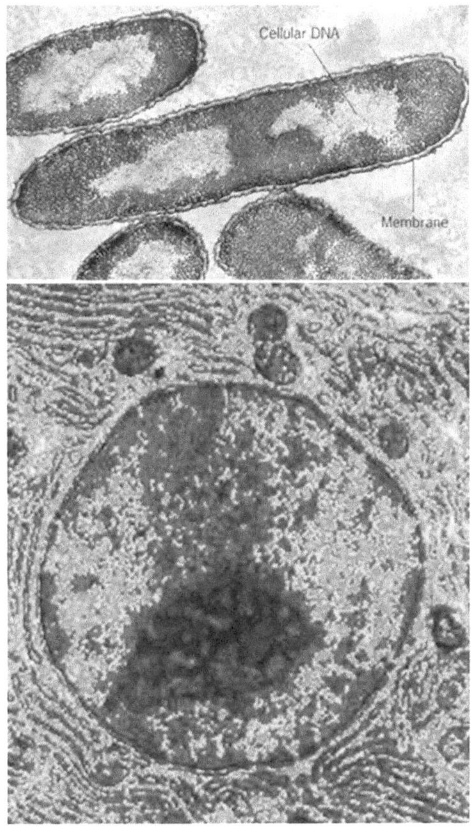

Fig. 3.1 Electron micrograph of *Escherichia coli* (**a**) and a eukaryotic hepatic cell (**b**) showing genetic material, which lies directly in the protoplasm in bacterial cell, whereas a distinct nuclear envelope surrounds the genetic material (chromatin) in the eukaryotic cell. The genetic material in the case of prokaryotic cell is in the form of an irregularly shaped region of low electron density containing naked DNA molecules. On the other hand, in eukaryotic cell in (**b**), a well-defined nucleus contains chromatin consisting of (*i*) darkly stained and (*ii*) a lightly stained, less dense region, known as euchromatin (Taken from Snustad and Simmons 2003)

During anaphase stage, when the DNA in the centromere region has replicated, the two daughter chromosomes get separated and then move toward the poles of the spindle. The chromosome at the same time starts decondensing, becoming gradually thinner and elongated until they pass on to the interphase stage where distinct chromosomal structures are not discernable.

Imagine 23 pairs of chromosomes in human cells having undergone decondensation after metaphase stage lying within the boundary of nuclear envelope, which will give an appearance of a compact and condensed apparently structureless chromatin mass during the interphase stage (Figs. 3.1 and 3.2).

3.2 Structure of a Metaphase Chromosome

Typically a metaphase chromosome (Fig. 3.4) is a thick darkly stained rodlike structure comprising chromatids and chromomeres and marked by specific constrictions, telomeres, and satellites. There are usually two constrictions, namely, primary constriction, also known as centromere, and secondary constriction.

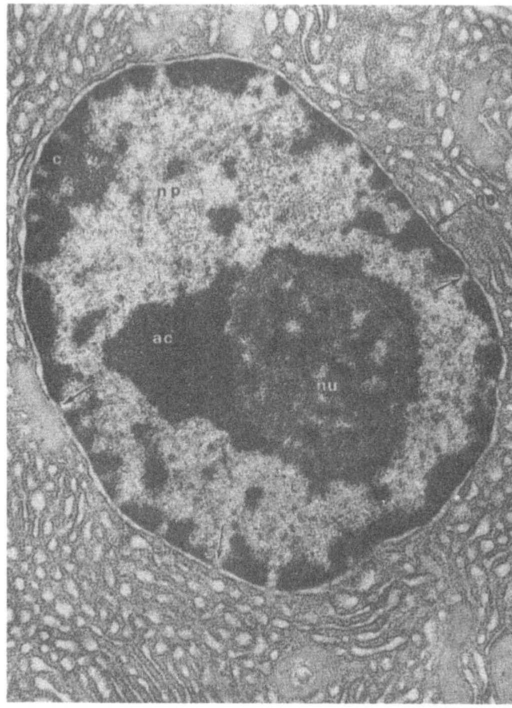

Fig. 3.2 Electron micrograph of a pancreatic cell. Staining with uranyl acetate enhances the density of DNA- and RNA-containing parts of the cell. The chromatin is mainly associated with the internal part of the nuclear membrane (*c*) except at the nuclear pores. The interchromatin channels are indicated by *arrows*; *ac* indicates chromatin associated with the nucleolus (*nu*); *np*, nucleoplasm. Note that nuclear envelope (*ne*) is double membrane with pores (Taken from De Robertis and De Robertis 2001)

Table 3.1 DNA content and chromosome components

Cells	Mean DNA Feulgen content	Presumed chromosome set
Spermatid	1.68	Haploid (n)
Liver	1.16	Diploid (2n)
Liver	6.30	Tetraploid (4n)
Liver	12.80	Octaploid (8n)

Table 3.2 Composition of chromatin

Components	Liver	Pea embryo
DNA	31 %	31 %
RNA	5 %	17.5 %
Histone	36 %	33 %
Nonhistone	28 %	18 %

3.2.1 Chromatids

At metaphase each chromosome consists of two symmetrical structures, the chromatids. The chromatids are attached to each other only by centromere, which become separated at the start of anaphase.

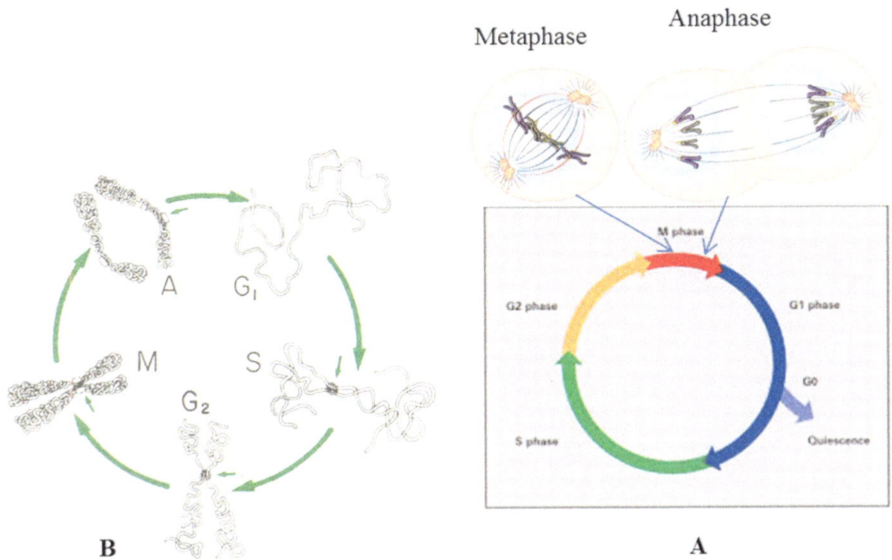

Fig. 3.3 The figure highlights condensation and decondensation cycle of chromatin fibers during different phases of cell cycle. The chromatin fibers are thin and elongated in the G1 period (**b**). They then duplicate during the S phase of cell cycle, but the two daughter molecules remain joined to each other in the region of centromere. The chromosomes are gradually thickened and shortened (see G2 and M phases in (**b**) and metaphase stage in (**a**)). The thick chromosomes, which lie at the equator of the spindle, split at the region of centromere, and the two daughter chromosomes thus formed move toward the poles of the spindle. This stage is known as anaphase (see anaphase in (**a**) and A in (**b**)). Now the mitotic cell will gradually move out of the anaphase into telophase followed by cytokinesis. After the chromosomes have reached the poles of the spindle, they start uncoiling, becoming thinner and elongated, and then ultimately are converted into chromatin that is normally seen in interphase period of the cell cycle. *G1*, chromosomes are completely dispersed; *S*, duplication occurs; *G2*, condensation starts; the condensation is maximum at metaphase (*M*) and anaphase (*A*), and the centromeres are clearly visible (Taken from De Robertis and De Robertis 2001)

Each chromatid comprises thin filament also known as chromonemata, which represent chromatid in early stages of condensation. Chromatid and chromonemata, therefore, are two names of the same structure.

3.2.2 Chromomeres

Chromomeres are bead-like structures that are sometimes visible during interphase chromosomes. Chromomeres are specially obvious in polytene chromosomes (Fig. 3.5), where they align with each other to constitute characteristic bands. In detailed analysis, the chromomeres are actually artifacts produced by close coiling of chromonemata giving an appearance of a granule.

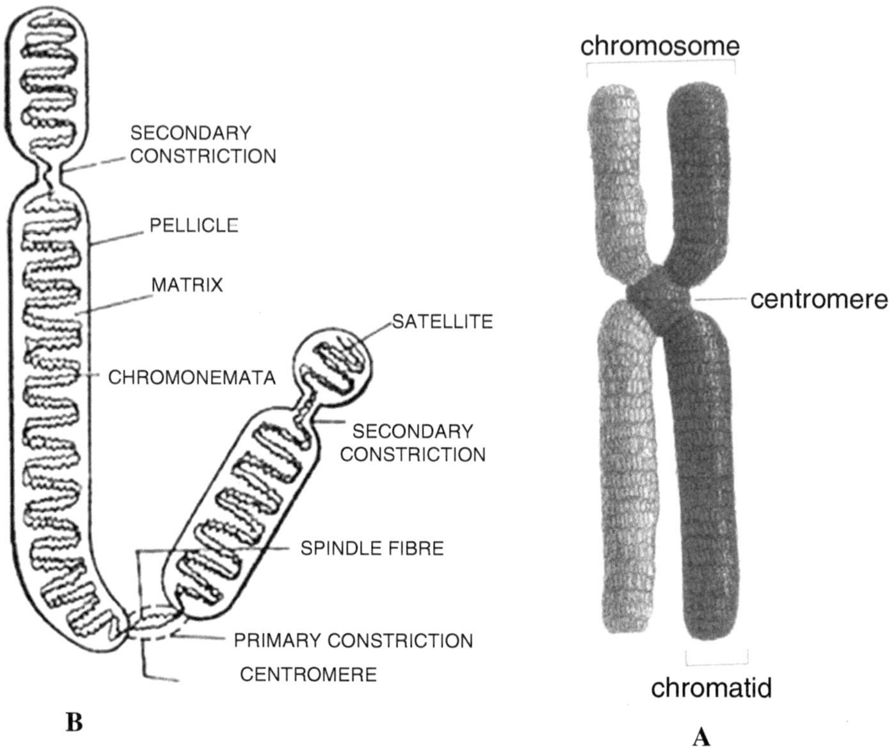

Fig. 3.4 Structure of a metaphase chromosome. In (**a**) the chromosome comprises two chromatids, which are joined to each other in the region of centromere. Each chromatid contains one of the two identical daughter DNA molecules generated earlier in the cell cycle at S phase by DNA replication. (**b**) shows typical structure of a eukaryotic chromosome showing primary and secondary constrictions, satellite, and chromonemata. The chromonemata is a 30 nm fiber, which is variously coiled and looped to give thickness (0.6–0.8 μm) to the chromosomes

3.2.3 Centromeres (CEN Region)

Centromere is also known as primary constriction or non-stainable gap of chromosome. The chromatin fiber which is densely coiled in the metaphase chromosome is not coiled in the centromere region (Fig. 3.4b). The chromatin fiber because of being uncoiled in nature in this region is not darkly stained and hence appears to be unstained region, though the fact is that it is also stained just like the rest of the chromosome, but the stain is not visible in this region, because of its extreme thinness.

The chromosomal spindle fibers (discontinuous spindle fibers) are attached to the chromosomes in the region of centromere at the metaphase plate (Fig. 3.3). The chromosomes attached to the spindle fibers are incorporated in the daughter cells and hence ensure genetic continuity. A chromosome in which centromere has been

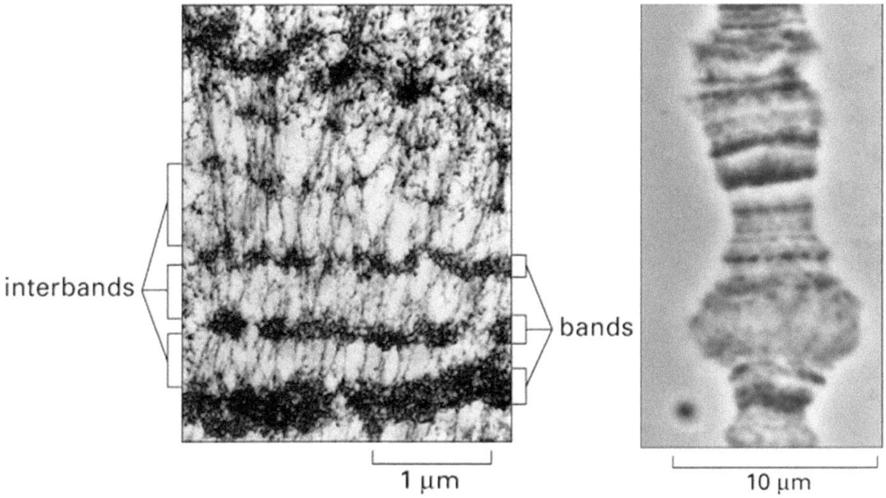

Fig. 3.5 Polytene chromosomes. Light micrograph of a portion of polytene chromosome from *Drosophila* salivary gland is shown on the *right* side of the figure. The bands are regions of increased chromatin condensation during interphase chromosomes. The figure on the *left* side is an electron micrograph of a small portion of *Drosophila* polytene chromosome. Bands of variable thickness alternate with interbands with less condensed chromatin. It is generally assumed that about 1000 chromatin fibers, produced by successive replication of DNA, but not followed by cytokinesis (a process also known as endomitosis), are lying parallel to each account for the diameter of giant chromosomes. Bands are regions where these chromatin fibers are coiled over themselves giving an appearance of a granule. When several granules, also known as chromomeres, lie close to each other, they give an appearance of a band. In interband regions, the chromatin fibers remain straight and do not coil to form chromomeres (Taken from Alberts et al. 2002)

lost will not be included in the daughter cells. The organisms with such cells will suffer from genetic syndromes.

The number and position of centromeres are highly variable. Usually a chromosome has only one centromere and hence are known a monocentric. Dicentric and tricentric chromosomes have also been reported in some animal and plant cells. Some chromosomes have diffused centromeric activity and hence are attached to the spindle fiber along their entire length. Such chromosomes are called holocentric chromosomes. In some chromosomal abnormalities, the chromosome may break, and the pieces of chromosomes may fuse with each other, producing chromosomes without centromere (acentric) or with two centromeres (dicentric).

The position of centromere is variable, and hence they are classified as telocentric, acrocentric, submetacentric, and metacentric (Fig. 3.6). The chromosomes acquire different shapes during anaphase when they are moving from equator to poles. Acrocentric and telocentric chromosomes acquire "i" shapes, submetacentric acquire "j" shapes, while metacentric chromosomes acquire "v" shapes.

Centromeres of all the chromosomes of *Saccharomyces cerevisiae* have been isolated and characterized. The CEN regions of different chromosomes are

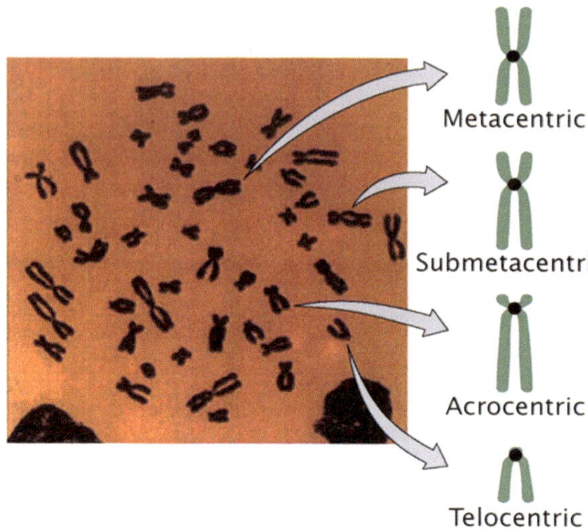

Fig. 3.6 Karyotype of a eukaryotic cell showing different types of chromosomes. The chromosomes have been classified on the basis of position of centromere on the chromosomes (Taken from Pierce (2012)

interchangeable without having any detectable effect on the host cell or its capacity to undergo normal cell division. Molecular studies have shown that *S. cerevisiae* centromere is 110–120 nucleotide pairs in length and has three essential regions. Regions I and III are short conserved boundary sequences, while region II is an AT-rich central segment about 90 nucleotide pairs long (Fig. 3.7). The length and AT-rich nature of region II are probably more important than its actual nucleotide sequence, whereas regions I and III contain specific sequences that serve as binding sites for proteins involved in spindle fiber attachment.

3.2.4 Secondary Constriction

Secondary constriction is an indentation in the chromosomes (Fig. 3.4). Usually nucleolar organizing region (NOR) is located in this constriction, though every secondary constriction does not necessarily contain NOR. NOR contains genes coding for 45S RNA, which is a precursor of 18S rRNA, 28S rRNA, and 5.8S rRNA. In human the NOR region is located in secondary constrictions of chromosomes 13, 14, 15, 21, and 22. All these chromosomes are acrocentric and have satellites.

The chromosomes which have NOR are known as nucleolar chromosomes. Usually there are two nucleolar chromosomes in a diploid cell.

3.2.5 Telomeres

The ends of eukaryotic chromosomes are known as telomeres. These are responsible for (i) preventing DNases from degrading the ends of the linear DNA molecules without loss of material, (ii) preventing fusion of ends with other DNA molecules,

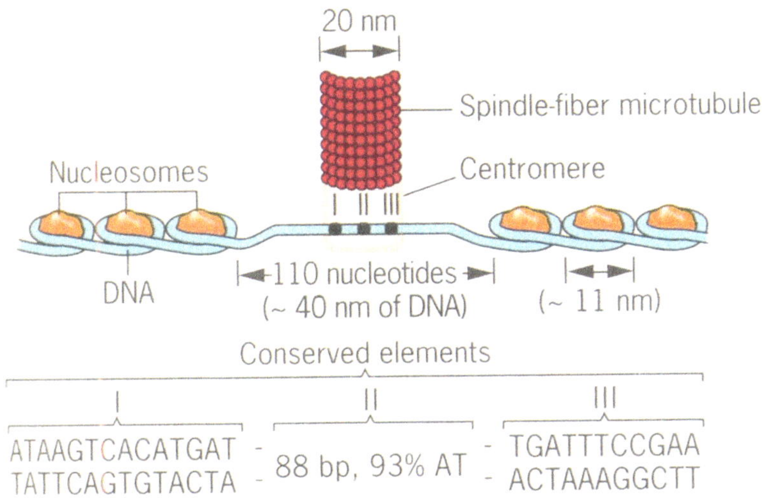

Fig. 3.7 Conserved structure of the centromere in *Saccharomyces cerevisiae* (*top*) and the sequence of the CEN region of chromosome 3 of *S. cerevisiae*

and (iii) facilitating replication of the ends of the linear DNA molecule without loss of material.

The telomeres of eukaryotic chromosomes have unique structures that include short nucleotide sequences present as tandem repeats (Fig. 3.8). Although the sequences vary somewhat in different species, the basic repeat unit has the pattern 5′-T_{1-4} A_{0-1} G_{1-8} -3′ in all but a few species. For example, the repeat sequence in humans and other vertebrates is TTAGGG, that of the protozoan *Tetrahymena thermophila* is TTGGGG, and that of the plant *Arabidopsis thaliana* is TTTAGGG. In most species additional repetitive DNA sequences are present adjacent to telomeres, which are known as telomere-associated sequences. In vertebrates the TTAGGG repeat is highly conserved and has been identified in more than 100 species. In normal human somatic cell, telomeres usually contain 500–3000 TTAGGG repeats which gradually shorten with age.

Most telomeres terminate with a C-rich single-stranded region of the DNA strand with the 3′ end – so-called 3′ overhang. These overhangs are about 12–16 bases long in *Tetrahymena* and about 125–275 bases long in human. Telomere sequences are known to be added to chromosomes by a special RNA-containing enzyme called telomerase. Figure 3.8 shows two complexes, shelterin and telomerase, which are responsible for telomere end protection and telomere elongation, respectively. Three shelterin subunits, TRF1, TRF2, and POT1, directly recognize TTAGGG repeats. They are interconnected by three additional shelterin proteins, TIN2, TPP1, and Rap1, forming a complex that allows cells to distinguish telomeres from sites of DNA damage. The telomerase complex consists of human telomerase reverse transcriptase (TERT), telomerase RNA (TERC), and dyskerin. Proteins involved in

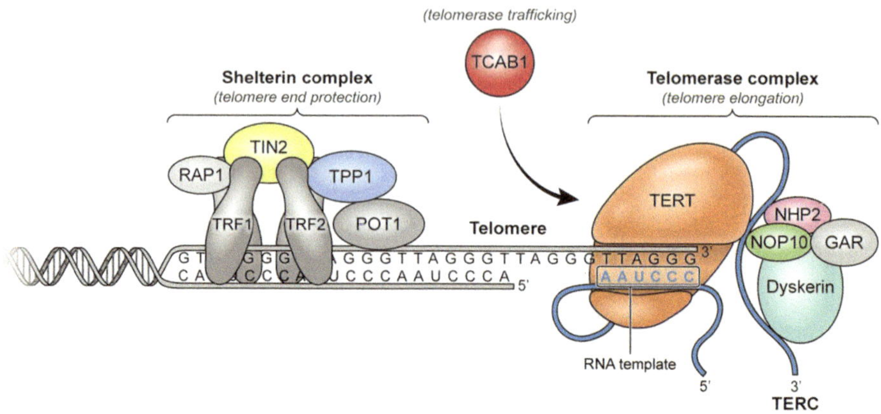

Fig. 3.8 Telomere: Shown here are the reactions involved in synthesizing the repeated G-rich sequences that form the ends of the human chromosomes (*telomeres*). The 3′ end of the parental DNA strand is extended by RNA-templated DNA synthesis. This allows the incomplete daughter DNA strand that is paired with it to be extended in its 5′ direction. This incomplete lagging strand is presumed to be completed by DNA polymerase α, which carries a DNA primase as one of its subunits. The figure also shows two complexes, shelterin and telomerase, which are responsible for telomere end protection and telomere elongation, respectively. The shelterin complex comprises TRF1, TRF2, and POT1 which directly recognize TTAGGG repeats and three additional proteins TIN2, TPP1, and Rap1. The telomerase complex consists TERT, TERC, dyskerin, NHP2, NOP10, GAR1, and TCAB1 in addition to a few other proteins (Taken from http://www.google.com.pk/url?sa=i&source=images&cd=&cad=rja&uact=8&ved=0CAcQjB0&url=http%3A%2F%2Fwww.bloodjournal.org%2Fcontent%2F124%2F18%2F2775&ei=pbM4Vc_WDIOdsAHYuoDACg&psig=AFQjCNGxmam0Uce7zEoyy3Tt-3PTyTvE7w&ust=1429865765273769)

human telomerase RNA processing and stability and in RNP trafficking and biogenesis include five box-containing RNPs, such as dyskerin, NHP2, NOP10, GAR1, and TCAB1 (telomerase Cajal body protein 1), in addition to a few other proteins.

3.3 Molecular Structure of Chromosomes

Interphase chromosomes are usually not visible with the light microscope. However, chemical analysis, electron microscopy, and X-ray diffraction studies of chromatin have provided valuable information about the structure of eukaryotic chromosomes.

The chemical analysis of isolated chromatin from interphase nuclei shows that it contains DNA, protein, and RNA (Table 3.2). The proteins are of two types: (1) basic proteins called histones and (2) a heterogeneous, largely acidic group of proteins referred to as nonhistone chromosomal protein.

Table 3.3 Some properties of histone

Histones	Molecular weight KDa	No. of amino acids	Amino acid composition	Lys + Arg (%)	Species variation	No. of molecules /200 bp of DNA
H1	22.5	244	Lysine rich	30.8	Wide	1
H2A	14.3	129	Moderately lysine rich	20.2	Fairly well conserved	2
H2B	13.8	125	Moderately lysine rich	22.4	Fairly well conserved	2
H3	15.3	135	Arginine rich	22.9	Highly conserved	2
H4	11.3	102	Arginine rich	25.5	Highly conserved	2

3.3.1 Histones

Histones play a major structural role in chromatin. They are present in the chromatin of all eukaryotes in amounts equivalent to the amount of DNA. In all animals and plants, the histones consist of five classes of proteins. These five major histone types called H_1, H_2a, H_2b, H_3, and H_4 are present in almost all cell types (Table 3.3). In some sperms, the histones are replaced by another class of small basic protein called protamines.

Four of these five proteins, viz., H_2a, H_2b, H_3, and H_4, are present in duplicate which is specifically complexed with DNA to produced ellipsoidal beads called nucleosomes. These nucleosomes comprise octamer (2 x H_2a, 2 x H_2b, 2 x H_3, and 2 x H_4) which is 11 mm in diameter and about 6 nm high with DNA wrapped two times around it (Fig. 3.9). The positively charged side groups on histone are important in their interaction with DNA, which is polyanionic because of the negatively charged phosphate group.

There are about 200 nucleotides per nucleosome, of which 146 are bound to the histone, while the remaining 64 (32 nucleotide on each side) become part of linker DNA.

3.3.2 Nonhistone Chromosomal Proteins

High-mobility-group (HMG) protein is a family of low molecular weight nonhistone proteins which are relatively present in high amount in chromatin. About one-tenth of all nucleosomes in the cells are associated with HMGs. The nonhistone chromosomal protein probably does not play a central role in the packaging of DNA into chromosome. Instead they are likely candidates for regulation of gene expression and DNA-dependent processes such as replication, DNA repair, transcription, and DNA recombination.

Table 3.4 lists some characteristics of some HMG proteins. These HMGs are known to be associated with active chromatin.

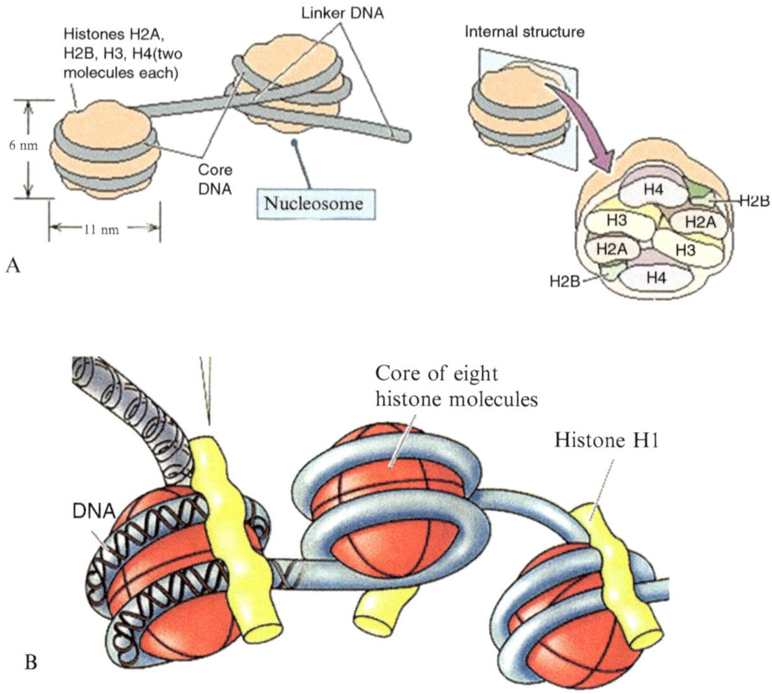

Fig. 3.9 Structure of a nucleosome. (**a**) Nucleosome which has dimensions of 11 nm × 6 nm consists of octamer core of eight histone protein molecules (two of H2A, two of H2B, two of H3, and two of H4), with 200 nucleotide pairs of DNA, of which 146 are wrapped twice around the histone core and the remaining 64 (32 nucleotide pairs on each side) are parts of the linker DNA. (**b**) This figure shows that the nucleosomal structure is stabilized by one molecule of H1 histone per nucleosome (Taken from and modified http: //www.cbs.dtu.dk/courses/genomics_course/roanoke/fig6_08.jpg; http://biology.kenyon.edu/courses/biol114/Chap01/chrom_struct.html)

Table 3.4 Characteristics of some abundantly present nonhistone proteins in calf thymus chromatin

Proteins	Number of amino acids	Chemical properties	Location in chromatin
Ubiquitin	74	Can bind covalently to a lysine side chain of histone H2A	Nucleosomal core
HMG1	270	Soluble in 2 % trichloroacetic acid	Internucleosomal linker DNA
HMG2	270	Soluble in 2 % trichloroacetic acid	Internucleosomal linker DNA
HMG14	100	Soluble in 10 % trichloroacetic acid	Nucleosomal core
HMG17	89	Soluble in 10 % trichloroacetic acid	Nucleosomal core

3.3.3 DNA Structure

3.3.3.1 Brief History

Johann Friedrich Miescher in 1868 (Fig. 3.10) described a white-colored acidic material in the pus cells of wounds of soldiers. He named this material as "nuclein." This discovery was almost forgotten when Frederick Griffith performed an interesting experiment in 1928 and postulated a transforming principle, responsible for transmission of genetic information from one generation to the other.

Frederick Griffith worked with *Streptococcus pneumoniae* (pneumococcus) that causes respiratory disease in mammals, viz., rats and man. These pneumococci exhibit genetic variably that can be recognized by different phenotypes. Pneumococci with polysaccharide capsule when grown on blood agar medium in Petri dishes form large smooth colonies, designated as type S. These encapsulated pneumococci are virulent and cause pneumonia in rats. The virulent type S pneumococci have a mutant which are without a polysaccharide capsule and are non-virulent. When grown on blood agar plate, they produce small rough-surfaced colonies. These are designated as type R. When capsule is present, it may be of several different antigenic types (such as types I, II, III, etc.) depending on the specific molecular composition, which ultimately depends on the genotype of the cell. The different types of the pneumococci can be identified immunologically. Antibodies raised against Type II cells in a rabbit will agglutinate only type II pneumococci and not type I or III.

In his experiment (Fig. 3.11), Griffith injected live type III S pneumococci in rats. The rats died. The rats did not die when he injected heat-killed type III S pneumococci. He then injected live II R cells in the rats, since these were non-virulent cells, and the rats did not die. Griffith's unexpected discovery was that the rats died when he injected heat-killed type III S and live II R pneumococci in the rats. He recovered type III S (virulent) bacteria from the carcasses. It appears some

Fig. 3.10 Johann Friedrich Miescher. J. F. Miescher, a young Swiss medical student, isolated an acidic substance "nuclein" in 1868 from the pus cells of human wound. This acidic substance "nuclein" was actually DNA, which Miescher was unaware of at that time (Taken from http://www.google.com.pk/imgres?imgurl=http://upload.wikimedia.org/wikipedia/commons/b/bc/Friedrich_Miescher.jpg&imgrefurl=http://en.wikipedia.org/wiki/Friedrich_Miescher)

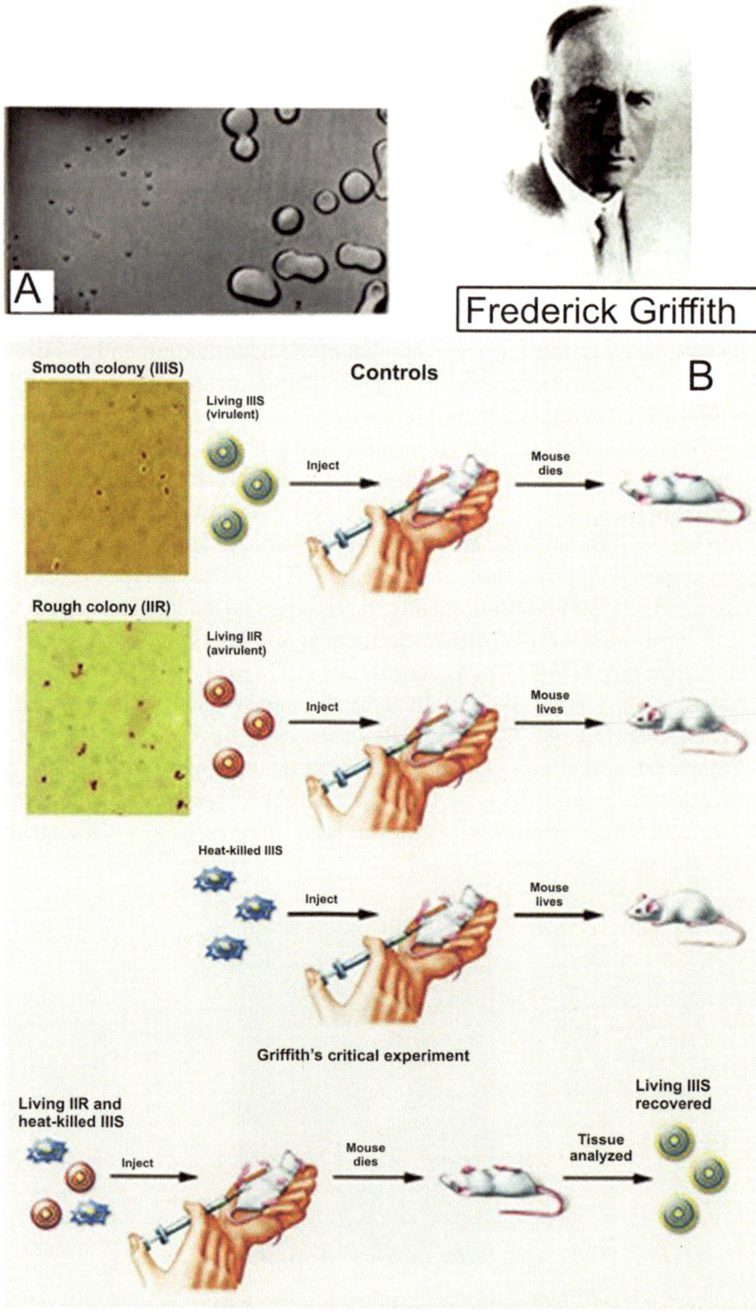

Fig. 3.11 Frederick Griffith's experiment on pneumococcus (*Streptococcus pneumonia*) proved that genetic characters are transmitted from one generation to the other through a "transforming principal." He injected live type III S bacteria in rats. The rats contracted disease and they died. The pneumococcus type III S forms smooth colonies and are virulent because of specific structure of

component of the dead type III S must have converted live type II R to type III S. Griffith called that component as the transforming principle.

Subsequent experiments showed that the phenomenon described by Griffith is now called transformation. The genetic information responsible for forming smooth type III capsule shifted from dead bacterium to the live type II bacterium and was ultimately taken by type II pneumococcus by genetic recombination. Later, as the bacterial growth progressed, the genetic information received from the dead type III S cells expressed itself in the non-viril live type II R cells thus converting them into virulent type III S cells, which killed the rats.

3.3.3.2 Evidence that DNA Is the Genetic Material

Oswald Avery, Colin MacLeod, and Maclyn McCarty in 1944 provided the first convincing evidence that DNA is the genetic material (Fig. 3.12). They showed that if highly purified DNA of pneumococcus of type III S is mixed up with type II R cells, some of the type II R cells were transformed to type III S. They also proved that if the purified DNA was treated with DNase, the transforming activity was completely eliminated.

Alfred Hershey and Marsha Chase provided conclusive evidence for DNA as the genetic material in 1952, when they showed that DNA in bacteriophage T_2 carries the genetic information. T_2 phage infects the common colon bacillus, *Escherichia coli*, and injects its DNA into the bacterial cell leaving protein coat outside. The viral DNA replicates in the bacterium and then provides information to the bacterial system for synthesis of viral proteins. The viral proteins then arrange themselves to form specific shape of viral coat and enclose DNA in its head in that process. This virus is now ready to move out of bacterium and invade other bacteria.

Hershey and Chase labeled the viral DNA with radioisotope ^{32}P, by growing the bacterium for several generations in a medium containing ^{32}P instead of normal isotope of phosphorus ^{31}P. In another set of experiment, they also labeled protein coat of the virus, by growing the bacteria for several generations in medium containing radioactive ^{35}S in place of the normal isotope ^{32}S (Fig. 3.13).

Fig. 3.11 (continued) the cell wall, which is smooth (see *big-sized* colonies of pneumococcus on the *right* side of (**a**)). On injecting live II R strain of pneumococcus, the rats did not die. Type II R is a mutant of type II S, which forms rough colonies, because of mutation in the protein components of the cell wall, and hence these bacteria give rough-surfaced colonies (see *small-sized* colonies of pneumococcus on the *left* side of (**a**)). Because of the rough surface, the bacteria have also lost their virility. In another experiment, Griffith injected heat-killed III S bacteria in the rats. The rats did not die, since the smooth surface was damaged because of heat, and bacteria have lost their virility. In the critical fourth experiment, Griffith injected live type II S and heat-killed type III S together to the rats. The rat surprisingly died. On analysis the tissues of the dead mouse were found to harbor live type III S pneumococci. Griffith postulated that something from the type III S heat-killed bacteria entered into the liver of type II R bacteria and then transformed type II R bacteria into type III S. That "something" was called "transforming principal" at that time (Taken from and modified: http://www.mun.ca/biology/scarr/Fg10_03smc.gif)

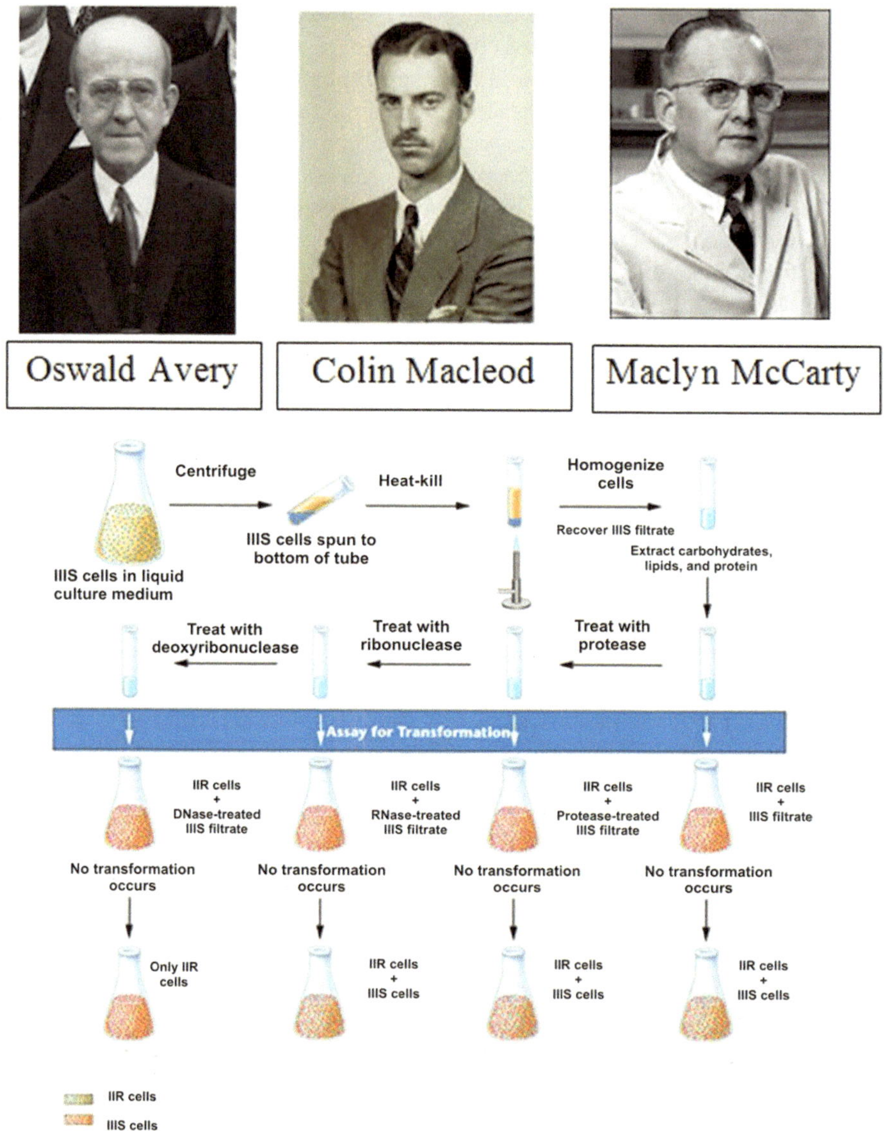

Fig. 3.12 Experimental demonstration that DNA is the genetic material. These experiments carried out in the 1940s by Oswald Avery, Colin McLeod, and Maclyn McCarty showed that adding purified DNA to a bacterium changed its properties and that this change was faithfully passed on to the subsequent generation. They extracted carbohydrates, lipids, and proteins from the homogenates of heat-killed type III S pneumococci and mixed them up with liquid culture of type II R cells. They obtained live type II S bacterial cells. They also mixed the above homogenate of heat-killed type III S pneumococci with the type II R bacterial culture after treatment with proteases, ribonuclease (RNase), and deoxyribonuclease (DNase). The first two cultures got transformed, while the third culture with DNase-treated heat-killed type III S was not transformed. This experiment showed that the DNA carried the genetic information (Taken from and modified: http://www.slideshare.net/purakichha/dna-as-genetic-material)

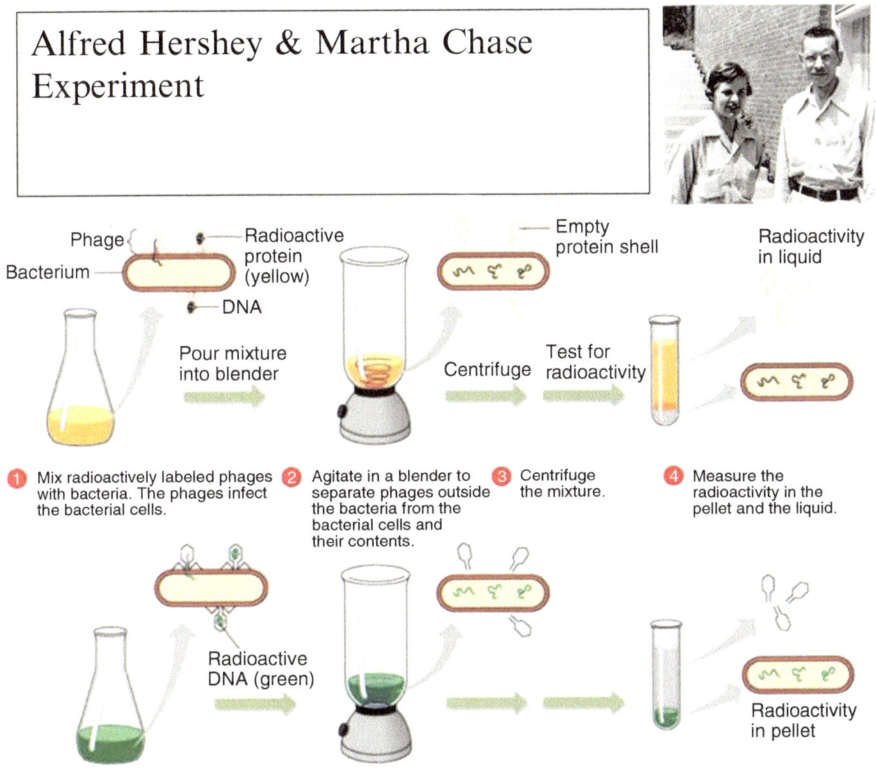

Fig. 3.13 Alfred Hershey and Martha Chase provided conclusive evidence that DNA is the genetic material. They labeled protein coat of the bacteriophage with radioisotope ^{35}S and then infected *E. coli* cells with these ^{35}S-labeled bacteriophages. They then spun the culture in a waring blender in order to remove the bacteriophages that were attached to the surface of the bacteria. The blended culture was centrifuged and the radioactivity detected both in the supernatant and the pellet (cells). The radioactivity was detected in the supernatant. No radioactivity was detectable in the cells. In the second set of experiment, Hershey and Chase labeled the DNA with ^{32}P and repeated the same procedure as was done after the ^{35}S labeling. This time no radioactivity was detected in the supernatant; instead it was detected in the cells (pellet). This experiment shows that DNA is transferred to the bacteria after infection and not the protein coat. The DNA therefore has all the genetic information for synthesis of components of new viruses (Taken from and modified: http://www.mun.ca/biology/scarr/Fg10_03smc.gif)

When T_2 phage particles labeled with ^{35}S were mixed with *E. coli* cells for a few minutes and the phage-infected cells were then subjected to shearing forces in a waring blender, most of the radioactivity (and then the proteins) could be removed from the cells without affecting progeny phage production after mild centrifugation. The radioactivity was present in the supernatant and not in the pellet. On the other hand, when T_2 particles, in which the DNA was labeled with ^{32}P, were used, essentially all the radioactivity was found inside the cells. The radioactivity was detected in the pellet and not in the supernatant.

Hershey and Chase results indicated that genetic information for synthesis of both the DNA and protein coat of the progeny virus must be present in the parental DNA.

3.3.3.3 Structure of DNA and RNA

The genetic information of all living organisms except for RNA viruses is stored in DNA. Nucleic acids are macromolecules composed of repeating subunits called nucleotides. Each nucleotide comprises (i) a five-carbon sugar or pentose, (ii) a cyclic nitrogenous base, and (iii) a phosphoric acid. In DNA the pentose sugar is 2-deoxyribose, hence the name deoxyribonucleic acid, and in RNA, the sugar is ribose, hence the name ribonucleic acid.

Four different bases commonly found in DNA are adenine (A), guanine (G), thymine (T), and cytosine (C). RNA also usually contains adenine, guanine, and cytosine, but has a base uracil (U) instead of thymine. Adenine and guanine are double-ringed bases called purines; cytosine, thymine, and uracil are single-ringed bases called pyrimidines.

In a polynucleotide chain, the successive nucleotides are linked together through phosphodiester linkages. Phosphoric acid is linked through an oxygen bridge with carbon number 3 of deoxyribose sugar in front and with carbon 5 of sugar behind (Fig. 3.14).

3.3.3.4 Double Helical Structure of DNA

One of the most exciting breakthroughs in the history of biology occurred in 1953 when James D. Watson and Francis Crick proposed double helical structure of DNA (Fig. 3.15). This structure was based on the following two major developments:

1. Erwin Chargaff and his colleagues analyzed the composition of DNA from different organisms and found that the concentration of thymine was always equal to the concentration of adenine and the concentration of cytosine was always equal to the concentration of guanine. Their data also showed that the total concentration of pyrimidine (thymine + cytosine) was also equal to the total concentration of purine (adenine + guanine). In contrast the (thymine + adenine)/(cytosine + guanine) ratio varied widely in DNAs of different species.
2. Watson and Crick used X-ray diffraction data on DNA structure provided by Maurice Wilkins, Rosalind Franklin, and their coworkers (Fig. 3.15). These data indicated that DNA was a highly ordered two-stranded structure with repeating substructures every 0.34 nm along the axis of the molecule.

On the basis of Chargaff's chemical data, Wilkin's and Franklin's X-ray diffraction data, and the inferences from model building, Watson and Crick proposed that DNA exists as a right-handed double helix in which the two polynucleotide chains are coiled around one another in a spiral (Fig. 3.16). Each polynucleotide chain consists of a sequence of nucleotides linked together by phosphodiester bonds, joining adjacent deoxyribose moieties. The two polynucleotide strands are held together in their helical configuration by hydrogen bonding between bases in opposing strands; the resulting base pairs stacked between the two chains perpendicular to the

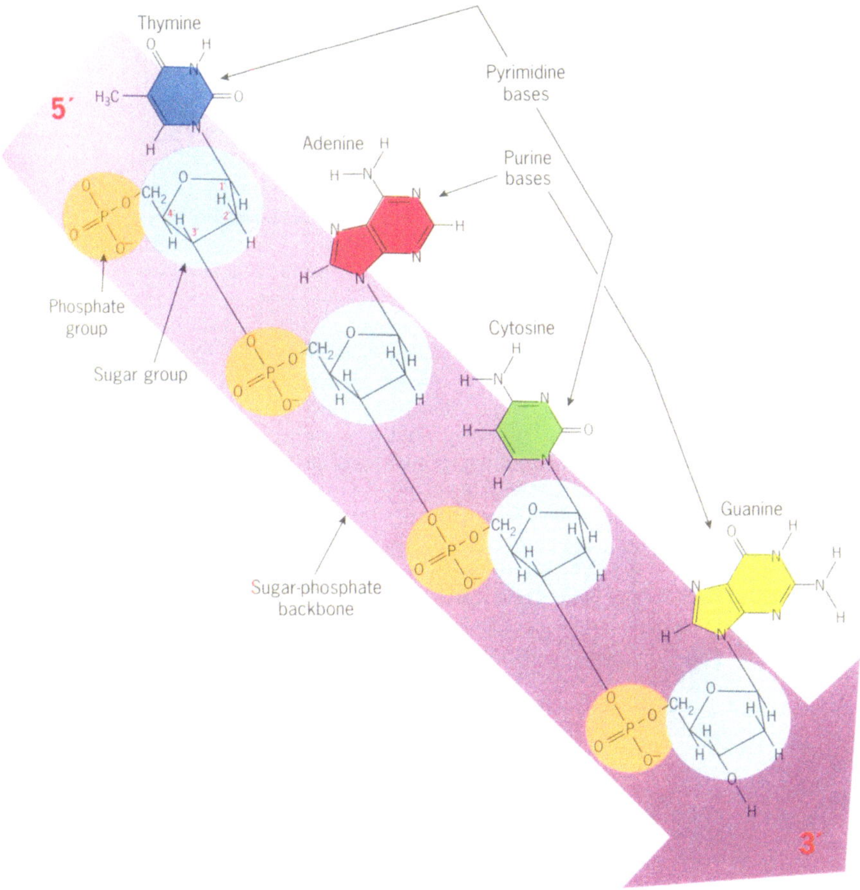

Fig. 3.14 A polynucleotide chain comprising 5′ thymidylic acid-adenylic acid-cytidylic acid-guanylic acid 3′, which can also be shown as 5′TACG 3′. In this chain the nucleotides consisting of a deoxyribose sugar, phosphoric acid, and a nitrogenous base are linked with one another through phosphodiester linkages. These linkages are present between phosphoric acid and carbon 3 of sugar in front and carbon 5 of the sugar behind. These linkages have established a polarity of the polynucleotide chain which goes from 5′ end to 3′ end. The nitrogenous bases are of two types, viz., double-ringed purines (adenine, A; guanine, G) and single-ringed pyrimidines (thymine, T; cytosine, C; uracil, U)

axis of the molecule, like the steps of a spiral stair case. The base pairing is specific: adenine is always paired with thymine, and guanine is always paired with cytosine (Fig. 3.17). Thus all base pairs consist of one purine and one pyrimidine. The specificity of base pairing results from the hydrogen binding capacities of the bases in their normal configurations. In their common structural configuration, adenine and thymine form two hydrogen bonds, and guanine and cytosine form three hydrogen bonds. Once the sequence of bases in one strand of a DNA double helix is known,

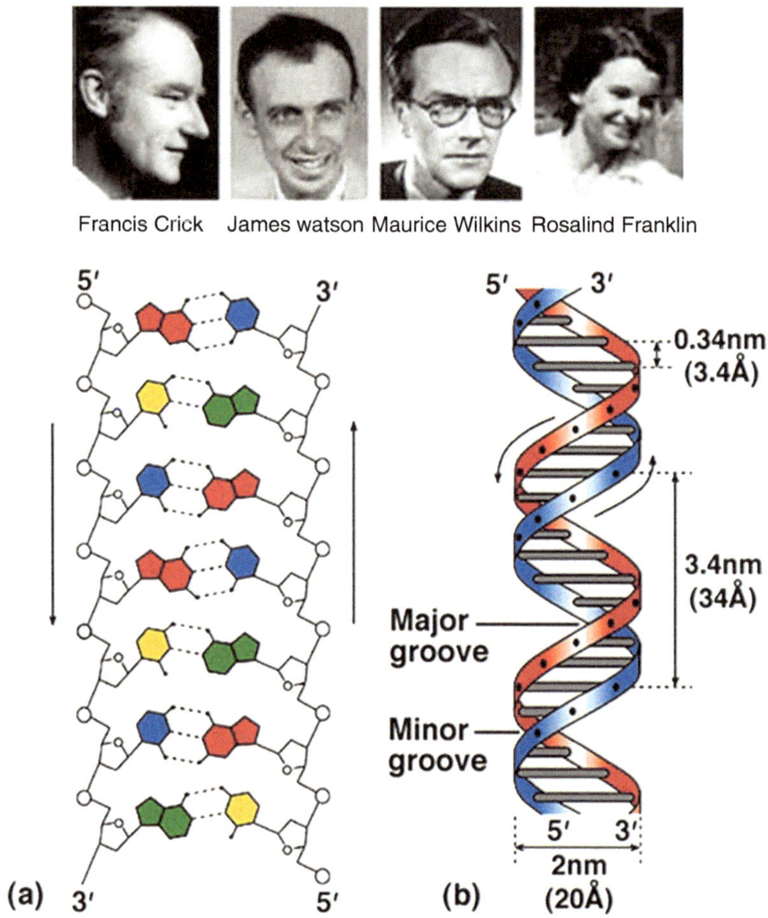

Fig. 3.15 On the basis of X-ray diffraction studies of Maurice Wilkins and Rosalind Franklin, Watson and Crick proposed the double helical structure of DNA (Taken from and modified http://pratclif.com/biologiemoleculaire/dna/helix.html)

the sequence of bases in the other strand can be easily predicted because of the specific base pairing. The two DNA strands are therefore said to be complementary to each other.

The base pairs in the DNA are stacked about 0.34 nm apart, with ten base pairs per turn of the double helix. The sugar-phosphate backbones of the two complementary strands are antiparallel. The phosphodiester bonds in one strand go from a 3′ carbon of one nucleotide to a 5′ carbon of the adjacent nucleotide, whereas those in the complementary strand go from a 5′ carbon to a 3′ carbon. The opposite polarities of the complementary strands of DNA double helix play an important role in DNA replication, transcription, and recombination.

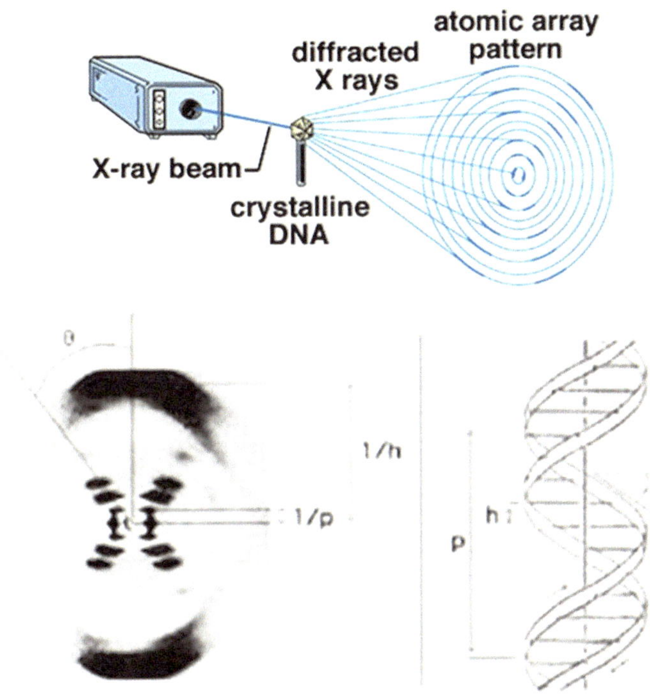

Fig. 3.16 The DNA double helix. The phosphate-ribose backbone is indicated as *ribbons* (**a** and **b**). The base pairs are flat structures stacked one on top of another perpendicular to the long axis of DNA, and therefore they are represented as horizontal lines in (**b**). Note that the two strands are antiparallel and that the molecule has a minor groove and a major groove. The double helix gives one complete turn every 10 base pairs (3.4 nm). The DNA molecule is 2 nm broad, and the distance between two consecutive nucleotides is 0.34 nm (Taken from and modified http://home.sandiego.edu/~cloer/bio482/dna_xray_newmed.gif)

3.4 Chromosome Structure in Prokaryotes

The contour length of the circular DNA molecule present in the chromosome of the bacterium *Escherichia coli* is about 1500 μm, which contains 4.7×10^6 nucleotide pairs of DNA. Since *E. coli* has a diameter of only 1-2 μm, the DNA must exist in a highly condensed (folded or coiled) centrifugation. This structure, known as the folded genome, is the functional state of a bacterial chromosome.

Within the folded genome, the large DNA molecule in an *E. coli* chromosome is organized into 50–100 domains or loops, each of which is independently negatively supercoiled (Fig. 3.18).

RNA and proteins are both components of the folded genome, which can be partially relaxed by treatment with either DNase or RNase. Treatment with DNase cleaves DNA at internal site and will relax the DNA only in the nicked domains. All unnicked loops will remain supercoiled.

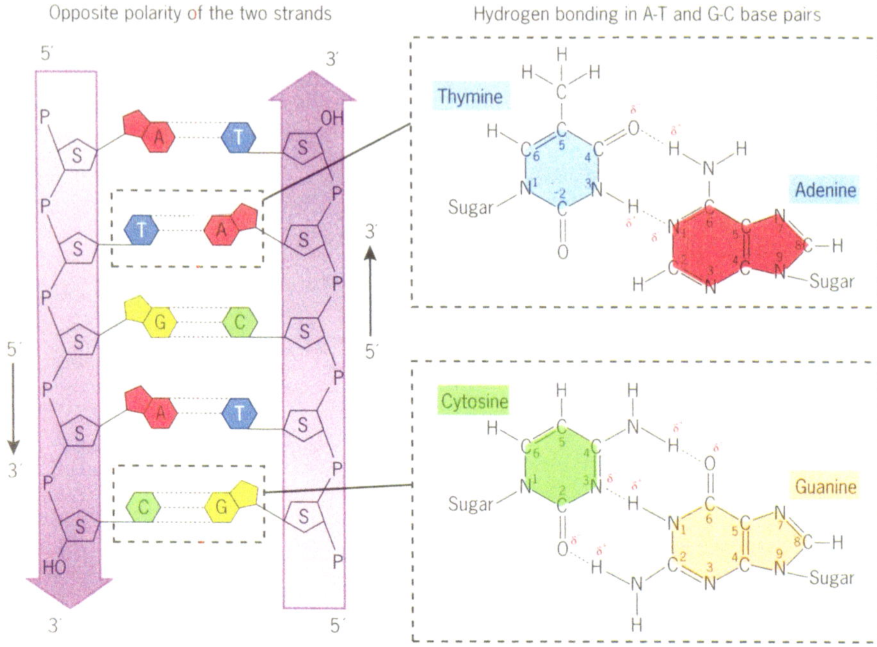

Fig. 3.17 The arrangement of bases on the two opposite strands is complementary; adenine (A) always lies in front of thymine (T) and cytosine (C) lies in front of guanine (G). In other words if the sequence of bases is known for one strand, those on the other can be easily predicted. Moreover, there are two hydrogen bonds between A and T and three hydrogen bonds between C and G

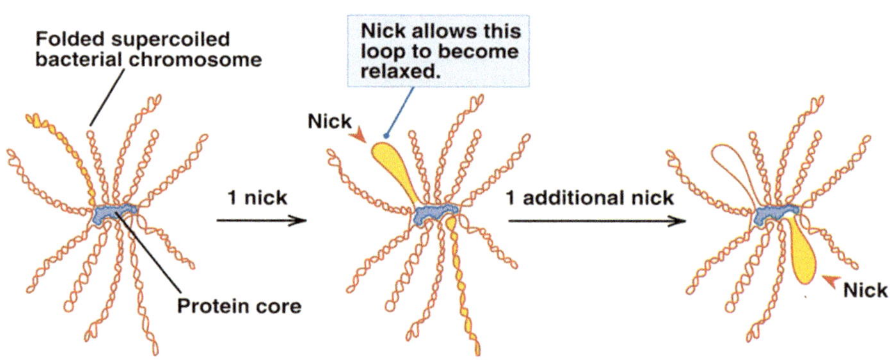

Fig. 3.18 Structure of functional state of *E. coli* chromosome. It is a supercoiled folded chromosome which have about 40–50 loops. These loops become relaxed after nicking

Destruction of DNA connectors by RNase will unfold the folded genome partially by eliminating the organization of the DNA molecule into 50–100 loops. RNase treatment however will not affect the supercoiling of the domains of the chromosomes.

3.5 Chromosomal Structure in Eukaryotes

Eukaryote chromosome is much more complex compared to the prokaryotic chromosome. In contrast to prokaryotes, eukaryotes are diploid, having two complete sets of genes, one from each parent. Although eukaryotes have only about 2–15 times as many genes as in *E. coli*, they have orders of magnitude more DNA. This DNA is packaged into several chromosomes, and each chromosome is present in two or more copies.

In contrast to 1.5 mm of contour length of chromosome in *E. coli*, the human genome contains about 1000 mm of DNA, which is about 2000 mm per diploid cells. This DNA is subdivided among 23 chromosomes of variable size and shape, with each chromosome containing 15–85 mm of DNA. When considered on the basis of number of nucleotides, the human genome contains about 3×10^9 nucleotide pairs, which is divided in 24 chromosomes (22 autosomes and 2 sex chromosomes), each containing 50×10^6–250×10^6 nucleotide pairs of DNA.

A typical eukaryotic chromosome contains 1–20 cm (10^4–2×10^5 μm) of DNA. During metaphase of mitosis and meiosis, this DNA is packaged in a chromosome with a length of only 1–10 μm. To account for the thickness of the chromosome, multineme and unineme theories have been debated. Considerable evidence however now indicates that each chromosome contains a single giant molecule of DNA that extends from one end of the chromosome through centromere all the way to the other end of chromosome.

As already mentioned on the previous pages, in the largest human chromosome, the 85 mm (85,000 μm or 8.5×10^7 nm) of DNA is packaged into a metaphase chromosome which is 0.5 μm in diameter and 10 μm in length – a condensation of about 10^4 fold in length.

After partial digestion of the DNA in chromatin with an endonuclease, DNA approximately 200 nucleotide pairs in length is associated with each nucleosome (Fig. 3.19). After extensive nuclease digestion, a 146 nucleotide-pain-long segment of DNA remains present in each nucleosome. The nuclease-resistant structure is called as nucleosome core. Its structure consists of a 146-nucleotide pair length of DNA and two molecules each of H_2A, H_2b, H_3, and H_4. The histones protect the segment of DNA in the nucleosome core from cleavage by endonucleases. X-ray diffraction studies show that DNA is wound as two turns of a superhelix around the outside of histone octamer (Fig. 3.9). The complete chromatin subunit consists of the nucleosome core, the linker DNA, and the associated nonhistone chromosomal proteins all stabilized by binding of one molecule of histone H1 to the outside of the structure (Fig. 3.9).

The 10 nm fiber is formed by direct juxtaposition with each other without detectable linker region. If this 10 nm fiber, in turn, is wound in a solenoidal coiling, a 30 nm fiber can be generated. How are these 30 nm fibers further condensed to form a metaphase chromosome? Unfortunately there is no clear cut answer to this question. Figure 3.20 shows degrees of coiling involved in the formation of 10 nm fiber (11 nm in the figure), which after solenoidal coiling is converted into 30 nm fiber. Further coiling of 30 nm fiber into 300 nm and then condensation into 700 nm thick

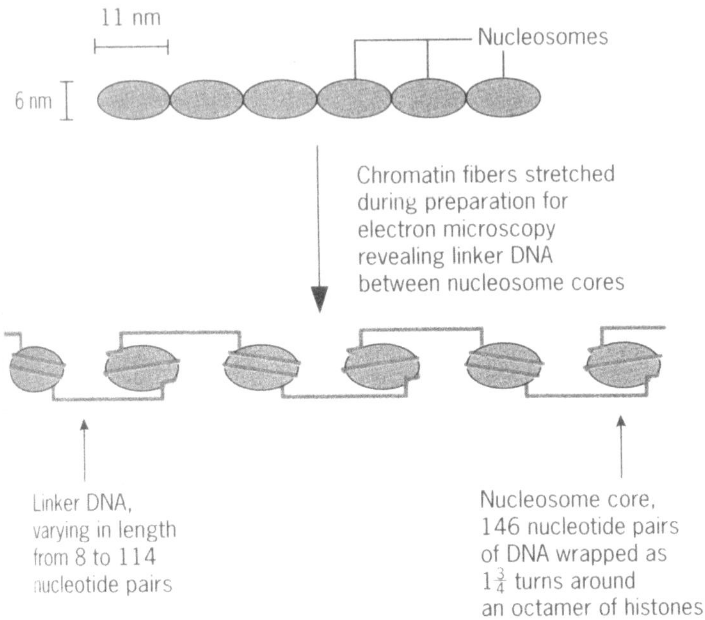

Fig. 3.19 Nucleosome structure of chromosome

structure corresponds to the diameter of the chromatid of the metaphase chromosome.

3.5.1 Structure of Metaphase Chromosome After Histone Depletion

Electron micrographs of isolated metaphase chromosome from which the histones have been removed reveal a scaffold which is surrounded by a huge pool or halo of DNA (Fig. 3.21). The scaffold must be composed of nonhistone chromosomal proteins, which retains the general shape of the metaphase chromosome.

In the histone-depleted chromosome, the DNA, which was packed 40-fold in the 30 nm chromatin fiber, becomes extended and produces loops with an average length of 25 μm (75,000 bp). In each loop the DNA exits from the scaffold and returns to an adjacent point. On the basis of these observations, a model of chromosome structure has been proposed in which the DNA is arranged in loop anchored to the nonhistone scaffold (Fig. 3.21). Because the lateral loop has 25 μm of DNA, after contracting 40-fold in 30 nm fiber, they would be only about 0.6 μm long, a length consistent with the diameter of metaphase chromosome.

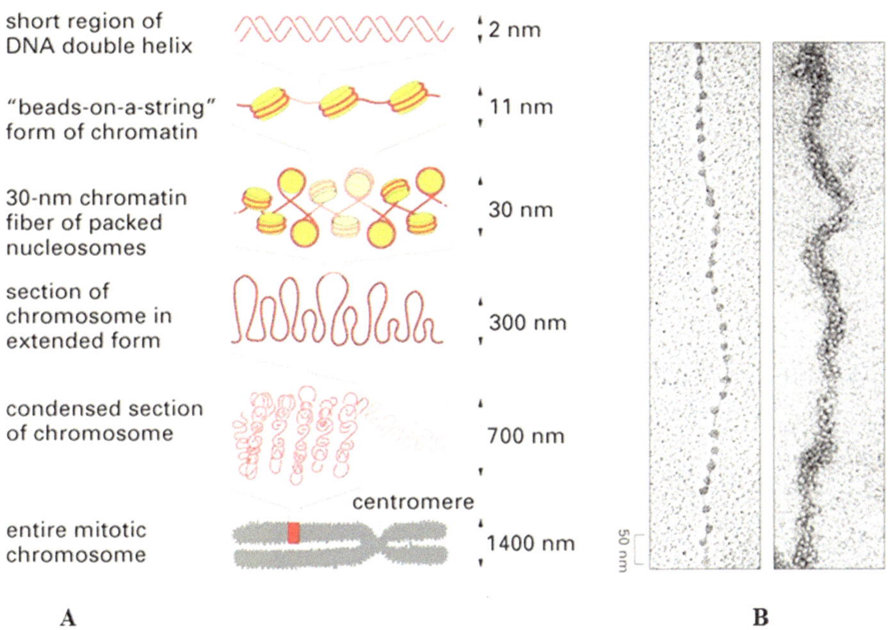

Fig. 3.20 (**a**) The packaging of chromatin fiber. The model shows many levels of chromatin packing to give rise to the highly condensed metaphase chromosome. The 11 nm fiber (also known as 10 nm fiber) is converted into 30 nm fiber after solenoidal coiling. The 30 nm fiber is then coiled and condensed into 300 nm fiber which is further condensed because of further coiling into 700 nm fiber which corresponds to the thickness of a chromatid. Figure (**b**) shows electron microscopic structure of nucleosome and 30 nm fiber. The chromatin isolated directly from an interphase nucleus shows 30 nm thick thread (*right*). This length of chromatin has been experimentally unpacked or decondensed after isolation to show the nucleosomes (*left*) (Taken from Alberts et al. 2002, http://oregonstate.edu/instruction/bi314/fall11/figure_05_21.jpg)

3.5.2 Laemmli's Radial Loop Model of Chromosome Without Histone Depletion

Paulsen and Laemmli described the radial loop model in 1979, in which they proposed that the nucleoprotein fiber is folded into loops which are arranged in the chromatid in radial fashion, in such a way that their bases become the central axis of the chromatid.

Paulsen and Laemmli based their model on the following observations:

(i) Electron micrographs obtained from chromosomes swollen by chelation of the divalent cation showed that the thick nucleoprotein fiber (200–300 Å) relaxed to the thin fiber (100 Å), which is probably a linear array of nucleosomes. Cross sections show a central area from which the fibers emerge in a radial fashion, often forming loops which were 3–4 μm long. Figure 3.22 shows central dense core from which loop of 10 nm fibers emerge. (ii) Chromosomes fixed in the presence of 1 mM $MgCl_2$ were more compact, showing thick fibers (200–300 Å) with an average

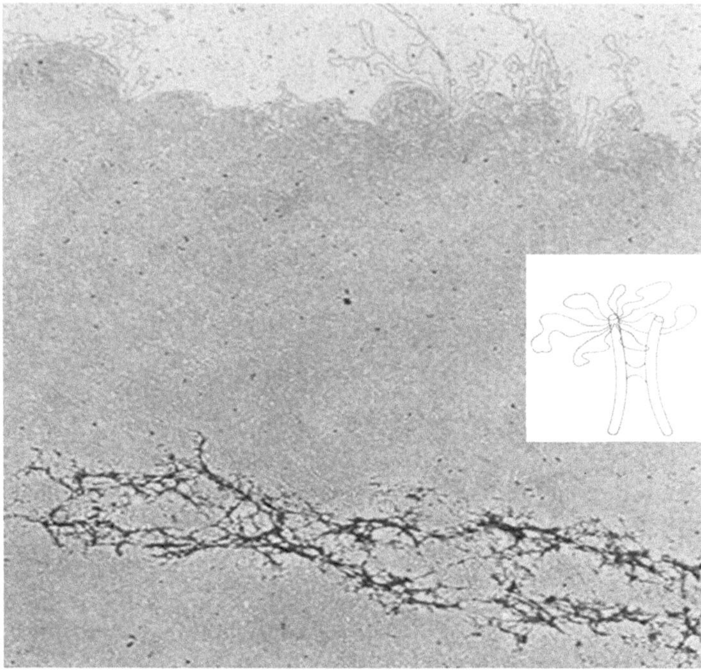

Fig. 3.21 Electron micrograph of a histone-depleted human chromosome. The nonhistone proteins form two scaffolds, one per chromatid, which are joined at the centromere. The scaffold maintains the shape of an intact chromosome, while the naked DNA fibers form a halo around it. The inset shows an interpretation of the model, which proposes that chromosomes are organized in loop of DNA that emerge from the nonhistone protein scaffold (Taken from and modified: http://mol-biol4masters.masters.grkraj.org/html/Gene_Expression_II13-Chromosomal_Nature_Before_During_and_After_Gene_Activation.htm

chromatid diameter of about 1 μm. Radial loops of about 0.6 μm could be observed. (iii) Chromosomes isolated with the help of hexylene glycol were extremely compact (500 Å thick fiber with diameter about 0.6 μm) with projections which appeared to be loops of the thick fiber (200–300 Å), possibly shortened by twisting into a short supercoil. The chromatin loops observed in the intact chromosomes were thought to be structurally related to the DNA loops observed previously in the histone-depleted chromosomes.

When fully condensed chromosomes are observed under scanning electron microscope (SEM), compact chromatin projections can be seen protruding radially on the chromatid surface (Fig. 3.23). Each projection is thought to be a loop of the 30 nm thick chromatin fiber.

Acknowledgments The assistance of my wife Dr. Farah Rauf Shakoori and Mr. Qadeer Ahmad in preparing the figures for this chapter is gratefully acknowledged.

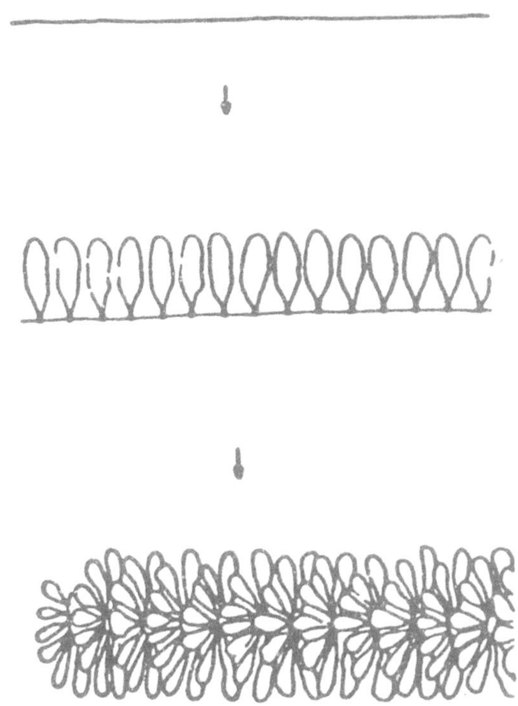

Fig. 3.22 This shows Laemmli's radial loop model of chromosome structure. In this model the nucleoprotein fiber is in the 100 Angstrom configuration instead of thick nucleoprotein fiber (200–300 Å) which is probably a linear array of nucleosomes. The model proposes that the nucleoprotein fiber is folded into loops which are arranged in the chromatid in radial fashion, in such a way that their bases become the central axis of the chromatid (Permission required from McGraw Hill)

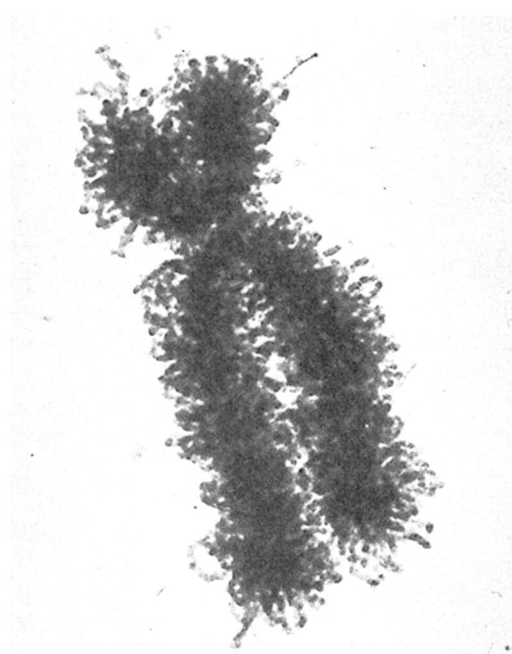

Fig. 3.23 A low-power electron micrograph of a whole mount of human chromosome 12, showing that the two chromatids are composed of 30 nm thick chromatin fiber. The two chromatids are joined at the centromere (Taken from: http://sciencelearn.org.nz/Contexts/You-Me-and-UV/Sci-Media/Images/Electron-micrograph-of-a-human-chromosome)

Glossary

A

Acentric Chromosome segment lacking a centromere.

Acrocentric Chromosome or chromatid that has its centromere near the end.

Anaphase A stage in cell division during which chromosomes are moved toward opposite ends of the spindle.

Autosome One of any pair of the chromosomes that do not carry the genes determining the sex.

B

Bacteriophage A virus that attacks bacteria.

C

Cell cycle The sequence of stages that a dividing eukaryotic cell goes through from the time it is created to the time it undergoes mitosis. It includes M, G1, S, and G2 phases.

Centromere The region joining the two sister chromatids during cell division.

Chromatid One of the daughter strands of a chromosome that has recently been copied during the cell division and which is still connected to the other daughter strand. Upon separation from each other, the chromatids become chromosomes.

Chromatin The entire complex of protein and DNA that makes up a eukaryotic chromosome.

Chromatin fibers A basic organizational unit of eukaryotic chromosomes that consists of DNA and associated proteins assembled into strands of average diameter 30 nm.

Chromomeres Small bodies that are identified by their characteristic size and linear arrangement along a chromosome.

Chromonema An optically single thread forming an axial structure within each chromosome.

Chromosome A single long molecule of DNA and any associated protein. Each chromosome carries a linear array of genes.

D

Dalton A unit of molecular mass approximately equal to that of a hydrogen atom or 1.66×10^{-24} g. Alternatively, there is approximately 1 mole of daltons per gram.

Deoxyribonuclease Any enzyme that hydrolyzes DNA.

Dicentric chromosome One chromosome having two centromeres.

Diploid Containing two sets of homologous chromosomes and thus two copies of each gene.

E

Euchromatin Genetic material that is not stained so intensely by certain dyes during interphase and that comprises many different kinds of genes.

Eukaryote A member of the large group of organisms that have nuclei enclosed by a membrane within their cells.

F

Folded genome The condensed intracellular state of the DNA in the nucleoid of bacterium. The DNA is segregated into domains, and each domain is independently negatively supercoiled.

G

G1 phase The phase of a cell cycle that constitutes the first part of the interphase before DNA synthesis.

G2 phase The phase of the cell cycle between synthesis of DNA (S phase) and mitosis (M phase), the last part of interphase.

Genome All of the hereditary genetic information in an organism. A complete set of chromosomes (n) inherited as a unit from one parent.

H

Haploid Containing a single set of chromosomes. Mature eggs and sperm and bacteria are examples of haploid cells.

Helix Any structure with a spiral shape. The Watson and Crick Model is in the form of double helix.

Histones Group of proteins rich in the basic amino acids. They take part in the formation of nucleosomes.

Hydrogen bond A weak electrical attraction between two molecules, due to the partially positive hydrogen atoms of one molecule attracting particularly negative atoms of the other molecule. This is a bond between hydrogen and an electronegative element, and the above hydrogen is attached to another electronegative element.

I

Interphase The part of the cell cycle during which no cell division occurs. It includes G1 phase, the S phase, and G2 phase.

K

Karyotype A systematized arrangement of the chromosomes from a cell.

Kinetochore The attachment that forms during the cell division between the microtubules of a spindle fiber and a chromosome. It contains motor proteins that move the chromosomes along the microtubules.

M

M phase The phase of the cell cycle during which mitosis occurs. It includes karyokinesis and cytokinesis.

Metacentric A chromosome structure in which the centromere is roughly equidistant between the two ends.

Metaphase A stage in cell division during which chromosomes line up in the equator of the spindle.

Metaphase plate The imaginary plane along which the chromosomes line up during metaphase of cell division.

Mitosis Nuclear division in eukaryotes producing two daughter nuclei that are genetically identical to the parent.

Mitotic spindle An array of microtubules formed during cell division that moves chromosomes to opposite sides of the cell.

N

Nonhistone chromosomal proteins All of the proteins in chromosomes except the histones.

Nuclear envelope The double-layered membrane enclosing the nucleus of eukaryotic cell.

Nuclear pore An opening in the nuclear envelope that connects the inside of the nucleus with the cytoplasm and through which molecules can pass.

Nucleolus A structure in the eukaryotic nucleus where ribosomal RNA processing occurs and ribosomal subunits are assembled.

Nucleosome A repeating, bead-like structure of eukaryotic chromosome, consisting of about 200 nucleotides of DNA wrapped twice around eight molecules of histone proteins.

Nucleotide A monomer that can be polymerized to form the nucleic acids DNA or RNA. One nucleotide consists of a five-carbon sugar (ribose or deoxyribose), a phosphate group, and one of the nitrogenous bases.

P

Phosphodiester bond The bond formed in the polymerization reaction between the phosphate bond of one nucleotide and the hydroxyl group on the sugar component of another nucleotide.
Polynucleotide A linear sequence of joined nucleotides in DNA or RNA.
Polytene chromosomes Giant chromosomes produced by interphase replication without division and consisting of many identical chromatids arranged side by side in a cable-like pattern.
Prokaryotes A member of a large group of organisms that lack true nuclei in their cells including bacteria and blue-green algae.
Prophase The first stage of cell division during which chromosomes become visible, the mitotic spindle forms, and the nuclear membrane breaks down.
Protamines Small basic proteins that replace the histones in the chromosomes of some sperm cells.
Purine A double-ringed nitrogen-containing base present in nucleic acids; adenine and guanine are the two purines present in most DNA and RNA molecules.
Pyrimidines A single-ringed nitrogen-containing base present in nucleic acids; cytosine and thymine are commonly present in the DNA, whereas uracil replaces thymine in RNA.

R

Radioactive isotope An isotope that spontaneously decays to form a different isotope or element, by emitting radiation or a particle.
Recombination Any change in the combination of genes or alleles found on a chromosome or in a given individual. The production of gene combinations not found in the parents by the assortment of nonhomologous chromosomes and crossing over between homologous chromosomes during meiosis.
Ribosomal RNAs The RNA molecules that are structural components of ribosomes.

S

S phase The phase of the cell cycle during which DNA is synthesized and the chromosomes are duplicated.
Sex chromosomes One of the pair of chromosomes that carries the gene(s) that determine gender.

Sister chromatids The paired strands of a recently replicated chromosome that has not yet divided.

T

Telomerase An enzyme that adds telomere sequences to the ends of eukaryotic chromosomes.

Telomere The unique structure found at the end of eukaryotic chromosomes.

Telophase The last stage in cell division in which chromosomes are assembled at the poles of the mitotic spindle.

Transformation Genetic alteration of an organism brought about by the incorporation of foreign DNA into cells.

V

Virulence The ability to cause disease and death.

References and Additional Reading

Adolph KW (ed) (1988) Chromosomes and chromatin, vol 1–3. CRC Press, Boca Raton

Alberts B, Johnson A, Lewis J, Raff M, Roberts K, Walter P (2002) Molecular biology of the cells, 4th edn. Garland, New York

Avery OT, MacLeod CM, McCarty M (1944) Studies on the chemical nature of the substance inducing transformation in pneumococcal types. J Expt Med 79:137–158

Blackburn EH, Szostak JW (1984) The molecular structure of centromeres and telomeres. Annu Rev Biochem 53:163–194

Blakburn EH (1991) Telomeres. Trends Biochem Sci 16:378–381

Chargaff E, Davidson JN (eds) (1955) The nucleic acids. Chemistry and biology, vol 1. Academic, New York, p 576

De Robertis EDP, De Robertis EMF (2001) Cell & molecular biology. Lippincott Williams & Wilkins, Philadelphia

De Lange T (2005) Shelterin: the protein complex that shapes and safeguards human telomeres. Genes Develop 19:2100–2110

Felsenfeld G (1985) DNA. Sci Am 253(4):58–67

Georgiev GP, Nedospasov SA, Bakayev VV (1978) Supranucleosomal levels of chromatin organization. In: Busch H (ed) The cell nucleus, vol 6. Academic, New York, pp 3–34

Griffith F (1928) The significance of pneumococcal types. J Hyg 27:113–159

Gross DS, Garrard WT (1988) Nuclease hypersensitive sites in chromatin. Annu Rev Biochem 57:159–198

Hershey AD, Chase M (1952) Independent functions of viral protein and nucleic acids in growth of bacteriophage. J Gen Physiol 36:39–56

Isenberg I (1979) Histones. Annu Rev Biochem 48:159–191

Kornberg RD, Klug A (1981) The nucleosome. Sci Am 244(2):52–64

Lodish H, Berk A, Zipursky SL, Matsudaira PO, Baltimore D, Darnell J (2000) Molecular cell biology, 4th edn. W.H. Freeman, New York

Mancini J, Autexier C (2012) The telomerase complex: an overview. In: Lue N, Autexier C (eds) Telomerases: chemistry, biology and clinical applications. Wiley, New York, pp 1–21

Marsden M, Laemmli UK (1979) Metaphase chromosome structure: evidence for a radial loop model. Cell 17:849–858

McGhee JD, Felsenfeld G (1980) Nucleosome structure. Annu Rev Biochem 49:1115–1156

Pierce BA (2012) Genetics: a conceptual approach, 4th edn. W.H. Freeman, New York

Snustad DP, Simmons MJ (2003) The principals of genetics. John Wiley & Sons, New York

Watson JD (1980) (orig. 1968). The double helix: a personal account of the discovery of the structure of DNA (critical edition). Norton

Watson JD, Crick FHC (1953) A structure for deoxyribonucleic acid. Nature 171:737–738

Watson JD, Baker TA, Bell SB, Gann A, Levine M, Losick R (2004) Molecular biology of the gene, 5th edn. Benjamin Cumming, Menlo Park

Zhimulev IF, Belyaeva ES (1991) Chromomeric organization of polytene chromosomes. Genetica 85:65–72

4 Laboratory Techniques of Studying Plant Chromosomes

Mahpara Kashtwari, Showkat A. Zargar, and Aijaz Ahmad Wani

Abstract

Study of chromosomes helps in revealing the basic number, structure, and behavior of chromosome complement of a particular species which can then be compared with that of another species. The properties of chromosomes remain constant within a particular species and change only upon hybridization and/or evolution. As such, by studying the chromosomes, the parental lineage or evolutionary history can be elucidated. Also, we can find a relation between the chromosome complement and the phenotype of the species. On the basis of the properties of the chromosome complement, various taxonomic rearrangements have been made. Several of the plant species are polyploid, having a multiple set of basic chromosomes. The sterility and fertility of a species can be ascertained to the odd or even number of chromosome complement, respectively. The biggest challenge in chromosomal study is to obtain deeply stained, condensed chromosomes with well-defined primary and secondary constrictions. For this, every step is optimized according to the type of tissue and the type of species being studied. For this a variety of metallic and nonmetallic chemicals have been used in combination for fixation with varied time durations accompanied with different types of stains. Sometimes the use of mordants becomes necessary in order to enhance the staining properties. The traditional method of chromosome study by sectioning has largely become outdated and replaced by fast, easy, and low-cost squash and smear techniques. The meiotic chromosomes are studied through anther smear, whereas the somatic chromosomes are commonly studied through root tip squashing. Lately, chromosomal banding techniques have

M. Kashtwari (✉) • S.A. Zargar
Cytogenetics and Reproductive Biology Laboratory, Department of Botany, University of Kashmir, 190 006 Srinagar, Jammu and Kashmir, India
e-mail: m.k.045@hotmail.com

A.A. Wani
Department of Botany, University of Kashmir, Srinagar, Jammu and Kashmir, India

T.A. Bhat, A.A. Wani (eds.), *Chromosome Structure and Aberrations*,
DOI 10.1007/978-81-322-3673-3_4

proved to be an additional tool in the hands of cytogeneticists for detailed chromosomal studies. In this chapter an attempt has been made to revisit the available procedures for studying plant chromosomes through squash and smear techniques and the fixatives and stains used therein in light of the available literature.

Keywords
Chromosome • Fixative • Pretreatment • Stain

4.1 Introduction

Study of plant chromosomes reveals the number and morphology of chromosomes for the construction of karyotype and the study of recombination through meiosis. During the chromosomal analysis, several factors like high chromosome number, smaller size of chromosomes, high stickiness/low spreading on metaphase plates, and odd number of chromosome complement hinder the optimal observation. The first step in the road of chromosomal analysis is the collection of tissue for mitotic and meiotic analysis. Fresh anthers from large flowers and whole inflorescences in case of smaller flowers are collected and fixed before anthesis for carrying out meiotic studies. For somatic analysis root tips or shoot tips are collected and fixed. Several fixatives are available that carry out various functions like coagulating the cytoplasm, precipitating or cross-linking the proteins and DNA, and increasing the efficiency of stain linkage to the chromatin. Some of the fixatives are metallic like mercuric chloride, osmium tetroxide, potassium dichromate, and others, while others are nonmetallic like ethanol, acetic acid, formaldehyde, propionic acid, and chloroform. The choice of fixative depends upon several factors like the reaction with nucleic acid, penetration into tissue, formation of artifacts, easy washing, and the effect on stainability, toxicity, and cost. Some fixatives coagulate the proteins (ethanol, mercuric chloride, chromium trioxide), while as others do not, i.e., they cross-link the proteins (osmium tetroxide, formaldehyde, potassium dichromate, acetic acid). The somatic tissues can be pretreated before fixation to increase the number of metaphase cells. In addition pretreatment enhances chromosome condensation and facilitates their morphological identification (Singh 2003). In 1929, Kagawa used chloral hydrate as a pretreatment agent for *Triticum* and *Aegilops* species. Pretreatment increases the number of mitotic cells by arresting the cells in the dividing phase. This mitotic arrest is brought about by chemicals that disrupt spindle polymerization and/or depolymerization, as such the metaphase chromosomes are not separated and are not pulled toward the two poles only after which the cytokinesis usually occurs. Colchicine is the most frequently used mitotic poison, others being *p*-dichlorobenzene, 8-hydroxyquinoline, α-dichlorobenzene, and ice cold water (Conagin 1972). Once the cells are pretreated and fixed, the tissue is then subjected to staining. In 1921, Belling described a technique for studying meiosis in plant species that involved the squashing of anthers. By the early 1940s, squashing technique had completely replaced the method of microtome sectioning for chromosomal studies using somatic and meiotic tissues (Aase 1935; O'Mara 1939). For smear preparations the anthers (Razaq et al. 1994) or young flower buds are placed on the slide and crushed in the drop of stain. The crushing causes the separation of cells and aids easy observation. Once the developmental stage of anther/bud is correlated with the stage present in the pollen mother cells (PMCs), the procedure becomes easily reproducible. The stains that are most often used in smear preparations are acetocarmine and aceto-orcein with varied concentrations from 1 to 2 %. The chromosomes are stained red against a clear cytoplasm after the excess stain is removed by processing. In case of mitotic analysis, the large number of cells in root and shoot tips hinders easy penetration of stain. Also, the fixatives somewhat harden the tissue, and as such, squashing becomes difficult. For this reason an important step is included in the procedure known as acid hydrolysis, in which the tissue is immersed in warm acid of a particular normality for a fixed duration of time in order to soften the tissue and dissolve the pectic layer which makes the tissue soft for squashing. Standardizing the acid hydrolysis is an important step in mitotic chromosomal study. Fuchsine is the most frequently used stain after hydrolysis. Fuchsine, when prepared for staining, is a straw yellow-colored liquid that turns magenta colored on oxidation. It stains the nucleic acid red with a clear background or can be counterstained by acetocarmine. Processing of the slide helps in chromosome spread and easy observation. Karyotypes can be prepared once well-spread and nonoverlapping chromosomes with well-defined constrictions are obtained. Another challenge, at the final stage of the studies, is making those slides permanent which contain valuable stages. This can be done using butanol: acetic acid schedule in appropriate concentration and volume. Several protocols for chromosome analysis are available in the literature; however, the application of any of these protocols depends upon the type of plant material one is dealing with. Besides, some minor laboratory modifications are required for mitotic and meiotic studies; it is generally necessary to adapt them for different plant materials (Ostergren and Heneen 1962). An overview of various techniques and procedures involved in chromosome analysis is described in detail below.

4.2 Pretreatment

The goal of all the efforts that we put in for making squash and smear preparations is to achieve well-spread, clear, and condensed chromosomes that can be easily studied for their structure and behavior. For achieving this, the plant tissue is subjected to pretreatment with certain chemicals before its fixation. The pretreatment also clears the cytoplasm and separates the middle lamella. La Cour (1935) considered pretreatment important for study of spiral structure of chromosomes. Pretreatment is usually followed in case of mitotic studies; however, for meiotic studies, the flower buds are directly put in the fixative. In general pretreatment causes the following changes in the mitotic cells:

1. Increases the number of metaphase cells
2. Constricts the chromosomes longitudinally

3. Clarifies the constrictions
4. Increases the viscosity of the cytoplasm
5. Allows easy penetration of fixative

During the cell division, chromosomes are arranged in a specific manner which facilitates their easy assortment into the daughter cells, but this arrangement hampers the observation of individual chromosomes. Thus, the main target of pretreatment becomes dissolution of this organization which is attained by changing the viscosity of the cytoplasm that collapses the spindle-forming mechanism. This releases chromosomes from the binding force of the spindle, and thus they become free. A slight pressure applied during slide preparation results in scattering of the chromosomes throughout the cell surface. Change in viscosity of the cytoplasm also clarifies the constrictions. Moreover, the disintegration of spindle and rapid cell death during pretreatment and fixation leaves many cells arrested in the metaphase stage which is highly beneficial in karyotyping. These microtubule-inhibiting pretreatment substances include colchicine, 8-hydroxyquinoline, *p*-dichlorobenzene, ice cold water, and α-bromonaphthalene (Osalou et al. 2013). Colchicine is by far the most commonly used pretreatment agent although *p*-dichlorobenzene is also widely used. Almost all the pretreatment agents effectively increase the number of metaphase cells and condense the chromosomes during mitosis.

4.2.1 Colchicine

Colchicine, having the formula $C_{22}H_{25}NO_6$ (Fig. 4.1) and molecular weight of 399.43 g/mol, is a water-soluble alkaloid which is derived from the dried corms and seeds of the autumn-flowering crocus (*Colchicum autumnale*, also known as meadow saffron). It was first isolated in 1820 by Pelletier and Caventou. In 1833,

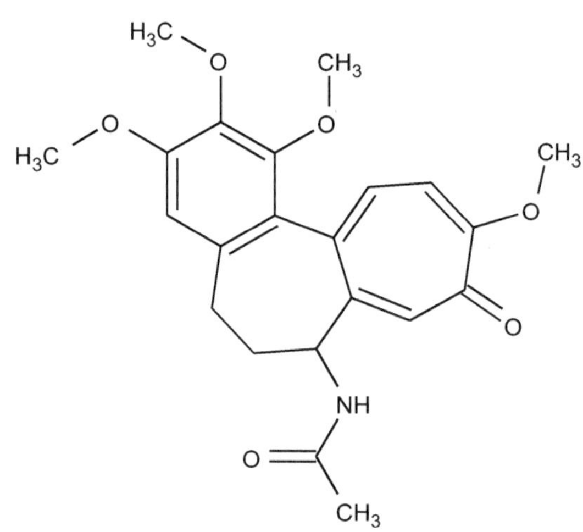

Fig. 4.1 Structure of colchicine

Geiger purified and named it, but was first used by O'Mara in 1939. Colchicine dramatically disrupts the pattern of mitosis (Dixon 1905) and thus is an inducer of the mitotic block. It was first considered as an active agent in increasing the number of mitotic cells (Dustin 1934; Lits 1934), but later Ludford in 1936 concluded that the increase in the number of mitotic cells is not due to stimulation but rather due to the accumulation of arrested mitoses. The effect of colchicine on the plant cells was first studied by Nebel and Ruttle in 1938 on the stamen hairs of *Tradescantia.* With the advance in scientific techniques, it was shown that colchicine acts on microtubules where it combines with tubulin and prevents its polymerization (Molad 2002). Microtubule assembly forms the mitotic apparatus which is responsible for separating chromatids during division (Palevitz 1993). According to Bhaduri (1939), it was first considered to be a catalyst that could change the colloidal state of the cytoplasm. During division, the metaphase stage gives an ideal state to view chromosomes as they are in a highly condensed form, and thus any treatment that increases the number of metaphase stage cells will be highly useful; colchicine fulfills this requirement. Besides its use as pretreatment agent, when used in very low concentrations, colchicine is also used as a polyploidizing agent, when applied in higher concentrations on actively growing meristems.

4.2.2 α-Bromonaphthalene

It is a simple bromine derivative of naphtalene with the formula $C_{10}H_7Br$ (Fig. 4.2) also known as 1-naphthyl bromide and has the molecular weight of 207.07 g. At 20 °C it is an oily liquid with a pungent order. It is slightly soluble in water but miscible in alcohol, benzene, ether and chloroform.

α-Bromonaphthalene was first used by Schmuck and Kostoff in 1939. Like colchicine, α-bromonaphthalene also prevents the formation of the spindle and arrests the dividing cells at metaphase. It is used in a saturated aqueous solution, which is obtained by adding a few milliliters to about 500 ml water, shaking thoroughly and letting it to settle. The undissolved α-bromonaphthalene appears as transparent pearls at the bottom of the flask which can be separated and reused. The duration of pretreatment can vary from 30 min to 3 h in some species like wheat (Mirzaghaderi 2010) to about 24 h as in the case of orchids (Pridgeon et al. 1999). Longer duration of exposure to pretreatment has been recommended especially for species having longer chromosomes. Besides wheat and barley, α-bromonaphthalene has been

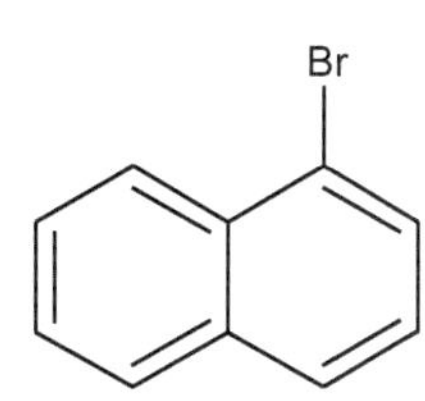

Fig. 4.2 Structure of α-bromonaphthalene

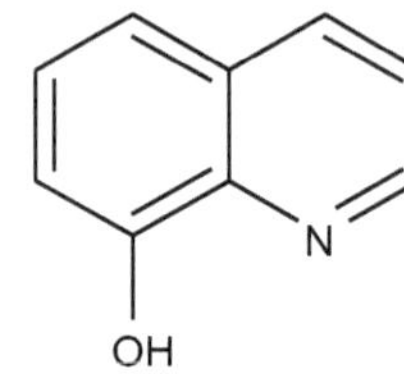

Fig. 4.3 Structure of 8-hydroxyquinoline

effectively used in *Lolium* (Ahloowalia 1965), strawberry (Owen and Miller 1993), *Voanioala gerardii* (Johnson 1989), *Piper* (Samuel 1987), some *Momordica* species (Roy et al. 1966), *Crocus* species (Candan 2007), and almond (Kazem et al. 2010).

4.2.3 Ice Cold Water

First used by Hill and Myers in 1945 as a pretreatment agent, ice cold water is generally recommended for cereals, temperate grasses and for *Arabidopsis* at 0 °C for 24 h (Maluszynski 2003). It was used for Mexican *Solanum* species (24 h at 4 °C) where it gave good results over 8-hydroxyquinoline (Chen and Li 2005). According to Tsuchiya (1971), the tissue should be kept in vials having cold water, and to maintain the low temperature, the vials are dipped in cold water topped with a layer of thick ice and kept as such in refrigerator. It shrinks the chromosome longitudinally but may also lead to stickiness so that the separation during slide preparation becomes difficult.

4.2.4 8-Hydroxyquinoline

It has the formula C_9H_7NO (Fig. 4.3) with a molecular mass of 145.16 g/mol. The usefulness of 8-hydroxyquinoline was first demonstrated by Tijo and Levan (1950). Like other pretreatment agents, it also affects the spindle formation and results in metaphase arrest due to an increase in the cytoplasmic viscosity resulting in immobilization of the chromosomes. It is generally effective for plant species with small chromosomes. It is good for resolving primary and secondary constriction positions.

Different concentrations of this chemical with varying time durations have been adopted in different plant species, e.g., 0.2 % solution for 30 min duration in aloe vera (Vig 1968) and 0.002 molar solution for 4–8 h in orchid species (Pridgeon et al. 1999). In general, a 0.0001 molar solution of 8-hydroxyquinoline for 90 min at 10–12 °C has proved good for several mangrove species (Sidhu 1968), whereas 2 millimolar solution for 5 h duration was found effective in the case of strawberry (Owen and Miller 1993).

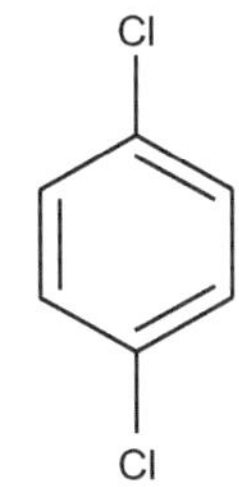

Fig. 4.4 Structure of *p*-dichlorobenzene

4.2.5 *p*-Dichlorobenzene (*p*-DCB)

p-Dichlorobenzene is a colorless solid having the formula $C_6H_4Cl_2$ (Fig. 4.4) and the molecular weight of 147 g/mol. It has a strong odor and is now frequently used as a pesticide in mothballs replacing naphthalene. Meyer in 1945 was the first person to use *p*-dichlorobenzene for pretreatment in root tip mitosis of several plants. According to him, the saturated solution gave good results for *Parthenium argentatum*, *Crepis*, *Allium*, *Pisum*, *Lycopersicon*, *Tradescantia*, and others. In pretreated material, it can cause early separation of chromatids, chromosome fragmentations, and chromosomal bridges and even can also cause polydiplodization (Carey and McDonough 1943; Sharma and Bhattacharya 1956; Oehlkers 1952; Srivastava 1966). Chromosome breaks have been observed in *Lens esculenta*, near secondary constrictions and satellites (Sarbhoy 1980). However, it has not been found to be a mutagen (Loeser and Litchfield 1983). It is equally effective for long as well as short chromosomes. Due to its very low water solubility, a saturated solution is often used. For plants with higher chromosome numbers, Sharma and Mookerjea (1955) treated the root tips in a saturated aqueous solution for 3 h at 12–16 °C for temporary smear preparation used for chromosome counting. *p*-Dichlorobenzene has been effectively used in several plant species such as *Parthenium* (Meyer 1948), tea (Bezbaruah 1968), *Amaranthus* (Madhusoodanan and Nazeer 1983), soya bean (Palmer and Heer 1973), *Vicia*, *Lathyrus*, and *Pisum* (Srivastava 1966). It was also used for shortening the chromosomes and spreading out the chromatid arms in succulent plant tribe Aloineae (Riley and Majumdar 2015).

4.3 Fixation

Fixation is a process by which all activities in a cell or tissue (either separated from the body of the organism or the organism as a whole) are ceased, and each constituent is fixed at the point where it was at the time of fixation. In short, fixation is freezing the time frame of the cell more or less accurate to its original living form. Since we cannot witness what is happening inside the living cell in real time, we study the fixed specimens of the cell across a time range. When this is done, we unite the data from different stages and thus try to understand a particular process of the cell.

The type of process to be studied determines the selectivity of fixation at a particular stage. In cytological studies the main focus remains the unraveling of chromosomal structure and behavior without the interference of the other cellular components. It aims at finding out the original position as well as presence of chromosomes, that is faithful to the living cell, without the formation of the artifacts. With the advancement in cytological techniques, we have moved from mere visibility of chromosomes toward analyzing the details of the chromosome structure such as chromatic and heterochromatic regions and primary and secondary constrictions. As we move across the taxon, cell and tissue composition varies considerably. This composition is the basis for differential response toward the fixatives. Such variation is evident in the fixation processes of monocots, dicots, and cryptogams. The presence of secondary metabolites and the wide variation in chromosome number and behavior are some of the factors that add to the complexity.

The process of fixation brings into picture certain metallic and nonmetallic compounds (Table 4.1) which either alone or in combination penetrates the tissue, stabilizes the cell, and makes it sensitive to further treatments. These compounds are grouped on the basis of their oxidation potential, and desirable combinations are made (Table 4.2). In general a strong reductant is not mixed with a strong oxidant and vice versa, but exceptions being present in special mixtures. Since our goal is to study a cell in its original state, a good fixative should have all the characteristics such as rapid penetration power, freezing the cell state, but neither dissolving nor distorting, shrinking or swelling the cell, be able to kill the pathogens, and inactivate the autolytic enzymes. One of the important features of a good fixative is that it sensitizes different constituents of the cell to take up specific stains so that they can be easily studied. A single fixative with all these properties is not known; thus to be effective, one or more compatible chemicals are mixed together. In such a mixture, the limitation of one chemical is fulfilled with the beneficial property of the other, e.g., a fixative that penetrates rapidly can be mixed with a fixative that precipitates chromatin. The balancing action of fixatives is important to make an effective fixation mixture. Baker (1944) suggested that the best fixation mixture will be one with simple non-precipitant chemical combined with a non-fixative salt. In general, different fixatives behave differently; some cross-link proteins and form a meshwork (Telleyesniczky 1905); others clump them and form a coagulum, while others dissolve them (Table 4.1). Their reaction with carbohydrates and lipids is also different. Some cause shrinking, while others swell tissues. Baker (1958) on the basis of

Table 4.1 Showing some commonly used fixatives grouped on the basis of reaction with albumin and subgrouping on the basis of metallic nature

Coagulants		Non-coagulants	
Metallic fixatives	Nonmetallic fixatives	Metallic fixatives	Nonmetallic fixatives
Mercuric chloride ($HgCl_2$)	Ethanol (C_2H_5OH)	Osmium tetraoxide (OsO_4)	Formaldehyde (HCOH)
Chromium trioxide (CrO_3)		Potassium dichromate ($K_2Cr_2O_7$)	Acetic acid (CH_3COOH)

Table 4.2 Showing some common fixatives grouped on the basis of their oxidation potential

Reductants	Oxidants
Ethanol (C_2H_5OH)	Osmium tetroxide (OsO_4)
	Chromium trioxide (CrO_3)
	Potassium dichromate ($K_2Cr_2O_7$)
	Formaldehyde (HCOH)
Mercuric chloride ($HgCl_2$)	Acetic acid (CH_3COOH)

reaction with albumin divided fixatives into two groups: coagulants and non-coagulants. Coagulants when reacting with albumin form a coagulum, while non-coagulants form a meshwork. A fixative may alter the nature of proteins without the addition of atoms to them and cause denaturation (Wolman 1955), known as nonadditive fixatives, or may add atoms to the protein backbone and interlink chains (Baker 1966), known as additive fixatives; some precipitate and others do not.

According to Levan (1949), precipitation of the protein causes a marked change in the refractive index of the chromosome and thus rendering them visible. Those fixatives which precipitate the chromatin are used in chromosomal studies. Some of the most commonly used and the most widely tested fixatives are described below. However, it is important to note that these fixatives are usually used in squash and smear preparations; some of them might even work for microtome sectioning.

4.3.1 Mercuric Chloride ($HgCl_2$)

It is an additive coagulant fixative, first used by Lang in 1878. In older literature it was referred to as "corrosive sublimate" as it even reacts with stainless steel and turns aluminum foil to gray dust. Mercury in a saturated aqueous solution (7 %) at moderate temperatures precipitates protein in an insoluble form (Baker 1966). This property resembles that of the many cationic fixatives. The reactions are dependent on the pH at which the fixative is used (Haarmann 1943). Its reaction with –COOH, $–NH_2$, and –SH groups makes it a strong protein precipitant. It penetrates moderately (Baker 1958), shrinks slightly (Baker 1966) and fixes chromatin weakly (Pischinger 1926; Fischer 1899) by reacting with phosphate groups. Chromosomes take up dyes well after fixation with mercuric chloride which is compatible with many fixatives, but one of the serious limitations is that it leaves behind metallic needle-like precipitates (Mayer 1918) which can be removed by washing with iodine in alcoholic solution (1 g iodine and 2 g PI in 100 cc of 70 % ethanol) or more quickly by 0.2 % sodium thiosulfate (Khasim 2002).

4.3.2 Potassium Dichromate ($K_2Cr_2O_7$)

First used in biological studies by Muller in 1859, it is slightly soluble in water. At pH greater than 4.6, it maintains the structure of chromosomes, where it reacts like chromic acid and also as a coagulant, whereas at a lower pH, it behaves as a non-coagulant and only the cytoplasm remains preserved (Zirkle 1928a). Un-acidified $K_2Cr_2O_7$ dissolves DNA, docs not fix proteins, but renders them insoluble in water (Baker 1966). In acidified condition, chromosomes are well fixed; cytoplasm and chromatin are precipitated. It is best used to study mitochondrion (Johansen 1940) because it reacts with unsaturated lipids and renders them insoluble. The fixation reactions are thought to involve the oxidation of proteins with the interaction of reduced chromate ions forming some cross-links. Chromium ions are reported to react with carbonyl and hydroxyl side chains of proteins. Wolman (1955) suggested that the primary reaction is formation of chromium bridges between proteins leaving amino groups available which favor staining with acid dyes, but at low pH, the response shifts toward basic dyes. When further processed with alcohol, reaction with chromate salt leads to the formation of insoluble chromate suboxide.

4.3.3 Chromic Acid (H_2CrO_4) or Chromium Trioxide (CrO_3)

Chromium trioxide is the anhydride of chromic acid. Since chromic acid is a strong acid, it significantly dissociates in water to give pH of 1.2 to a 1 % solution (Casselman 1955a, b). Molecular chromic acid could in principle be made by adding chromium trioxide to water, but in practice the reverse reaction occurs when molecular chromic acid is dehydrated. This is what happens when concentrated sulfuric acid is added to a dichromate solution. Chromic acid contains chromium in an oxidation state of +6 which makes it a strong and corrosive oxidizing agent, the strongest of all fixatives. Chromic acid should not be used with organic reducers like ethanol or formalin with which it reacts violently. However, some useful fixatives do contain chromic acid and formaldehyde.

It is a strong coagulant of protein (Baker 1966), the coagulation occurring due to blocking of –OH and –COOH groups (Bowes and Kenten 1949), and thus makes the proteins less basophilic. Berg (1927) describes the reaction with proteins as a two-step process: first step is coagulation and precipitation due to the primary effect of chromic acid and the second step is hardening due to $HCrO_4^-$ ion. It precipitates DNA in an insoluble form (Fischer 1899). Overall the cytoplasm is coarsely coagulated, nucleus is well fixed, but the chromosomes show excellent fixation. The fixative has to be thoroughly washed from the tissue as free chromium trioxide is reduced to green chromic oxide and is difficult to remove. This may prevent staining and or may obstruct visibility.

4.3.4 Osmium Tetroxide (OsO_4)

Schultze, one of the proponents of the cell theory, in 1864 first used osmium tetroxide in cytology where he observed that the treated cells remained in a lifelike condition. In an aqueous solution, it takes up a molecule of water to form hydrogen per-perosmate (H_2OsO_5) (Yost and White 1928) which weakly ionizes to H^+ and $HOsO_4^-$ ions. Because of its low ionization, osmium tetroxide is a weaker acid than acetic acid; thus, it undergoes oxidation to lower oxides even in the presence of light. This is the reason why the solution of osmium tetroxide is kept in an amber bottle away from light.

Osmium tetroxide is an additive fixative as it intercalates across double bonds (Boeseken and Giffen 1920; Criegee 1936) and can also cross-link aromatic or aliphatic double-bonded ring compounds together (Criegee 1936; Criegee et al. 1942). It also oxidizes –OH, $–NH_3$, and –SH groups (Wolman 1955). Although proteins lack appreciable amount of double bonds, some amino acids like tryptophan and histidine contain double bonds in their rings and thus react strongly with osmium tetroxide forming dark precipitates (Bahr 1954). Thus, the more double-bonded-ringed amino acids are in the protein, the more the protein reactive toward osmium tetroxide (Porter and Kallman 1953). Reaction of proteins with osmium tetroxide renders them nonreactive to acidic dyes (Baker 1958). It does not coagulate proteins and also renders them non-coagulable by heat or ethanol (Berg 1927). Osmium tetroxide is soluble in lipids and blackens when the tissue is placed in ethanol. It also blackens unsaturated lipids due to the reaction with double bonds (Altmann 1894). Since it does not precipitate DNA (Fischer 1899), it is an effective pretreatment agent although it cannot be used for studying the interphase chromosomes. The extent of fixation depends upon temperature (Sjöstrand, 1956, 1959) and pH (Baker 1966). One of the serious limitations, apart from being costly, is the blackening of the tissues. It reacts with ethanol and formaldehyde to be reduced to black osmium dioxide. Washing in running water is strongly recommended before immersing in such reducing solutions. Hydrogen peroxide can be used to bleach the tissue, but it affects the stainability of the chromosomes.

4.3.5 Ethanol (C_2H_5OH)

It is the most commonly used chemical in fixation. It is soluble in water in all concentrations; however, in fixation, 70–100% ethanol is used. It is a nonadditive fixative, i.e., it does not react with the macromolecules but acts as a dehydrating agent. It removes water molecules that stabilize the proteins and thus renders them insoluble. The insoluble proteins then aggregate and form a precipitate. It is a reductant and undergoes oxidation to acetaldehyde and then to acetic acid. Thus it should not be used with oxidizing fixatives such as metallic oxides but is compatible with many other fixatives. Ethanol also precipitates DNA (Fischer 1899) in the same way it precipitates proteins. Since it does not react with the macromolecules like DNA, it does not change their staining ability, i.e., it does not increase or decrease their affinity toward basic or acidic dyes. Main limitation with ethanol as a fixative is that it shrinks the tissue excessively and hardens it significantly. Due to these reasons, ethanol has to be used with other fixatives that reverse the shrinkage, e.g., acetic acid. The tissue, once fixed, has to be treated with acids to remove the hardness in order to be able to study them. The penetration property of ethanol is moderate (Baker 1960).

4.3.6 Acetic Acid (CH_3COOH)

Acetic acid also known as ethanoic acid in IUPAC nomenclature is the second simplest carboxylic acid after formic acid. When undiluted, it crystallizes to form ice-like crystals just below the room temperature at 16.6 ° C; this is the reason why its anhydrous form is known as "glacial" acetic acid. Presence of 0.1 % water lowers the freezing point further to 0.2 °C (Armarego and Chai, Christina 2009). When diluted to about 4–18 %, it is commonly known as vinegar. Acetic acid is a weak monoprotic acid (only one proton is released). It is dissociated to a small extent in water (Lassek 1950). In aqueous solution, it has an acid dissociation value of 4.76 (Goldberg et al. 2002). A one molar solution has a pH of about 2.4 indicating dissociation of only 0.4 %. An important property of acetic acid is that it is a wide solvent similar to ethanol and water. It dissolves not only polar compounds such as inorganic salts and sugars but also nonpolar compounds such as oils and elements such as sulfur and iodine. This dissolving property and miscibility of acetic acid make it a widely used chemical for dissolving dyes. It is an important component of staining mixtures like acetic-carmine, acetic-orcein, etc. As a fixative, it precipitates nucleic acid and dissolves histones but does not fix proteins (Pischinger 1937). This property may be due to the action of acetate ions on DNA. Baker (1966) found that it precipitates nucleoproteins from a solution but does not precipitate albumin. In spite of these facts, it is an ideal fixative for chromosomes. It has the property to maintain the chromosome structure intact presumably by not causing any distortion of the nucleoprotein. An interesting fact about this fixative is the excessive swelling of the cellular components including chromosomes. For chromosomal behavior studies, the swelling of chromosomes is not a problem, but for karyotyping and fine structural studies, acetic acid is necessarily used in combination with fixatives that tend to shrink the tissues, e.g., ethanol and some metallic oxides. It ideally fixes pachytene chromosomes, but the metaphase and anaphase chromosomes are also well fixed. Since it induces swelling, it leaves the tissue soft after treatment. It does not form any artifacts; therefore, no special washing is required. Fixation by acetic acid depends upon the pH. When pH is more acidic than 4 or around, the tissues are characteristically fixed, but above it fixation does not occur (Zirkle 1928b). It does have an effect on the stainablity of the cells. Cytoplasm tends to react more toward acidic dyes, whereas the metaphase and anaphase chromosomes react more toward the basic dyes.

4.3.7 Formaldehyde (HCHO)

It is the simplest aldehyde, also known by its IUPAC name as methanal, and is an anhydride of formic acid. It is an unusual compound as it acquires different forms in different states. As a gas at room temperature, it occurs as simple individual molecule, but when condensed, it mostly occurs as a cyclic trimer $(CH_2O)_3$ and a linear polymer. In water it forms a hydrate methanediol, $H_2C(OH)_2$, apart from forming oligomers. A 40 % saturated solution of formaldehyde by volume in water is known as 100 % formalin. A small amount of methanol, 10–12 %, is added to the solution of formaldehyde as a stabilizer which suppresses oxidation to form formic acid, but it can also be reduced to methanol (CH_3OH) and thus can be used as an oxidizing agent; however, its oxidation potential is lowest among the fixatives, and the speed of oxidation is very low. It can also be used at neutral pH (Lillie 1954) by adjusting magnesium or calcium carbonate. It is specially used as an excellent tissue fixative chiefly because of its ability to react with proteins. It reacts with the terminal and secondary amino groups (NH_3) to cross-link proteins without librating water (Woodroffe 1941; Baker 1966). It also reacts with immino groups (NH).

Bridges can also be formed between –CO and –NH groups (Pearse 1968). These reactions occur in standard solutions having pH of 4, but in highly alkaline solutions of more than 8 pH, it can reduce sulfur bridges (–S–S–) to –SH groups and further react to form methylene bridge ($–S–CH_2–S–$) (Middlebrook and Phillips 1942). Since it only reacts with basic groups, it leaves the proteins reactive to acidic dyes, and the basophilic nature is highly reduced. The property of bridge formation hardens the tissue remarkably. Due to this reason, the tissue becomes hard to smear impeding cytological analysis. It does not fix or precipitate DNA and therefore is not considered as a good fixative for the chromosome studies. Formaldehyde is more widely used as a specimen preserver than a fixative although it can only delay decay but not avoid it.

4.3.8 Propionic Acid (C_2H_5COOH)

Similar to acetic acid in structure and having similar solvent properties, propionic acid thus can be used as an effective substitute. It has an added advantage that it does not swell the chromosomes to the extent acetic acid does and thus used extensively in the fixation of chromosomes. It penetrates rapidly but to a lesser extent than acetic acid. Propionic-carmine and propionic-orcein are staining solutions in which propionic acid is used as a solvent.

4.3.9 Chloroform ($CHCl_3$)

It is slightly soluble in water but dissolves easily in alcohol, ether, or acetone. In plants where fixation is hampered due to a waxy coating on the tissue or presence of elevated level of fatty acids, chloroform can be used as an option as it is a good solvent for fats. It dissolves away the upper layers and thus aids in the penetration of the fixative. Since chloroform-treated tissue becomes brittle, it can be used in combination with other fixatives that are useful in smear preparation. It slowly converts into carbonyl chloride in the presence of light or air. Mixing ethanol with chloroform checks its decomposition into carbonyl chloride which is highly toxic. Also, a prolonged exposure of chloroform or an overdose can be toxic.

4.4 Fixing Mixtures

Since none of the fixatives can be used alone because each has several drawbacks, it always remains fruitful to mix two or more of them to get effective fixing and uninterrupted downstream effect. Some of the classic mixtures of fixatives that proved to be effective in one or more plants for chromosomal studies are described below.

4.4.1 Flemming's Weak Fluid (1882) and Flemming's Strong Fluid (1884)

Flemming's Weak Fluid (1882)		Flemming's Strong Fluid (1884)	
Osmic acid	0.1 %	Chromium trioxide, 5 % aq	0–3 ml
Chromic acid aqueous	0.25 %	Osmium tetroxide, 2 % aq	0–4 ml
Glacial acetic acid	0.1 %	Acetic acid, 20 % aq	0–5 ml
		Distilled water	0–8 ml

Walther Flemming was the person who gave the term "chromatin" in 1879. A great deal of his work on chromosomes was the result of this fixative. Flemming pointed out that acetic acid cleared nuclear details and assisted in staining. In this fluid the shrinkage caused by osmic acid is opposed by the swelling action of acetic acid. In the tissues that are multilayered, there is differential staining of cells where thin sections are nicely stained. Since osmic acid has abrupt action and causes intense blackening, the tissue should not be kept in the fixative for long. Lower temperature slows down the action of osmic acid and prevents blackening of tissue. The constituents are kept separate and mixed just before use. The weak fluid is used for small tissue samples, whereas the strong fluid is used for thick samples for better penetration, e.g., root tips. An important modification was prepared by Benda in 1902, where the proportion of acetic acid is highly reduced to 2–3 drops, used for study of mitotic chromosome.

4.4.2 Carnoy's Fluid (1886)

Glacial acetic acid	1 part
Ethanol	3 parts

Also known as Carnoy's fixative I, it consists of one part by volume glacial acetic acid and three parts by volume ethanol. It is a simple and effective fixative for nuclear and mitochondrial organelles. It is one of the oldest known fixative mixtures first used by Clarke in 1851 (first published by Carnoy in 1886) in neurological microtechnique but is now used for almost all plant samples. Acetic acid and ethanol as such are not good fixatives, but once we mix them together in appropriate proportions, the drawbacks of the two are removed. Ethanol causes shrinkage of tissue, fixes cytoplasm but leaves nucleoproteins unfixed; acetic acid causes swelling, does not fix cytoplasm but fixes nucleoproteins. This mutually collaborative nature of the two fixatives gives this mixture a wide application both in plants and animals. Both squash and smear preparations can be studied using this fixative. Depending upon the thickness and composition of the material, the time of fixation varies greatly from 15 min to 24 h. It is rather effective in the study of chromosomes for both meiotic and mitotic works.

Several modifications have been published where mordents have been used, e.g. ferric ammonium sulfate (1–3 %) (Lesins 1954) which is used 3–12 h after fixation and 0.5–5 % aq. solution (Austin 1959) which is used after fixation. Apart from this, several modifications in basic Carnoy's fixative have been proposed that work for different plants and even different parts of the same plant. Some of them are 1:1 and 1:2 (Van Beneden and Neyt 1887), 1:4 (Zacharias 1888), and 3:2 (Burns and Yang 1961). The fixative has to be always prepared afresh.

4.4.3 Carnoy's Fluid II (1886)

Glacial acetic acid	1 part
Chloroform	3 parts
Absolute ethanol	6 parts

It is widely used for floral bud fixation. In Asteraceae and other families, 1:3:4 and 1:1:3 proportions are used (Turner 1956). Metzer and Lang (1955) modified it for legumes by saturating it with $HgCl_2$.

4.4.4 Navaschin's Fluid (Navaschin 1925)

Solution A	
Chromic anhydride	1.5 g
Glacial acetic acid	10 ml
Dist. water	90 ml
Solution B	
Formaldehyde aq. sol.	40 ml
Dist. water	60 ml

The solution A is separated from solution B because the former consists of chromic anhydride which is an oxidizer, and the latter consists of formaldehyde which is a reducer. When mixed, an oxidation reaction occurs that disrupts the fixation capacity of the fluid. It seems unreasonable to mix these fluids together, but the penetration of the liquids is so fast that oxidation does not get a chance to be complete. A 3–4 h treatment seems enough for the fixative to act on the tissue, there being no need for further storage. For soft tissues time period of fixation is further reduced, e.g., for smear preparations one hour treatment is enough. The original formula was published by S. Navaschin in 1912; later M. Navaschin in 1925 modified the formula in proportion especially for root tip fixation. For meiotic chromosome analysis, Belling's modification is important. For the study of chromosomal details, the proportion of acetic acid should be kept low as it causes swelling of chromosomes; however, for gross morphology and behavior, higher concentration of this fluid is strongly recommended.

4.5 Stain and Staining

Living cells have an inherent property of being transparent, and thus distinguishing different components of a cell under the microscope becomes difficult. Dyes or stains are usually used to overcome this barrier to visibility. The dyes that are used most often react with the components of the cells, be it cell wall, proteins, or other cellular components. Their attachment to the component does not affect its gross morphology, and thus the component becomes visible and distinct during analysis. According to Baker (1960), three factors control the staining ability of cellular components, viz., (i) affinity between the dye and the component, (ii) density of the component, and (iii) permeability of the dye in the same component. An important property of a dye is that it should not lose the ability to color when dissolved in an aqueous solution.

For all dyes, the coloring ability is attributed to interatomic interactions such as resonance of double bonds in aromatic rings. All the dyes are based fundamentally on the structure of an aryl ring having delocalized electrons, i.e., benzene. Benzene itself is colorless because it absorbs wavelength in UV region. This colorless property of aryl ring is changed by interaction with a group of atoms, known as chromophore. Chromophore has alternating single and double bonds also known as conjugated double bonds. They function by altering the energy of the delocalized electrons of the parent compound, and in result the compound absorbs and reflects from the visible range. Sometimes whole of the modified molecule is called the chromophore. Apart from the chromophore, dyes also contain additional groups attached to the parent compound or the chromophore that increases the intensity of color; these groups are known as auxochromes. They modify the ability of the chromophore to absorb light but themselves fail to produce any color. Auxochromes have a lone pair of electrons. These groups have a property that they essentially ionize and thus are responsible for the charge of the dye. Auxochromes are of two types, having positive charge such as the amino group or having negative charge such as the carboxyl group. Both types of groups can be present on a chromophore. These dyes can be cationic (containing positive charge) and thus are attracted toward acidic components or anionic (containing negative charge) and thus are attracted toward basic components. The cellular components react differently to different dyes, i.e., some components react to a particular dye, while others do not. When more than one stain is used and each stain reacts with different components, the procedure is known as counterstaining.

Staining of chromosomes is of two kinds: vital and nonvital. Vital staining is carried out in live cells, and as such those stains are required which are nontoxic and do kill the cell. But, in case of chromosomal studies on plants, the cells are first killed in fixatives and then stained. This kind of staining is known as nonvital staining. Baker (1958) claimed that the process of staining is that of absorption, but it can be a mixed contribution of both physical adsorption and chemical reaction. The basic objective behind the process of differential staining is compromised if all the cellular components are stained equally because of excessive soaking in the dye. In order to reach the desired staining level of the components, we need to limit the period of dying. It is done either by progressive or regressive dying. In progressive dying the strength of the color is gradually increased till desired. In regressive dying the tissue is initially overstained, and then the dye is gradually removed till the specific cellular component attains a high contrast. The regressive stage of this process is therefore called differentiation. Sometimes staining is not enough as the stain may start to fade after some time or by the action of succeeding treatments. To prevent this, another technique is implied known as mordanting. By mordanting, we try to make sure that the color of the dye attained by the tissue is not removed by differentiation, dehydration, or counterstaining. A mordant is a polyvalent metal ion or a metal-containing compound that is used before dying and which forms a complex with certain dyes and with the biomolecules through covalent and coordinate bonds. Mordants help in the attachment of the dye to the biological molecule and also help enhance the contrast of the cellular component. Enhancing of the color follows the same general principle as that of the auxochromes. Metals have relatively low energy levels, so their incorporation into a delocalized system results in lowering of the overall energy. The absorbance of the mordant-dye complex thus is positioned further in the visible range, and what is reflected is related to the filtered color from the white light. The mordant-dye complex formed between the dye and the polyvalent metal ion is known as a "lake" and is actually a coordination complex. This assembly is based on the formation of two bonds by the metal ion with the dye, one covalent bond with hydroxyl oxygen and a coordinate bond with separate double-bonded oxygen atom. In parallel to this, the metal atom also attaches itself to the biological molecule, thereby acting as a bridge. In case of nucleic acids, the important reacting component is the phosphate backbone, not the bases, as it provides the ideal combination of phosphate hydroxyl and nearby electron-donating atoms. The chief mordants are salts of aluminum, iron, and chromium, and among the dyes that are used with mordants, carminic acid and hematein are most common. All the dyes that are used with mordants can also be used without them, but they give varied color.

4.5.1 Carmine

It is one of the most frequently used dyes in the study of chromosomes also known by the name C.I. Natural Red 4. It is not a single chemical but a mixture of complex compounds derived from an insect known as *Coccus cacti* living on the plant *Opuntia coccinellifera.* The female bodies of this insect are dried and ground to a fine powder which forms the crimson-colored dye. The major constituent of the mixture is carminic acid ($C_{22}H_{20}O_{13}$) having the molecular weight 492.38. The chemical structure of the carminic acid consists of a core of anthraquinione (Fig. 4.5) linked to a glucose unit. The color is because of the quinonoid ring. It possesses a ketonic (=O) and a phenolic (–OH) group placed in such a relation to each other that makes it reactive to mordants, but its color (crimson) does not undergo much change when it links itself to aluminum. Initially it was prepared from the cochineal extract by reaction with lead acetate followed by sulfuric acid (Gatenby and Beams 1950); later it has been synthetically produced by organic chemists (Allevi et al. 1991). Carminic acid (Fig. 4.6) dissolves in distilled water in all proportions (Gurr 1960) and also in ethanol, whereas the crude extract does not dissolve in distilled water. Carminic acid is not soluble at its isoelectric point, pH 4–4.5 (Baker 1950), but in acidic pH, it behaves as a basic dye and stains chromatin, whereas at alkaline pH, it behaves as an acidic dye. This is the reason why carmine is used in a solution of acetic acid for staining plant cells (Godward 1948).

Since for the staining of chromosomes carmine is dissolved in a solution of acetic acid, the derived dye is therefore known as acetocarmine (Belling 1921). It was originally used as a 1 % solution in boiling 45 % acetic acid solution. It is best to

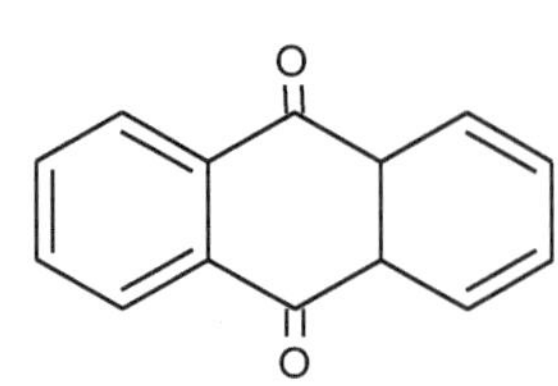

Fig. 4.5 Structure of anthraquinone

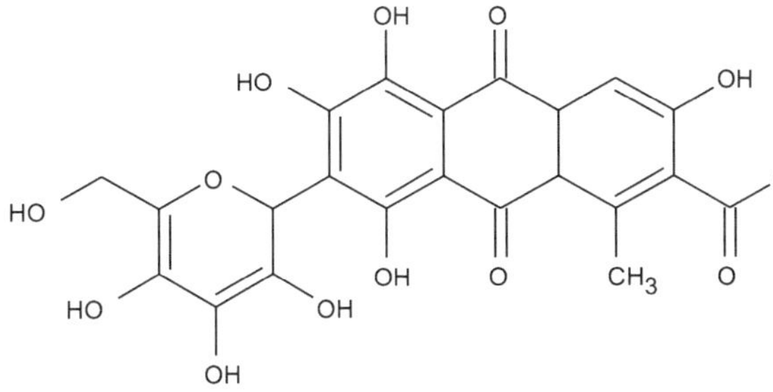

Fig. 4.6 Structure of carminic acid

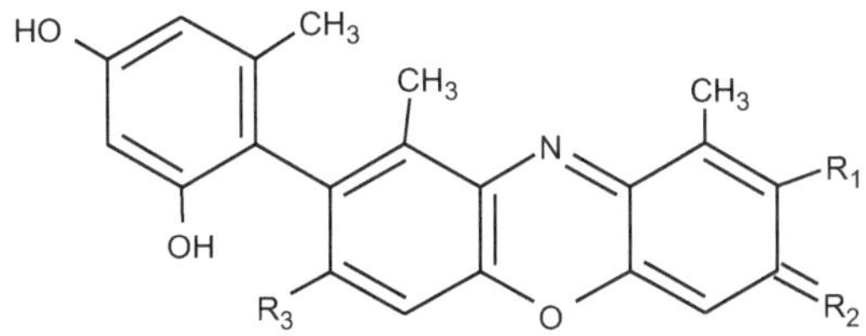

Fig. 4.7 Structure of α-amino orcein, one of the components of orcein

filter the solution before use as the natural carmine contains many impurities. Belling suggested that the addition of few drops of ferric hydroxide as a mordant will improve the color. For some plant materials, a 2 % acetocarmine can also be used. For some hard or bulk tissues, warm acetocarmine in HCl can fulfill a double purpose of softening and staining. Although it also has a property of fixation as it contains acetic acid, it should not be used for this purpose and should be used as a stain only. In order to offset the swelling action of acetic acid on the tissue, it is replaced by propionic acid; in that case the prepared stain is called propiono-carmine.

4.5.2 Orcein

Orcein, also known as and C.I. Natural Red 28, is the dye extracted from several species of lichen, commonly known as "orchella weeds" major source being the archil lichen, *Roccella tinctoria*, and also from *Lacenora parella*. Orcinol, the parent compound, is extracted from lichens and is then converted to orcein by ammonia and air. Orcein is a mixture of phenoxazone derivates (heterocyclic compounds containing one oxygen and one nitrogen atom), such as hydroxy-orceins, amino-orceins, and amino-orceinimines, the dominant component being α-amino orcein (Fig. 4.7) The chemical components of orcein were elucidated only in the 1950s (Musso 1961 in Henwood 2003). Orcein is less soluble in water but easily dissolves in ethanol and is reddish brown in color. In 1941, it was first used in the study of chromosome by La Cour in the form of aceto-orcein. It has an advantage over acetocarmine that the use of mordant is not required. Although synthetic orcein is available, the natural orcein gives better contrast (Conn 1953). This solution is unstable and should be prepared fresh before use.

A solution of 2 g of natural orcein dissolved in 100 ml of a mixture of lactic and propionic acids (1:1) and diluted to 45 % with water was used by Dyer in 1963 for studying fresh pollen mother cells. It was also found suitable for the rapid preparation of root tip chromosomes for studying detailed morphology (Dyer 1963). In root tip and shoot tip studies, acetic-orcein is mixed with HCl for hydrolysis before mounting in acetic-orcein stain (Tijo and Levan 1950; Sharma and Sharma 1957). As in case of carmine, acetic acid can be replaced by other acids such as propionic acid, lactic acid, or others to prepare the stain. Overheating the acetic-orcein can induce chromosomal breakages (Sharma and Roy 1955).

4.5.3 Crystal Violet (Methyl Violet 10B)

Popularly known for its role in Gram staining for the identification of bacteria, crystal violet (Fig. 4.8) was first used by Newton in 1926. It was first synthesized in 1883 by Alfred Kern. The dye is so named because it combines with six molecules of water to form large blue crystals. It is a basic dye and is readily soluble in ethanol but less soluble in water. It is usually used as a 1 % solution. Baker (1950, 1958) explained that the solubility of crystal violet is reduced by iodine, and this renders chromosomes more visible against a clear background. This effect is called Gram effect based on staining in bacteria.

Pure crystal violet belongs to the triarylmethyl group where three methyl attached to nitrogen behave as auxochromes and the aryl rings are chromophores. In total three aryl rings are bonded to a central carbon atom. This structure closely resembles that of rosalanine. The color of crystal violet is bluish purple resembling that of *Gentiana* flower near neutral pH. In alkaline solution the two amines loose the protons to the water, and the central carbon atom is attacked by the hydroxyl nucleophile, rendering the compound colorless. At pH = 1, the color changes to green. At

Fig. 4.8 Structure of crystal violet in ionized form

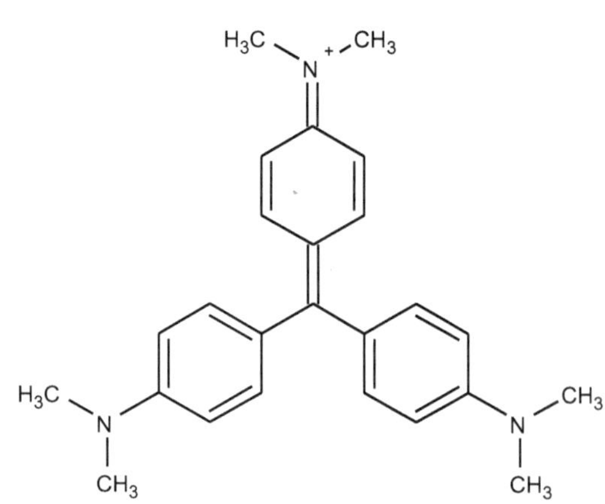

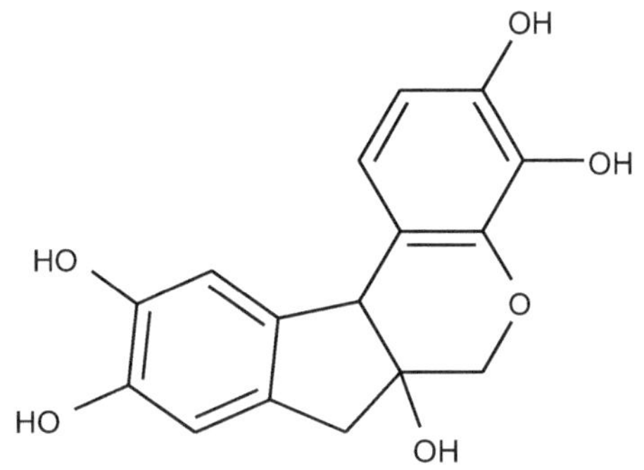

Fig. 4.9 Structure of hematoxylin

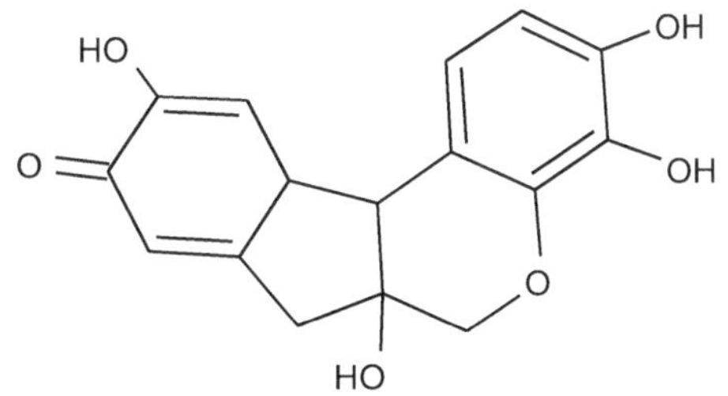

Fig. 4.10 Structure of hematein

further acidic pH, all the nitrogen becomes positively charged out of which two are protonated, and the dye becomes yellow. It cannot be effectively used for chromosomal study in the tissues that are being squashed after staining because the rate of penetration is low and also because it is nonspecific in nature.

4.5.4 Hematoxylin

Hematoxylin also known as "natural black 1" is a dye extracted from the heartwood of *Haematoxylon campechianum* native to Mexico. The heartwood releases a red dye when treated with water which turns yellow, then black on cooling. The water is evaporated leaving crude hematoxylin. Hematoxylin (Fig. 4.9) in itself does not stain, but its oxidation product hematein (Fig. 4.10) stains deep purple to violet because of the formation of a quinonoid ring (Perkin and Everest 1918; Baker 1950). During the preparation of the staining solution, hematoxylin is treated with oxidizing agents such as sodium iodate, hydrogen peroxide, potassium permanganate, mercuric oxide, and others. This process is known as ripening. A major drawback of hematoxylin is that it cannot be used without a mordant. Commonly used mordants are that of aluminum such as potassium aluminum sulfate (Mayer 1903), ammonium alum, and iron such as ferric chloride and iron alum (Benda 1896). The dye and the mordant should be mixed just before staining especially in the case of iron alum because of the possibility of heavy iron precipitation.

4.5.5 Fuchsine (Fuschin)

Belonging to the triphenylmethane, it gives a magenta red color when dissolved in water, and the dehydrated crystalline form is green. It is actually a mixture of several chemical dyes, viz.,

1. Magenta 0: *p*-Rosaniline chloride ($C_{19}H_{18}N_3Cl$) (Fig. 4.11) (MW. 323.4)
2. Magenta I: Rosaniline chloride ($C_{20}H_{20}N_3Cl$) (Fig. 4.12) (MW. 337.4)
3. Magenta II: ($C_{20}H_{21}N_3Cl$) (Fig. 4.13) (MW. 338.4)
4. Magenta III: New magenta/ new fuchsine ($C_{22}H_{24}N_3Cl$) (Fig. 4.14) (MW. 365.4)

The amount of methylation increases from Magenta I to Magenta III, where Magenta 0 does not contain any methyl group and Magenta III has three methyl groups. Similarly Magenta I consists of one and Magenta II consists of two methyl groups. However the composition of these components varies in the basic

Fig. 4.11 Magenta 0

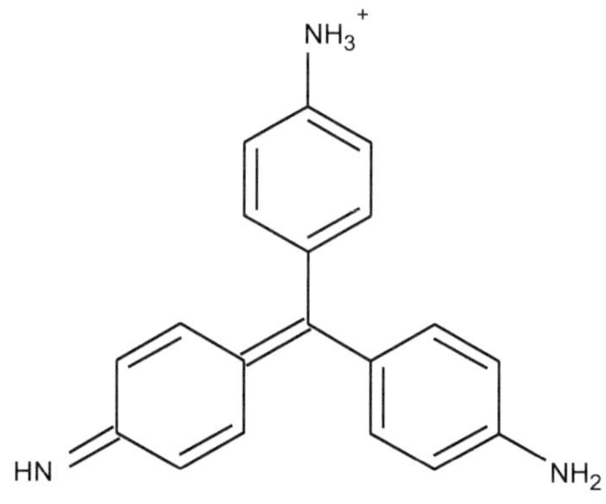

Fig. 4.12 Magenta I

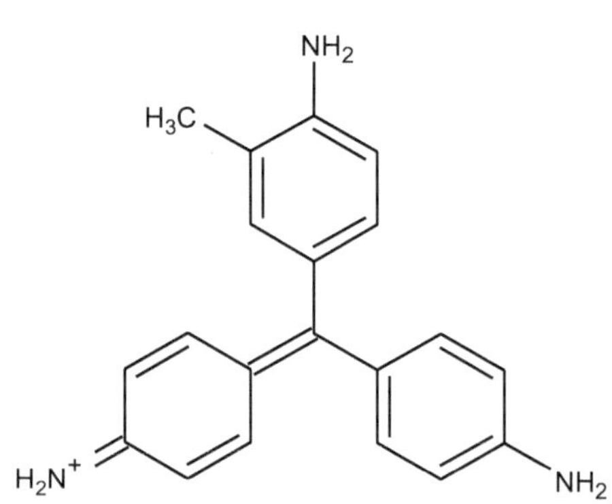

Fig. 4.13 Magenta II

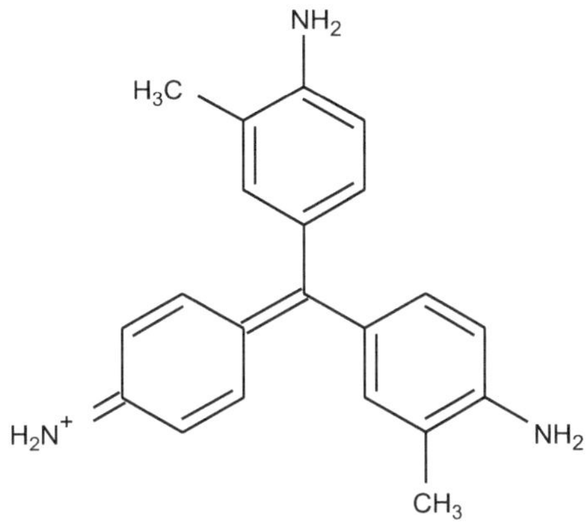

Fig. 4.14 Magenta III

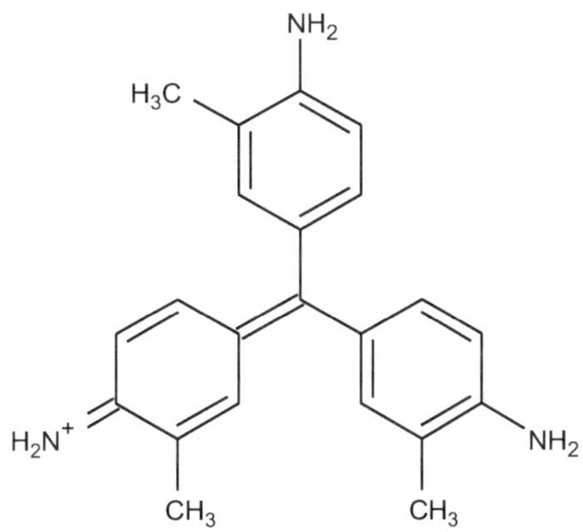

fuchsine. Magenta 0 being the most abundant and most important for producing the red color is the best Schiff's reagent. In order to get accurate results, Meganta 0 is specifically used alone, whereas for general work, fuchsine is highly recommended. All of these components have one quinonoid ring that is responsible for the change of color upon hydration.

Fuchsine can be used as basic fuchsine or as acidic fuchsine (Fig. 4.15). Basic fuchsine also sold under the name Fuchsine is the mixture of either all the four compounds from Magenta 0 to Magenta III or simply Magenta 0/p-rosaniline chloride. Acid fuchsine is produced by reaction of basic fuchsine with sulfonic groups to give acid derivative of Magenta 0–III compounds. Each compound can take up to three such groups, thereby making 12 possible chemicals, thus making acid fuchsine a variable colorless mixture.

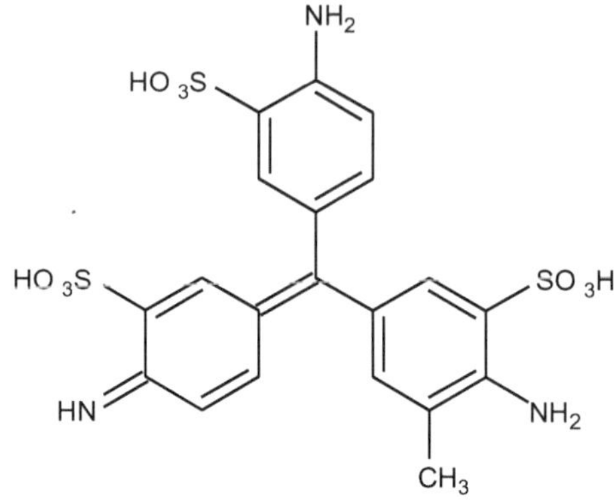

Fig. 4.15 Structure of one of the components of acid fuchsine

4.5.5.1 The Feulgen Reaction

Feulgen and Rossenbeck (1924) first used fuchsine solution with sulfurous acid to form a colorless liquid that stains chromosomes deep red to demonstrate the DNA component of the chromosomes. The principle behind the reaction is the combination of reactive or unmasked aldehyde groups of sugar bases of the DNA with the Feulgen sulfurous acid to yield the characteristic magenta color. The Feulgen sulfurous acid or lucosulfinic acid is derived from the addition of sulfurous acid to basic fuchsine or more specifically Magenta 0. This transformation leaves the solution colorless. The unmasking of sugar aldehydes is accomplished by treating with warm HCl, the process is known as hydrolysis. There is a considerable debate not only over the actual structure of the reaction product between fuchsine and sulfuric acid but also between that of the reaction product and aldehyde of DNA. Wieland and Scheuing (1921) were of the opinion that the reaction product is lucosulfinic and that one molecule of lucosulfinic combines with two adjacent aldehyde groups. This theory is known as N-sulfinic acid formula of aldehyde-Schiff's reagent, whereas the long forgotten sulfonic acid theory of Prud'homme (1900) suggested an entirely different structure for the Schiff's reagent. According to them, instead of lucosulfinic acid, alkyl sulfonic acid is formed. The alkyl sulfonic acid structure is more accepted and is supported by many authors (Hörmann et al. 1958).

In the studies on reactions of aldehydes with various amines, Schiff (1865a, b) noted replacement of hydrogen atoms by aldehyde residues and recommended this method for determination of the number of reactive hydrogen atoms in bases. Schiff (1865c) then extended these studies to basic fuchsine and found that it reacted with aldehydes in the same way as other bases which contained replaceable hydrogen atoms. This is the reason why the reaction product between fuchsine and sulfuric acid is known as Schiff's reagent. However, addition of alcohol and HCl greatly enhanced the reaction of basic fuchsine with aldehydes. This is essentially the method used in histochemistry for the preparation of aldehyde-fuchsine, a Schiff base. Hantzsch and Osswald (1900) isolated leuco-fuchsine-sulfonic acid in solid form and proved that it was identical with fuchsine-sulfurous acid, i.e., Schiff's reagent.

4.6 Preparations of Some of the Important Stains Used in Squash and Smear Techniques

4.6.1 Acetocarmine, Aceto-orcein

Materials required		
1.	Carmine, Orcein	1 g
2.	Glacial acetic acid	45 ml
3.	Distilled water	55 ml

Procedure

- Make 45 % glacial acetic acid solution.
- Heat to boil in a conical flask and add the dye slowly and stir well.
- Cool down to room temperature and filter.

Precaution

- The glacial acetic acid solution does not boil easily; thus, the flask needs to be properly covered.

4.6.2 Crystal Violet

Materials required		
1.	Crystal violet	1 g
2.	Distilled water	100 ml

Procedure

- Heat 100 ml of distilled water to boil.
- Dissolve the dye with constant stirring.
- Filter through Whatman Filter Paper No. 1.
- Allow to mature for 1 week before use.

4.6.3 Schiff's Reagent/Fuchsine Reagent

Materials required		
1.	Basic fuchsine	0.5 g
2.	Distilled water	100 ml
3.	1 N HCl	10 ml
4.	Potassium metabisulfite	0.5 g
5.	Activated charcoal	0.5 g

Procedure

- Bring to boil 100 ml of distilled water in a 200 ml beaker or flask.
- Add 0.5 g of basic fuchsine to it.
- Cool to 58 °C and filter through Whatman Filter Paper No.1 into an amber-/black-colored bottle or one wrapped in black paper.
- Cool down to 26 °C and add to it 10 ml of 1 N HCl and 0.5 g of potassium metabisulfite.
- Seal and store for 24 h; during this time the reaction between basic fuchsine and potassium metabisulfite leaves the solution colorless.
- If the solution is transparent and straw colored, it is ready to use.
- If not, cautiously bleach the solution with a small quantity of activated charcoal (0.25 g–0.5 g).
- Shake thoroughly and filter immediately.

Precautions

- Always keep the solution in a dark bottle and away from light. As the Schiff's reagent is unstable, it deteriorates in contact with light. Do not store beyond 6 months (Lhotka and Davenport 1949).

4.7 Squash and Smear Preparation

After all these years, the discipline of cytology has gone through several stages. The procedure of sectioning has mostly become outdated, but the procedures that find wide application even today are the squash and smear preparations. They have become popular because the study of chromosomes is now easy and fast. The most common tissues that are studied for chromosomes are root tips and anthers. The term "squash" and "smear" are loosely used but are totally different. Smears are formed when fresh tissue is spread over a slide and no treatment is necessary. Pollen mother cells are the ideal tissue for smear preparation. In case of squash, the tissue quantity is more, and as such some treatment with acids is necessary for the dissolution of the middle lamella which will leave the tissue softer. This tissue is then pressed under a cover slip till the desired spread is reached.

4.7.1 Smear Method

Smear method is invariably used for the study of meiosis in the pollen mother cells (PMCs). For this study, the chromosomes must be easily visible and decently countable. The haploid and the diploid numbers are established through this procedure along with the presence of any abnormalities in meiosis. Chromosome number being an important data for hybrids and newly found or rare species, this method still finds its place in the modern techniques. In addition the effect of several environmental factors such as pollution or use of chemicals on cultivated plants can be quantified by the degree of abnormalities they cause in meiosis and their general effect on plant vigor. The procedure involves either taking pollen mother cells out of the anther or using the anther, flower or whole inflorescence as such if they are minute. The pollen/anther is placed on the slide moistened by breath and quickly covered with a drop of acetocarmine. The stain prevents the tissue from drying up which will lead to chromosome clumping. The tissue is then either smashed with the back of the needle made flat or covered with a cover slip and then pressed to release the contents and/or for spreading the cells. The slide is then processed several times to attain the desired clarity and contrast. In this process it is important to observe every stage of the meiosis, for this we need to study the anthers in a gradual way from the smallest flower bud to the largest one or vice versa.

4.7.2 Squash Method

In the case of squash preparation, the main focus is the study of mitotic chromosomes and the karyotypic description. A good technique should permit the chromosomes to become well spread. Bringing out the details of chromosome morphology as distinctly as possible depends upon proper fixation that varies from plant to plant. Since, as already described, the tissue used in squash preparation is comparatively thicker and harder due to fixation, the primary requirement is the softening of the tissue before staining. This softening can be achieved by either enzymatic or chemical treatment which (i) digests the pectic materials in the middle lamella; (ii) loosens the cytoplasmic cross-linkage created by the fixatives, thereby clearing the cytoplasm; and (iii) allows the penetration of the stain. Once the tissue is soft, it is placed in a drop of stain between the slide and the cover slip. Pressure is then applied on the cover slip to form a uniform layer. The slide is then processed several times and, if found fruitful, is made permanent. In case of the chemical agents, warm dilute HCl used for a few minutes is the most commonly applied method. Usually 1N HCL is used for a period of 5–12 min at 50–60 °C. Before mounting in stain, the acid needs to be dried off, or it will not allow the stain to interact with the tissue. Prolonged treatment in acid causes depolymerization of DNA and breakage of nucleoprotein links which cause the chromosomes to appear blurry. Other chemical softening agents are trichloroacetic acid (Sharma 1956), 1 % chromic acid (Elliott 1956), and 2 % NaOH (Tandler 1959).

4.7.3 Making Slides Permanent

Once a good preparation has been made during mitosis or meiosis, one would like to make the slide permanent for future reference. The following chemical combination is used for making permanent slides:

1. Solution A: Acetic acid + butanol (1:1 ratio)
2. Solution B: Acetic acid (100 %)

Procedure

- Take 50–100 ml of solution A and B each in two separate petri plates.
- Wax the temporary slides on all sides of the coverslip and keep for 24 h at room temperature or in the refrigerator. The slides in the cover glass area are also marked with at least 3 ink dots on upper and lower sides acting as points of alignment between the cover glass and the slide (note that the upper and lower dots must superimpose each other).
- After 24 h, the wax is carefully removed with a sharp blade.
- The slide is then immersed first in solution A in such a way that the cover glass faces the solution and is completely immersed in it.
- Once the cover glass drops down in solution A, both slide and cover glass are carefully transferred to solution B for 1 min.
- The slide and cover glass are then taken out, and excess solution is removed with the help of blotting paper.
- A drop of Canada balsam or euparol is added on the slide near the cover glass area. The cover glass is then placed back on the slide in such a way that the ink dots match and superimpose again.
- The slide is then kept in an oven at 50 °C for 1–2 days.
- The slide is now permanent and can be stored in a wooden box.

Precautions

- The wax applied on the slide must be of a thin layer so that its removal becomes easy.
- There must not be any air bubbles formed under the cover glass area after the slide is waxed.
- The position and face of slide and cover glass is continuously monitored during their transfer in solutions A and B.

References

Aase HC (1935) Cytology of cereals. Bot Rev 1(12):467–496

Ahloowalia BS (1965) A root tip squash technique for screening chromosome number in Lolium. Euphytica 14(2):170–172

Allevi P, Mario A, Pierangela C, Alberto F, Antonio S, Steve B, Max M, John T (1991) The first total synthesis of carminic acid. J Chem Soc Chem Commun 18:1319–1320

Altmann R (1894) Die Elementarorganismen und ihre Beziehungen zu den Zellen. Veit, Leipzig

Armarego WL, Chai CLL (2009) Chapter 4-purification of organic chemicals purification of laboratory chemicals

Austin AP (1959) Iron-alum aceto-carmine staining for chromosomes and other anatomical features of Rhodophyceae. Stain Technol 34(2):69–75

Bahr GF (1954) Osmium tetroxide and ruthenium tetroxide and their reactions with biologically important substances: electron stains III. Exp Cell Res 7(2):457–479

Baker JR (1944) Memoirs: the structure and chemical composition of the Golgi element. Q J Microsc Sci 2(337):1–71

Baker JR (1950) Cytological technique. Methuen, London

Baker JR (1958) Principles of biological microtechnique: a study of fixation and dyeing. Methuen & Co Ltd, London, p 357

Baker JR (1960) Cytological techniques: the principles underlying routine methods. Methuen, London

Baker JR (1966) Cytological technique. The principles underlying routine methods, 5th edn. Methuen, London

Belling J (1921) On counting chromosomes in pollen-mother cells. Am Nat 55:573–574

Benda C (1896) Neuere Mitteilungen über die Histogenese der Säugetierspermatozoen. Verh d phys Ges zu Berlin, Jahrg, 1897

Benda C (1902) Akromegalie. Deutsche Klinik am Eingange Des Zwanzigsten Jahrhunderts 3:261

Berg WM (1927) Articles Chromsaure Chromsalze and Osmiumsaure in Enzyklopadie der Mikroskopischen Techinik Krause R. Berlin; Urban and Schawarzenberg

Bezbaruah HP (1968) An evaluation of preparatory procedures for leaf-tip chromosomes spreads of the tea plant (Camellia sinensis). Stain Tech 43(5):279–282

Bhaduri PN (1939) A study of the effects of different forms of Colchicine on the roots of *Vicia Faba* L. J R Microsc Soc 59(4):245–276

Böeseken J, Van Giffen J (1920) Sur les Cyclohexanediols 1, 2, et la Souplesse de L'Anneau Benzénique. Recl Trav Chim Pays-Bas 39(3):183–186

Bowes JH, Kenten RH (1949) The effect of deamination and esterification on the reactivity of collagen. Biochem J 44(2):142

Burns JA, Yang SJ (1961) Stain Tech 36:206

Candan F (2007) Morphological, anatomical, cytological and palynological investigations on C. ancyrensis, C. sieheanus, C. chrysanthus and C. flavus taxa of the genus Crocus L. – PhD thesis. Natural and Applied Sciences Institute, Celal Bayar University, Manisa, Turkey

Carey MA, McDonough ES (1943) On the production of polyploidy in Allium with paradichlorobenzene. J Hered 34:238–240

Carnoy JB (1886) La cvtodierê se de l'oeuf. Cellule 3:1–92

Casselman WGB (1955a) Cytological fixation by chromic acid and dicbromates. Quart J Micr Sci 96:203–222

Casselman WGB (1955b) The relative oxidizing properties of certain reagents and mixtures used for the fixation of tissues. Ibid 96:223–226

Chen Q, Li HY (2005) An improved technique for high resolution mitotic chromosome studies in Solanum. HortSci 40(1):54–56

Clarke JL (1851) Researches into the structure of the spinal cord. Philos Trans R Soc Lond 141:607–621

Conagin CHTM (1972) Colchicine effects on Arachis hypogaea L. Bragantia 31:187–198

Conn HJ (1953) Biological stains. A handbook on the nature and uses of the dyes employed in the biological laboratory, 6th edn. Williams & Wilkins, Baltimore

Criegee R (1936) Osmiumsäure-ester als Zwischenprodukte bei Oxydationen. Justus Liebigs Ann Chem 522(1):75–96

Criegee R, Marchand B, Wannowius H (1942) Zur kentniss der organischer osmium-verbinguen gen ½On organic osmium compounds. Ann Chem 550:99–133

Dixon WE (1905) A manual of pharmacology. E. Arnold, London
Dustin AP (1934) Action de la colchicine sur le sarcome greffe, type Crocker, de la souris. Bull acad med Belg 14:487–488
Dyer AF (1963) The use of lacto-propionic orcein in rapid squash methods for chromosome preparations. Biotech Histochem 38:85–90
Elliott CG (1956) Chromosomes in micro-organisms. Symp Soc gen Microbiol 6:279–295
Feulgen R, Rossenbeck H (1924) Mikroskopisch-chenischer Nachweis einer Nucleinsaiure vom Typus der Thymonucleinsafiure und die darauf beruhende elektive Farbung yon Zellkernen in mikroskopischen Priiparaten. Hoppe-Seyler's. Z Physiol Chem 135:203–252
Fischer A (1899) Fixierung Farbung and Bau des Protoplasmas. Fischer Verlag, Jena
Flemming W (1879) Über das Verhalten des Kerns bei der Zelltheilung. Virchows Arch [A] 77:1–29
Flemming W (1882) Zellsubstanz Kern und Zellteilung. FCW Vogel, Leipzig
Flemming W (1884) Z wiss. Mirk 1: 349
Gatenby JB, Beams HW (1950) The microtomist's vade-mecum, 11th edn. Churchill, London, p 753
Geiger PL (1833) Ueber einige neue giftige organische Alkalien. Annalen der Pharmacie 7(3):269–280
Godward MBE (1948) The iron alum acetocarmine method for algae. Nature 161(203):101–117
Goldberg RN, Kishore N, Lennen RM (2002) Thermodynamic quantities for the ionization reactions of buffers. J Phys Chem Ref Data 31(2):231–370
Gurr E (1960) Encyclopedia of microscopic stains. Leonard Hill, London
Haarmann W (1943) The complex affinity between heavy metals and proteins. V. Mercury binding by non-protein substances and the mechanism of the Takata reaction. Biochem Z 314:18
Hantzsch A, Osswald G (1900) Uber die Umwandlung von Farbbasen in Pseudoammonium-Hydrate -Cyanide und -Sulfonsauren. Ber Dtsch Chem Ges 33:278–317
Henwood A (2003) Current applications of orcein in histochemistry: a brief review with some new observations concerning influence of dye batch variation and aging of dye solutions on staining. Biotech Histochem 78:303–308
Hill HD, Myers WM (1945) A schedule including cold treatment to facilitate somatic chromosome counts in certain forage grasses. Stain Technol 20(3):89–92
Hörmann H, Grassmann W, Fries G (1958) Über den Mechanismus der Schiffschen Reaktion. Justus Liebigs Ann Chem 616(1):125–147
Johansen DA (1940) Plant microtechnique. McGraw-Hill Book Company, Inc., New York
Johnson MA (1989) An unusually high chromosome number in Voanioala gerardii (Palmae: Arecoideae: Cocoeae: Butiinae) from Madagascar. Kew Bull 44:207–210
Kagawa F (1929) A study on the phylogeny of some species in Triticum and Aegilops, based upon the comparison of chromosomes. J Coll Agr Inp Uni Tokyo 10:173–228
Kazem Y, Houshmand S, Madani B, MartínezGómez P (2010) Karyotypic studies in Iranian wild almond species. Caryologia 63(2):117–123
Kern A (1883) U.S. Patent No. 290,893. Washington, DC: U.S. Patent and Trademark Office
Khasim SM (2002) Botanical microtechnique; principles and practice. Capital Publishing Company, New Delhi, 41–52, 85–88
La Cour LF (1935) Technic for studying chromosome structure. Stain Tech 10:57–60
La Cour L (1941) Acetic-orcein: a new stain-fixative for chromosomes. Stain Tech 16:169–174
Lang A (1878) Ueber Conservation der Planarien. Zool Anz 1:14–15
Lassek AM (1950) A study of the precipitating effects of basic fixing solutions. Anat Rec 107(4):409–414
Lesins K (1954) Procedure to facilitate chromosome counts in difficult plant material. Stain Technol 29:261–264
Levan A (1949) The influence on chromosomes and mitosis of chemicals as studied by the Allium test. Hereditas 35(S1):325–337

Lhotka JF, Davenport HA (1949) Deterioration of Schiff's reagent. Biotech Histochem 24(4):237–239
Lillie RD (1954) Argentaffin and Schiff reactions after periodic acid oxidation and aldehyde blocking reactions. J Histochem Cytochem 2(2):127–136
Lits F (1934) Contribution à l'étude des réactions cellulaires provoquées par la colchicine. Compt rend Soc de biol 115(1421):15
Loeser E, Litchfield MH (1983) Review of recent toxicology studies on p dichlorobenzene. Food Chem Toxicol 21(6):825–832
Ludford RJ (1936) The action of toxic substances upon the division of normal and malignant cells in vitro and in vivo. Arch exp Zellforsch 18:411–441
Madhusoodanan KJ, Nazeer MA (1983) Comparative morphology of the somatic karyotypes of vegetable Amaranthus and its phylogenetic significance. Cytologia 48:237–244
Maluszynski M (2003) Doubled haploid production in crop plants: a manual. Springer Science & Business Media, Dordrecht
Mayer P (1903) Die Caprellidae der Siboga-expedition, vol 34. Buchhandlung und druckerei vormals EJ Brill, Leiden
Mayer P (1918) Ueber die Reinheit unserer Farbstoffe. Zeitsehr f wiss Mikr 34:305
Metzer RL, Lang EL (1955) Stain Tech 30:41
Meyer JR (1948) Modification of mitosis by chemicals. Science 108(2799):188
Middlebrook WR, Phillips H (1942) The action of formaldehyde on the cystine disulphide linkages in wool: the subdivision of the combined cystine into two fractions differing in their reactivity towards formaldehyde. Biochem J 36(3–4):294
Mirzaghaderi G (2010) Simple metaphase chromosome preparation from meristematic root tip cells of wheat for karyotyping or in situ hybridization. Afr J Biotechnol 9(3):314
Molad Y (2002) Update on colchicine and its mechanism of action. Curr Rheumatol Rep 4(3):252–256
Muller H (1859) Veber Glatte Muskelu und Nervenge flechte de chlorioidea un menschliche auge. Verhandlungen der physikalisch-medizinschen gesselschaft zu wurzurg 10:179
Navašchin SG (1912) 0 dimorfizme jader v somati6eskich kletkach u Gaetonia candicans. – Izv. Imp Akad Nauk Ser 6(6):373–385
Navaschin M (1925) Morphologische Kernstudien der*Crepis*-Arten in bezug auf die Artbildung. Z Zellforsch 2:98–110
Nebel BR, Ruttle ML (1938) The cytological and genetical significance of colchicine. J Hered 29(1):3–10
Newton WCF (1926) Chromosome Studies in Tulipa and some related Genera. J Linn Soc Lond Bot 47(316):339–354
Oehlkers F (1952) Chromosome breaks induced by chemicals. J Hered 6(Suppl):95–105
O'Mara JG (1939) Observations on the immediate effects of colchicine. J Hered 30(2):34–37
Ostergren G, Heneen WK (1962) A squash technique for chromosome morphological studies. Hereditas 48:332–341
Owen HR, Miller AR (1993) A comparison of staining techniques for somatic chromosomes of strawberry. HortSci 28(2):155–156
Palevitz BA (1993) Morphological plasticity of the mitotic apparatus in plants and its developmental consequences. Plant Cell 5(9):1001
Palmer RG, Heer H (1973) A root-tip squash technique for soybean chromosomes. Crop Sci 13:389–391
Pearse AE (1968) Histochemistry. Theor Appl 1:1–759
Pelletier PJ, Caventou JB (1820) Examen chimique de plusieurs végétaux de la famille des Colchicées, et du principe actif qu'ils renferment [Cévadille (veratrum sabadilla); hellébore blanc (veratrum album); colchique commum (colchicum autumnale)]. Ann Chim Phys 14:69–81
Perkin AG, Everest AE (1918) The natural organic colouring matters. Longmans/Green and Company, London

Pischinger A (1926) Die Lage des isoelektrischen Punktes histologischer Elemente als Ursache ihrer verschiedenen Färbbarkeit. Z Zellforsch Mikrosk Anat 3(2):169–197
Pischinger A (1937) Untersuchungen über die Kernstruktur besonders über die Beziehungen der Struktur im Leben und nach der Fixier-ung. Zsch Zellf 26:249–280
Porter KR, Kallman F (1953) The properties and effects of osmium tetroxide as a tissue fixative with special reference to its use for electron microscopy. Exp Cell Res 4(1):127–141
Pridgeon AM, Cribb PJ, Chase JM, Rasmussen FN (1999) Genera orchidacearum. General introduction, apostasioideae, cypripedioideae (Reprinted edition)
Prud'homme M (1900) Nouvelles matigres colorantcs a fonctions acides. Bull Soc Chim 23:456–457
Razaq ZA, Vahidy AA, Ali SI (1994) Chromosome numbers in compositae from Pakistan. Ann Mo Bot Gard 81:800–808
Rezaei Osalou A, Daneshvar Rouyandezagh S, Alizadeh B, Er C, Sevimay CS (2013) A comparison of Ice cold water pretreatment and α-bromonaphthalene cytogenetic method for identification of Papaver species. Sci World J 2013:608650
Riley HP, Majumdar SK (2015) The aloineae: a biosystematic survey. University Press of Kentucky, Lexington
Roy RP, Thakur V, Trivedi RN (1966) Cytogenetical studies in momordica. J Cytol Genet 1:30–40
Samuel R (1987) Chromosome numbers in piper. Kew Bull 42:465–470
Sarbhoy RK (1980) Karyological studies in the genus Phaseolus. Linn Cytologia 45(3):363–373
Schiff H (1865a) Action des aldehydes sur les amines. C R Acad Sci 69:32–33
Schiff H (1865b) Sur quelques amides de la serie toluique. C R Acad Sci 60:913–916
Schiff H (1865c) Note sur l'action des aldehydes sur la rosaniline. C R Acad Sci 61:45–75
Schmuck A, Kostoff D (1939) Brome-acenaphthene and bromenaphthaline as agents inducing chromosome doubling in rye and wheat. C R Acad Sci 23:263–266
Schultze M (1864) IJber den Bauder Leuchtorgane der Mfinnchen yon *Lampyris splendidula*. Arch Mikrosk Anat 1:124–137
Sharma AK (1956) Fixation of plant chromosomes. Bot Rev 22(10):665–695
Sharma AK, Bhattacharya NK (1956) Chromosome breakage through para-dichlorobenzene treatment. Cytologia 21:253–360
Sharma AK, Mookerjea MA (1955) Paradichlorobenzene and other chemicals in chromosome work. Stain Tech 30(1):1–7
Sharma AK, Roy M (1955) Orcein staining and the study of the effect of chemicals on chromosomes. Chromosoma 7(1):275–280
Sharma AK, Sharma A (1957) Permanent smears of leaf-tips for the study chromosomes. Stain Tech 32:167–169
Sidhu SS (1968) Further studies on the cytology of Mangrove species of India. Caryologia 21(4):353–357
Singh RJ (2003) Plant cytogenetics. CRC Press, Boca Raton, p 488
Sjöstrand FS (1956) In: Oster G, Pollister AW (eds) Physical techniques in biological research, vol 3. Academic, New York, p 241
Sjostrand FS (1959) Physical techniques in biological research 3C. Academia Press, New York
Srivastava LM (1966) Induction of mitotic abnormalities in certain genera of tribe. Vicieae by paradichlorobenzene. Ctytologica 31:166–171
Tandler CJ (1959) An alkali-formaldehyde squash technic for plant cytology and cytochemistry. Stain Techn 34:234–236
Telleyesniczky K (1905) Arch mikr Anat 66:367
Tijo JH, Levan A (1950) The use of oxyquinoline in chromosome analysis. Anales de la Estacion Expt de Aula Dei 2:21–64
Tsuchiya T (1971) Telocentric chromosomes in barley. In: Nilan RA (ed) Barley genetics II. Proc 2nd Int Barley Genet Symp. Washington State University Press, Pullman, pp 72–81
Turner B (1956) Amer Jor Bot 43:577

Van Beneden E, Neyt A (1887) Bull Acad Roy Belg Iii 14:215–295
Vig BK (1968) Spontaneous chromosome abnormalities in roots and pollen mother cells in aloe Vera L. Bull Torrey Bot Club 95:254–261
Wieland H, Scheuing G (1921) Die Fuchsin-schweflige Säure und ihre Farbreaktion mit Aldehyden. Berichte der deutschen chemischen Gesellschaft (A and B Series) 54(10):2527–2555
Wolman M (1955) Problems of fixation in cytology, histology, and histochemistry. Int Rev Cytol 4:79–102
Woodroffe D (1941) Fundamentals of leather science. Harvey Waddon, Croydon
Yost DM, White RJ (1928) The acid-forming properties of osmium tetroxide. J Am Chem Soc 50(1):81–84
Zacharias (1888) Anat Anz 3:24
Zirkle C (1928a) The effect of hydrogen-ion concentration upon the fixation image of various salts of chromium. Protoplasma 4(1):201–227
Zirkle C (1928b) Fixation images with chromates and acetates. Protoplasma 5:511–534

Flow Cytometry and Its Utility

5

Mihir Halder, Sayantani Nath, and Sumita Jha

Abstract

Flow cytometry is a laser-based biophysical technique to study optical properties of microscopic particles in fluid suspension. The technology is not only routinely used in medicine (transplantation, hematology, cancer, prenatal diagnosis, genetics, and sperm sorting for sex preselection) but also has many other applications in basic research, crop improvement, clinical practice, and clinical trials. The principles of flow cytometric chromosome analysis and sorting known as flow cytogenetics and research in this field have increased over the past four decades due to high sensitivity and precision of this technique. The use of nucleic acid-specific fluorochromes has semiautomated quantitative chromosome analysis, thus reducing subjectivity of preexisting slide-based methods. Flow cytometric classification of chromosomes (flow karyotyping) enables detection of chromosomal aberrations, while flow sorting permits isolation of single chromosome types in large quantities to be used for gene mapping, preparation of chromosome-specific gene libraries, and ultimately sequencing leading to chromosome genomics. Flow fluorescent in situ hybridization (FISH), i.e., combined *fluorescent in situ hybridization in suspension* (FISHIS) and flow cytometry using microsatellite DNA or gene-specific probes, is another approach to increase our knowledge on structure, function, and evolution of chromosomes. These techniques provide increased value for diagnosis and management of clinical diseases, especially cancer which is generally characterized by high chromosomal instability. Flow cytogenetics also has important applications in plant biology, especially in the study of genome evolution and crop improvement. This review

M. Halder • S. Nath • S. Jha (✉)
Center of Advanced Study, Department of Botany, University of Calcutta,
35, Ballygunge Circular Road, Kolkata 700 019, West Bengal, India
e-mail: mhalder16@gmail.com; snsayantani1@gmail.com; sumitajha.cu@gmail.com; sjbot@caluniv.ac.in

T.A. Bhat, A.A. Wani (eds.), *Chromosome Structure and Aberrations*,
DOI 10.1007/978-81-322-3673-3_5

outlines the utility of flow cytometry in chromosomal analysis, its applications, limitations, and future directions.

Keywords
Flow cytometry • Fluidics • Optics • Electronics • Flow karyotyping • Chromosome sorting • Chromosome genomics • Chromosome proteomics

5.1 Introduction

Flow cytometry (FCM), i.e., "measurement of cells in motion," is a laser-based biophysical tcchniquc to study optical properties of microscopic particles, usually cells, during flow in a continuous liquid current. This technology simultaneously measures and then analyzes multiple physical and chemical characteristics of up to thousands of particles or cells (0.2–150 μm in size) per second providing means of statistically accurate analysis. Thus, cells must be disaggregated from tissue before analysis. Flow cytometry is crucial for biological investigations, because it permits qualitative and quantitative analyses of intact cells and prepared cellular components (nuclei, chromosomes, and organelles) that have been labeled with a wide array of commercially available reagents, such as specific dyes or monoclonal antibodies. The main advantage of flow cytometry is the "one by one" measuring principle which makes possible the measurement of thousands of individual cell or particles in a matter of seconds. This ability of bulk particle measurement makes flow cytometers a widespread analytical tool for a broad spectrum of scientific disciplines both at research and commercial levels.

The advancement of flow cytometry was mainly driven by clinical purpose initially developed as an alternative tool to microscopy for automated blood cell counting. Flow cytometry has been progressively utilized in immunophenotyping over the last two decades and was proven to be of great clinical utility. Immunophenotyping alongside multiparameter flow cytometric approaches was proved to be of prognostic as well specific diagnostic tool for various diseases. Other important applications include examination of transplanted tissues, diagnosis of fetomaternal hemorrhage, sex preselection, etc.

The principles of chromosome analysis and sorting with the help of flow cytometric known as flow cytogenetics and research in this field have increased over the past four decades due to high sensitivity and precision of this technique. The chromosomal theory of heredity proposed by Walter Sutton in 1903 merged the separate histories of cytology and genetics and gave birth to cytogenetics as a new discipline. From that time cytogenetics has added to the in-depth exploration and understanding of chromosome structure, function, and evolution. The improvement in this field was mostly reliant on the development of progressively advanced techniques for staining, differentiation, and isolation of chromosomes. Cytogenetics was taken one step ahead by the correlation of molecular biology and chromosome studying techniques, thus forming the new field of molecular cytogenetics. This new approach allowed quantitative chromosome analysis (cytophotometry, cytofluorometry, and image analysis) which was later on automated by speedy progress of computer biology. All these techniques analyzed chromosomes fixed on a flat surface until Gray et al. (1975a) classified and sorted chromosomes from Chinese hamster cells based on dye content by flow cytometry initiating the new era of flow cytogenetics. Till date, flow cytometry is the most successful and widespread technique for chromosome separation on a preparative scale. This method of flow cytometric classification of chromosomes was termed flow karyotyping which represents the relative distribution of chromosomal DNA content of the entire genome. Flow karyotyping enables detection of chromosomal aberrations, while flow sorting allows separation of individual chromosome types in enough quantities to be used for gene mapping, preparation of chromosome-specific gene libraries, and sequencing or proteomic study of the entire chromosomes. Flow fluorescent in situ hybridization (FISH), i.e., combined *fluorescent in situ hybridization in suspension* (FISHIS) and flow cytometry using microsatellite DNA or gene-specific probes, is another approach to augment our knowledge on structure, function, and evolution of chromosomes. These techniques provide increased value for diagnosis and management of clinical diseases, especially cancer which is generally characterized by high chromosomal instability. Flow cytogenetics also has important applications in plant biology, especially in the study of genome evolution and crop improvement.

The most important characteristic of flow cytometry is its versatility, which explains its vast range of utilities, both in the past and for future. The only stringent prerequisite for successful flow cytometric analysis is particles in suspension tagged with an optically detectable dye/probe. It provides loads of information regarding cell viability, apoptosis, ploidy, genome structure, and cell cycle status. It has thus become an indispensible tool in pathology, immunology, oncology, and many other fields of biomedical research. Although less explored in plant biology, flow cytometry has many uses in fundamental research and industrial purposes, mainly in the estimation of genome size and ploidy level. The present review outlines the principle and history of flow cytometry, its utility, limitations, and future directions.

5.2 Flow Cytometric Measurement

Flow cytometry generally monitors absorbance and scattering of light, fluorescence, and probes depending on the sample nature and sample preparation. Different parameters (like cell size and shape, granularity of cytoplasm, internal pH, surface receptors, viability, etc.) of the cell can be measured by using flow cytometry (Rieseberg et al. 2001).

5.2.1 Absorbance and Scattering Monitoring

Split laser beams have long been used in flow cytometry. In modern cytometer, when particles or cells passing through the laser beam scatter light, it is detected as forward scatter (FSC) and side scatter (SSC). In flow cytometer, a laser light is focused on a point on the flow stream. When no cell is in the path of the laser beam, light beam uninterruptedly passes through the stream and falls on the obscuration bar, as a result no light reaches at the forward scatter detector. When a cell passes through the laser, light is refracted in all directions. The amount of light that is refracted in the forward direction (less than 10 °) passes around the cell along the same axis that the laser is traveling, reaching the forward scatter detector, called forward scattered light. FSC is very useful for the determination of cell size and shape (Ahlgrim et al. 2013). Side scattered light is the light scattered orthogonal (90°) to the incident laser beam as a cell passes through the laser beam and detected by the side scatter detector. This SSC light is affected by overall cell size and internal complexity of the cell.

5.2.2 Fluorescence Monitoring

Fluorescence monitoring forms the basis of majority of flow cytometric applications for the last few decades. Cellular parameters can be characterized by flow cytometer by measuring fluorescence, either intrinsic (if present) or extrinsic, depending on the need for reagents. Intrinsic fluorescence (autofluorescence) is the phenomenon where any cellular molecules itself exhibit fluorescence properties. For those cellular components which do not exhibit autofluorescence, specific fluorescent dyes are used for detection or staining of target molecule. Excitation of fluorochrome with light of specific wavelength results in fluorescence that is detected by photomultiplying tubes (PMT's) in flow cytometer. Fluorescent light is filtered by a set of filters and finally fluorescence of a specific wavelength is detected by each sensor. The discovery of different fluorochromes with distinct excitation and emission spectrum opens up the new horizon of flow cytometry that allows application of several fluorochromes simultaneously for the analysis of several parameters of the sample at any a time.

5.2.3 Probes

Fluorescent probes such as monoclonal antibodies, nucleic acid probes, and fluorochrome-conjugated antibodies are used in flow cytometry for various purposes. Fluorescent probes more specifically recognize target molecules and allow to measure different biological and biochemical properties of the cell. Application of fluorescent probes includes cell sorting, determining nucleic acid content, measuring enzyme activity, identifying cell surface receptors and specific antigens, and for apoptosis studies. Multiparametric analysis of the sample can be achieved at any one time using more than one fluorescent probe.

5.3 Utility of Flow Cytometry

Flow cytometers are utilized in a wide range of purposes ranging from routine clinical tests to international genome sequencing projects (Fig. 5.1). Such versatility makes it impossible to summarize all its applications and is beyond the scope of this chapter. However the most prominent uses, with a focus on chromosome study, are as follows.

5.3.1 Cell Size, Shape, and Granularity

Flow cytometry has been employed for assessing size, shape, and granularity of particles for more than 50 years now. Cell size and shape are major characteristics of a cell that play a central role in cell function and tissue structure. Flow cytometry can be used to determine relative size and shape of an individual cell as well as distribution of different cells types in a subpopulation on the basis of these parameters. In a flow cytometer, the FSC is measured and is analyzed for the calculation of relative size and shape of cells and their distribution in the sample. FSC is the quantity of the laser light beam that passes around the cell; thus higher FSC means higher cell volume. Absolute size can be determined by using a calibration curve prepared by passing polystyrene beads of known size. Flow cytometry has been used for deduction of size and shape of microbial cells and thus helps in species identification (Gasol and Giorgio 2000). It is also established that changed light scattering profiles of a particular cell type correlate with perturbed morphology and such data can be utilized for various purposes. Flow cytometry has been used as a valuable tool for speedy identification, isolation, and characterization of cell shape mutants of pathogenic microbes like *Helicobacter pylori* (Sycuro et al. 2010, 2013), thus accelerating research in their regulation. Flow cytometry has been utilized for detection of the extent of platelet activation based on shape changes (Ruf and Patscheke 1995) which play key role in patho-mechanisms of various arterial disorders like stroke, myocardial infarction, peripheral arterial disease, etc. It is also employed in the investigation of abnormalities in RBC rheology which is altered in diseases like terminal renal failure, sepsis, diabetes mellitus, and acute inflammatory states (Piagnerelli et al. 2007) Flow cytometry is also used for the analysis of cellular responses in nanoparticle-treated cells (Jiang et al. 2008). Although FSC is regularly used for measuring cell shape and size, it is not a monotonic function of cell size (Shapiro 2003) and can be influenced by damaged membrane; presence of strong absorbing, reflecting, or refracting materials; as well as nucleus and other organelles.

As discussed earlier the intensity of SSC is influenced by internal complexity of a cell like mass, protein content, refractive index of cytoplasm, number and type of organelles, vesicles or granules (Steen 2004), internal or surface irregularities, as well as different physical states like particle uptake, sperm decondensation, and mitosis (Giaretti and Nüsse 1994; Steen 2004). Initial studies on internal complexity of cell by flow cytometry were based on the discovery that different blood cells,

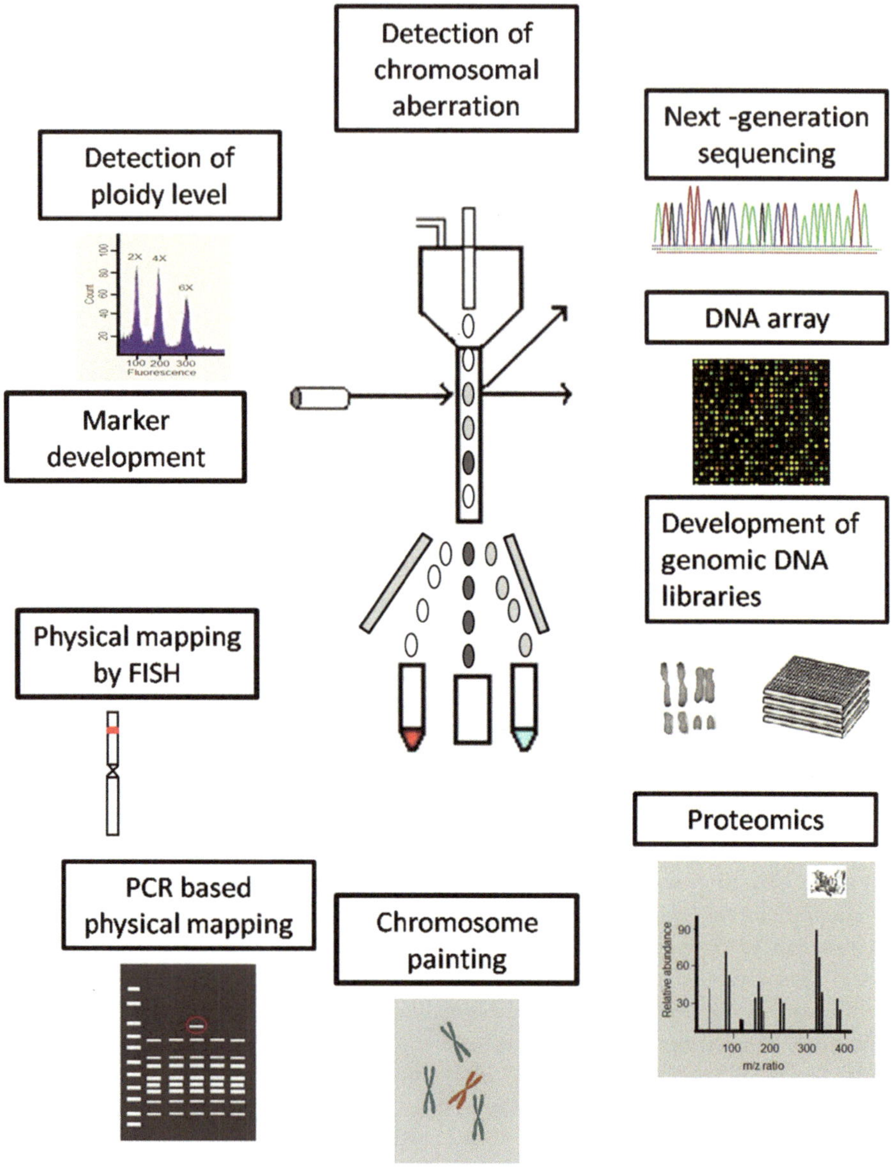

Fig. 5.1 Schematic diagram showing different applications of flow cytometry

viz., lymphocytes, monocytes, and granulocytes which differed in shape and granularity, could be discriminated by their distinctive FSC and SSC parameters (Salzman 2001). Flow cytometry has been shown to be useful in evaluation of nonmaterial-induced cytotoxicity based on alteration of light scatter properties of cells uptaking nanoparticle (Zucker et al. 2010).

Better analysis can be achieved by the application of light scatter measurements along with any other parameter. For example, different clusters of cells in a cytogram can be further correlated with other properties by using fluorescence-specific probes.

5.3.2 Viability and Apoptosis

Flow cytometry and various staining techniques are extensively used for cell viability assessment. Common dyes used for viability assay include rhodamine 123 (Evenson et al. 1982; López-Amorós et al. 1997), propidium iodide (PI) (López-Amorós et al. 1997; Shapiro 1995), carboxyfluorescein diacetate (cFDA) (Amor et al. 2002), bis-(1,3-dibutylbarbituric acid) trimethine oxonol (DiBAC4) (López-Amorós et al. 1997; Jepras et al. 1997), and ethidium bromide (EB) (Evenson et al. 1982). These are used in the assessment of microorganisms from different samples (Davey and Kell 1996; Breeuwer and Abee 2000; Bunthof et al. 2001) to analyze effects of antibiotics and chemical and physical agents and in turn help in their characterization and regulation. This process is also used in the assay of spermatozoa sample quality to estimate fertility (Evenson et al. 1982) accounted by cell motility and viability.

Apoptosis or controlled cell death of eukaryotes is operative during embryogenesis, tumor regression, immune response, etc., analysis of which is essential for basic research and clinical purposes. Apoptosis initiates an ordered series of cellular events (transcription of particular sequences, synthesis and localization of certain proteins, nuclease activation, DNA fragmentation, etc.) leading to certain morphological, biochemical, and molecular characteristics of cells. These changes can be easily investigated by various flow cytometric methods, the choice of which depends on the specific cell type and stage of apoptosis. Apoptosis results in altered light scattering properties of cells and thus can be investigated by studying FSC and SSC by flow cytometry (Darzynkiewicz et al. 1992). However the specific changes depend on the stage of apoptosis. For example, at initial stage, FSC decreases due to cell shrinkage, but SSC increases because of chromatin condensation. However at latter stages SSC also decreases because of secondary necrosis and cell leakage. Another method of flow cytometric analysis of apoptosis is by measurement of DNA content (using specific fluorochromes like propidium iodide, DAPI, etc.) which is reduced in apoptotic cell because of fragmentation and loss of nuclear DNA. Thus identification of cells with DNA content lesser than that of G1-cells is indicative of cell death by apoptosis (Darzynkiewicz et al. 1992). However this process involves certain chemical pretreatment of cells (permeabilization, ethanol fixation, etc.) which can lead to hyperfragmentation of DNA and overestimation of apoptotic cell percentage and thus should be carefully analyzed. The third method includes estimation of apoptotic DNA strand breaks by labeling with fluorochrome-conjugated dUTP using DNA polymerase or exogenous terminal deoxynucleotidyl transferase (TdT-assay) (Gorczyca et al. 1993). Although this method is more complex and costly as compared to others, it is more specific (necrotic cells show lesser breaks) and enables early identification (i.e., before other effects are visible) as well as identification of cell cycle stage when apoptosis initiates (by simultaneous analysis of DNA content). Flow cytometric analysis can also be done by identification and measuring of specific markers of apoptosis like phosphatidylserine after FITC annexin V staining (Fadok et al. 1992; Koopman et al. 1994). This method does not require any chemical pretreatment and allows early identification of apoptosis even before nuclear changes initiate. However time frame of apoptotic stages varies with cell and inducer types; thus the percentage of apoptotic population can differ with different methods for the same sample and hence should be carefully analyzed.

5.3.3 Nuclear DNA Content, Cell Cycle, and Ploidy

Nuclear DNA content or genome size (C-value) is a pivotal character in biodiversity, whose study offers a reliable unifying component in biology with practical and analytical significance. C-value represents "DNA content of unreplicated haploid nuclear chromosome complement" as described by Swift in 1950. Knowledge on genome size is increasingly useful in a phylogenetic study and in combination with other characters can contribute to taxa delimitation, intergeneric classification, or hybrid identification. Flow cytometry is the most precise method for quantifying genomic DNA (Crissman et al. 1979; Dolezel and Bartos 2005) and has been used to estimate the genome size of many plants (Bennett and Leitch 2005) and animal species (Hinegardner and Rosen 1972; Burton et al. 1989; Tiersch and Chandler 1989). Flow cytometric DNA content estimation is a relative method, where genome size of sample is quantified by comparison with a reference standard with known DNA amount. Trout red blood cells (TRBC) and chicken red blood cells (CRBC) are the most common internal standards for animals (Noguchi and Browne 1978; Vindeløv et al. 1983), whereas *Allium cepa* is the widely used one for plants (Bennett et al. 2000).

Flow cytometric DNA content estimation is used for determination of cell or sperm ploidy in fishes, amphibians, and reptiles (Bickham et al. 1985; Benfey et al. 1986; Echelle et al. 1988), sex identification and preselection in birds (Nakamura et al. 1990) and cattle (Johnson 1995), as well as for investigation of hybrid fishes (Goddard and Dawley 1990). Nuclear DNA content of plants has been extensively studied with the genome size of more than 4100 species of angiosperms published till date (Bennett and Leitch 2005). The study on plants has thus played a significant role to understand the origin, range, and effects of genome size variation of different taxa. Because of the rapid and accurate analysis, DNA content estimation by FCM analysis has important implications in basic research, breeding programs, and production. It is an easy, quick, and nondestructive tool for ploidy analysis which is useful in different crossing strategies for obtaining the desired ploidy level for many crops. It is an important tool for the assessment of ploidy stability of cultures and micropropagation of polyploids. FCM can also be employed for the detection of hybrids and aneuploids. FCM allows the analysis of DNA ploidy which is not observed by chromosome counting. Data on DNA content estimations can help resolve phylogenetic relationships or correlate with phenotypic characters and environmental factors. Estimation of genome size is still not possible for the majority of plant species mainly because of problems with sample preparation due to the presence of interfering cytoplasmic constituents and rigid cell wall. Improved sample preparation strategies can help solve the problem.

5.3.4 Flow Karyotyping and Chromosome Sorting

The use of nucleic acid-specific fluorochromes has semiautomated quantitative chromosome analysis and enabled flow cytometric classification of chromosomes termed as flow karyotyping. Flow karyotype represents the relative distribution of chromosomal DNA content of the entire genome. It enabled purification of different chromosomes into individual types in large numbers for further analysis establishing the new section of cytogenetic study, i.e., flow cytogenetics. Flow cytometry to analyze and sort metaphase chromosomes is applied in suspensions prepared from cell or tissue samples which were pretreated to achieve a high proportion of cells at this specific stage of mitotic division. Flow karyogram is prepared by the instrument based on fluorescence intensity (relative DNA content) of the DNA-specific fluorochrome used for staining. Theoretically, each chromosome forms a separate peak on the flow karyotype, whose position correlates with fluorescence intensity. Practically such results are very rare due to likeness in size and relative DNA content of chromosomes and peaks constituting different types of intergrades leading to poor resolution. This is mostly the case for monovariate analysis where DNA content is the only considered parameter. Monovariate analysis of human chromosomes by Gray et al. (1975b) resolved only seven peaks for the 24 chromosome types. This was improved by replacing the fluorochrome ethidium bromide with Hoechst 33258 by Carrano et al. (1979) which improved the resolution of human flow karyotypes by classifying 15 groups for 24 chromosome types. Bivariate flow karyotyping was proposed by Langlois et al. (1982) who analyzed human chromosomes stained with two separate fluorochromes with different base specificities—chromomycin A3 (GC specific) and Hoechst 33258 (AT specific). This improved the resolution of human flow karyotypes enabling the discrimination of all chromosomes in humans except chromosomes 9–12 and chromosomes 14 and 15. Since then flow karyotyping in man has been standardized to a level which allows semiautomated detection of chromosomal abnormalities (Boschman et al. 1992).

However, sample quality is of major significance for flow karyotyping and sorting of chromosomes. It should be optimized to obtain chromosomes in pure form without clumps free from all kinds of debris with unaltered morphology and structure (Bijman 1983; Telenius et al. 1993; Ng and Carter 2006). The progress in flow karyotyping of plants has been slowed down by problems in sample preparation due to the presence of rigid walls and difficulties in cell synchronization (Doležel et al. 1994). *Haplopappus gracilis* was the first plant with complete flow karyotypes reported by de Laat and Blaas (1984) who sorted its two chromosome types. Chromosome samples of plants were initially prepared from cultured cell (de Laat and Blaas 1984; Wang et al. 1992), but this approach was abandoned because of difficulties of initiation in some species, synchronization problems (Arumuganathan et al. 1991), and unstable karyotypes (Leitch et al. 1993). The leaf mesophyll protoplast (Conia et al. 1987, 1989) and root tip meristem from young seedling (Doležel et al. 1992) were suggested as useful alternative tissues for flow karyotyping of plants. Unlike animals and humans where bivariate flow karyotyping marked an immense advancement in flow cytogenetics, it did not cause any significant progress for plants (Lucretti and Doležel 1997; Lee et al. 2000). This setback was mainly because of the occurrence of homogenously dispersed repetitive DNA sequences in plants (Schubert et al. 2001). Thus flow cytometric analysis and sorting in plants have employed the use of only one DNA fluorochrome, typically DAPI, by many researchers (Vláčilová et al. 2002; Kubaláková et al. 2005). However the number of chromosomes which can be differentiated varies from plant to plant.

Flow cytometry is the current preferred approach for studying chromosomes because it allows quantitative, statistically accurate analysis over a short span of time. The sensitivity and precision of flow cytometers along with the advancement of nucleic acid-specific fluorochromes and probes widened the number of parameters that could be studied by this approach. Apart from characterizing the chromosome complement of a species, it is a prompt and accurate method for quantitative identification of structural and numerical chromosome aberrations (Otto and Oldiges 1980). Thus it is a particularly important tool for studying nuclear genome structure and analysis of specific and aberrant chromosomes. The development of *fluorescent in situ hybridization* in suspension (FISHIS) enables high-resolution mapping on sorted chromosomes. The DNA obtained from sorted chromosomes can be used for various purposes like PCR-based physical mapping, *in situ hybridization*, marker development, positional cloning, and preparation of chromosome-specific libraries. Flow sorting technology has allowed the reduction of nuclear genome to individual chromosomes and was proved to be revolutionary for genome sequencing particularly for species with large or complex genomes. Apart from reducing complexity in sequencing large genomes, it enables parallel work by independent team of researchers, each on a specific set of chromosome (s). This approach cross-linked cytology and genomics leading to the chromosome genomics era. Another novel but inadequately explored area is the proteomic analysis of flow sorted chromosomes which can throw light on structural aspects of chromosomes.

5.3.4.1 Chromosome Aberrations

Genetic aberrations resulting in changes in DNA content of a chromosome (deletions, insertions, tranlocations) of up to 3–5 Mbp (Trask et al. 1989) can be detected by flow karyotyping, while numerical chromosome aberrations remain unrecognized but can be detected by DNA content analysis. However translocations involving the same DNA content and inversions remain undetected by flow karyotyping. Aberration study is an important tool to analyze chromosomal rearrangements related to cancers and genetic diseases. FCM analyses and sorting can be used to identify breakpoints on aberrant chromosomes by hybridization with a large set of probes or spot blots. This method is so sensitive that it can be used to identify polymorphic variations within human populations. Amplification of chromosomal DNA permitted reverse chromosome painting where the paint or probe is developed from a flow-sorted aberrant chromosome of interest and hybridized to a normal karyotype to reveal the structure of the aberrant chromosome and location of breakpoints (Blennow 2004).

5.3.4.2 Physical Mapping

An important advantage of flow cytometric chromosome sorting is the preservation of normal morphology (Doležel et al. 1992). Thus, these isolated chromosomes can be used for physical mapping using PCR, FISH, etc. Initially specific genes were mapped on individual chromosomes by southern blotting or dot blots of sorted chromosomal DNA with labeled probes (Lebo 1982; Lebo et al. 1984; Arumuganathan et al. 1994). This was later replaced by the PCR-based method which is simpler and requires far less quantity of DNA. PCR-based mapping was widely used in humans, animals, and plants to localize sequences on specific chromosomes like sex chromosomes (Kejnovský et al. 2001; Matsunaga et al. 2005) and identify translocations (Kamnasaran et al. 2001) and breakpoints of chromosome deletions (Silverman et al. 1995). FISH was traditionally performed on metaphase chromosomes but can now be done on flow-sorted chromosomes entirely free from cytoplasmic debris (Lucretti et al. 1993) followed by mechanical stretching enabling higher-order spatial resolution of individual probes. This approach can help anchor specific linkage groups to chromosomes, ascertain order and orientation of sequences, and guess the length of contig gaps during the construction of physical maps (Szinay et al. 2010; Han et al. 2011). Clone-based physical mapping of flow-sorted chromosomes is a common strategy for producing reference sequence of cereal genomes as in the case of wheat by International Wheat Genome Sequencing Consortium (IWGSC) (Feuillet and Eversole 2007).

5.3.4.3 Genetic Marker Development

Markers are important sources for linkage map construction, trait inheritance study, and congregation of physical maps and shotgun sequences. Different types of markers include SNPs (single nucleotide polymorphisms), SSRs (simple sequence repeats), DArTs (diversity array technology), and ISBPs (insertion site-based polymorphisms) (Poczai et al. 2013). Marker development for chromosome-specific libraries is a useful strategy to lessen screening effort, specifically for polyploids. Marker development for flow-sorted chromosomes has been reported for field bean (Požárková et al. 2002), wheat (Michalak de Jimenez et al. 2013), barley (Mayer et al. 2011), and rye (Martis et al. 2013) and thus will enable prediction of chromosome specificity of sequences in the future.

5.3.4.4 Sequencing and Chromosome Genomics

In spite of the improvement of next-generation sequencing (NGS) over clone-by-clone sequencing (CBC), large genomes still pose a problem because of complexity and redundancies. Sequencing large genomes is costly, assembly of which is in turn hindered by a huge content of repetitive DNA, the problem being aggravated in allopolyploids, by the presence of homoeologous genomes in the same complement. Division of the genome into its constituent chromosomes offers a solution to these troubles. Apart from reducing complexity in sequencing large genomes, it enables parallel work by independent team of researchers, each on a specific set of chromosome(s), thus speeding up the entire process. Flow-sorted chromosomes thus have been proved appropriate for an array of applications which include physical and genetic mapping and whole genome sequencing giving rise to chromosome genomics. The first successful generation of DNA library of flow-sorted chromosomes was performed in wheat by Wang et al. (1992). This chromosome-based sequencing method is beneficial for the study of the complex genomes of cereals and presents considerable prospective for successful genome sequencing and gene cloning in these crops. Chromosome genomics thus can be used in genome sequencing projects to validate and support the precise sequence congregation obtained by NGS shotgun. Chromosome genomics has also opened up new avenues for the study of specialized chromosomes like supernumerary B chromosomes and sex chromosomes.

5.3.4.5 Chromosome Proteomics

A novel, almost unexplored area of flow cytogenetics is identification, study, and characterization of chromosomal proteins using proteomics approaches on sorted chromosomes collectively termed as chromosome proteomics. Such work was initiated by Uchiyama et al. (2005) in human chromosomes that resulted in detection of 107 proteins and prediction of probable model for metaphase chromosome structure (Takata et al. 2007; Fukui 2009). Some unattended areas for future research in this field include study of centromere structure, posttranslational modifications, etc.

5.3.5 Immunophenotyping and Clinical Uses

Flow cytometry has been progressively employed for immunophenotyping procedures over the last few decades and was proven to be of great clinical utility. Immunophenotyping by multiparameter flow cytometry is established as a prognostic as well as specific diagnostic tool for various hematological malignancies because neoplastic hematopoietic cells express aberrant phenotypes (Orfao et al. 1999). Another important application of immunophenotyping by flow cytometry is for inspection of quality and fate of transplanted tissues (both hematopoietic and solid organs) by counting human leukocyte antigens (HLA) like CD34+ hematopoietic progenitors (Sutherland et al. 1996) since its number in a transplant is the most reliable indicator of the fate of the graft and can predict chances of infections in patients. Also, flow cytometry crossmatch (FXCM) is very common before organ transplant which tests for IgG antibodies against B cells, T cells, or undifferentiated lymphocytes (Graff et al. 2009). The most widespread clinical application of flow cytometry is for counting of absolute and relative CD4+ T cells in peripheral blood for analyzing infection status of AIDS patients (Cassens et al. 2004). Some other diseases diagnosed by FCM include platelet dysfunction (Shattil et al. 1987; Michelson 1996), systemic mastocytosis (SM) (Escribano et al. 2001; Sotlar et al. 2004), and paroxysmal nocturnal hemoglobinuria (PNH) (Hall and Rosse 1996; Richards et al. 2000). It is also used for diagnosis and quantification of fetomaternal hemorrhage (Davis et al. 1998; Dziegiel et al. 2006). FACS can provide relatively pure subpopulations of X and Y chromosome bearing sperms for mammalian species which can then be used for sex preselection (Fugger 1999). This technique is now routinely used for various breeding purposes, particularly for cattle. This approach can also be used for reducing sex-linked and sex-limited disease risks in humans. Immunophenotypic by flow cytometry is also progressively commercialized for a wide array of procedures like platelet activation assay, basophil-activation-based allergy tests, diagnosis and monitoring of autoimmune disorders, identification and quantitation of autoantibodies, and detection of multidrug resistance in tumor cells.

5.4 Limitations and Future Prospects

Flow cytometry is the current preferred technique for both clinical and basic researches because of its capability to conduct speedy, precise, and large-scale analysis of a wide range of cellular or molecular properties. However improvement and utilization of flow cytometric techniques have been mainly driven by clinical purposes. This is important because assays which originate in basic research find quick clinical or commercial applications. Also utility of FCM in plant biology is still not fully explored and has important future implications. Flow karyotyping and chromosome genomics are the present emerging areas of research which will revolutionize approaches of studying chromosome and genome structure, respectively. Other new unexplored avenues with predictive novelty include the study of specialized chromosomes and chromosome proteomics which can unravel many mysteries of cell biology and genetics.

Glossary

Allopolyploid Polyploidy with more than one basic set of chromosomes

Apoptosis Biological event of program cell death observed in multicellular organism

Autofluorescence Intrinsic property of some biomolecules that emits fluorescence after excitation with light

B chromosome Extra-small chromosomes over standard set of chromosomes

Chromosomal aberrations Structural or numerical changes of chromosome

Chromosome painting Chromosomal in situ hybridization with multiple sequence-specific fluorescent probes

Cytogenetics Branch of genetics that deals with the study of structural and functional aspect of the cell, predominantly the chromosomes

Cytophotometry The analysis of cellular content with the help of an instrument called cytophotometer

Deletion Loss of a segment of a chromosome
DNA library Collection of DNA fragments of an organism that was cloned into vectors
Flow cytometry Laser-based biophysical technique to study optical properties of cells in liquid stream
Fluorescence in situ hybridization (FISH) Modern technique used for detecting and locating a specific DNA sequence on a chromosome using probe
Fluorochrome Chemical molecules that absorbed a particular wavelength of light, get excited and reemit light of higher wavelength in the form of fluorescent
Genetic marker Unique DNA sequence used in molecular mapping
GenomeComplete set of DNA in an organism
PCR Polymerase chain reaction
Probes Small nucleic acid sequence (DNA or RNA) or monoclonal antibodies or fluorochrome-conjugated antibodies used for detection of specific molecule
Sex chromosome Chromosomes responsible for the determination of sex of an organism
SSC The light scattered 90° to the incident laser beam as the cell flow through the flow cytometer
Translocation A chromosomal aberration that occurred due to the rearrangement of chromosomal parts in between two non-homologous chromosomes

References

Ahlgrim C, Pottgiesser T, Sander T, Schumacher YO, Baumstark MW (2013) Flow cytometric assessment of erythrocyte shape through analysis of FSC histograms: use of kurtosis and implications for longitudinal evaluation. PLoS One. doi:10.1371/journal.pone.0059862

Amor KB, Breeuwer P, Verbaarschot P, Rombouts FM, Akkermans AD, De Vos WM, Abee T (2002) Multiparametric flow cytometry and cell sorting for the assessment of viable, injured, and dead *Bifidobacterium* cells during bile salt stress. Appl Environ Microbiol 68(11):5209–5216

Arumuganathan K, Slattery JP, Tanksley SD, Earle ED (1991) Preparation and flow cytometric analysis of metaphase chromosomes of tomato. Theor Appl Genet 82:101–111

Arumuganathan K, Martin GB, Telenius H, Tanksley SD, Earle ED (1994) Chromosome 2-specific DNA clones from flow-sorted chromosomes of tomato. Mol Gen Genet 242(5):551–558

Benfey TJ, Solar II, de Jong G, Donaldson EM (1986) Flow-cytometric confirmation of aneuploidy in sperm from triploid rainbow trout. Trans Am Fish Soc 115(6):838–840

Bennett MD, Leitch IJ (2005) Plant genome size research: a field in focus. Ann Bot 95:1–6

Bennett MD, Johnston S, Hodnett GL, Price HJ (2000) *Allium cepa* L. cultivars from four continents compared by flow cytometry show nuclear DNA constancy. Ann Bot 85(3):351–357

Bickham JW, Tucker PK, Legler JM (1985) Diploid-triploid mosaicism: an unusual phenomenon in side-necked turtles *(Platemys platycephala)*. Science 227(4694):1591–1593

Bijman JT (1983) Optimization of mammalian chromosome suspension preparations employed in a flow cytometric analysis. Cytometry 3(5):354–358

Blennow E (2004) Reverse painting highlights the origin of chromosome aberrations. Chromosom Res 12(1):25–33

Boschman GA, Manders EM, Rens W, Slater R, Aten JA (1992) Semi-automated detection of aberrant chromosomes in bivariate flow karyotypes. Cytometry 13(5):469–477

Breeuwer P, Abee T (2000) Assessment of viability of microorganisms employing fluorescence techniques. Int J Food Microbiol 55(1–3):193–200

Bunthof CJ, Bloemen K, Breeuwer P, Rombouts FM, Abee T (2001) Flow cytometric assessment of viability of lactic acid bacteria. Appl Environ Microbiol 67(5):2326–2335

Burton DW, Bickham JW, Genoways HH (1989) Flow-cytometric analysis of nuclear DNA content in four families of neotropical bats. Evolution 43:756–765

Carrano AV, Gray JW, Langlois RG, Burkhartschultz KJ, Van Dilla MA (1979) Measurement and purification of human chromosomes by flow cytometry and sorting. Proc Natl Acad Sci U S A 76:1382–1384

Cassens U, Göhde W, Kuling G, Gröning A, Schlenke P, Lehman LG, Traoré Y, Servais J, Henin Y, Reichelt D, Greve B (2004) Simplified volumetric flow cytometry allows feasible and accurate determination of CD4 T lymphocytes in immunodeficient patients worldwide. Antivir Ther 9(3):395–406

Conia J, Bergounioux C, Perennes C, Muller P, Brown S, Gadal P (1987) Flow cytometric analysis and sorting of plant chromosomes from *Petunia* hybrida protoplasts. Cytometry 8(5):500–508

Conia J, Muller P, Brown S, Bergounioux C, Gadal P (1989) Monoparametric models of flow cytometric karyotypes with spreadsheet software. Theor Appl Genet 77(2):295–303

Crissman HA, Stevenson AP, Kissane RJ, Tobey RA (1979) Techniques for quantitative staining of cellular DNA for flow cytometry analysis. In: Melamed MR, Mullaney PF, Mendelsohn ML (eds) Flow cytometry and sorting. Wiley, New York, pp 243–261

Darzynkiewicz Z, Bruno S, Del Bino G, Gorczyca W, Hotz MA, Lassota P, Traganos F (1992) Features of apoptotic cells measured by flow cytometry. Cytometry 13(8):795–808

Davey HM, Kell DB (1996) Flow cytometry and cell sorting of heterogeneous microbial populations: the importance of single-cell analyses. Microbiol Rev 60(4):641–696

Davis BH, Olsen S, Bigelow NC, Chen JC (1998) Detection of fetal red cells in fetomaternal hemorrhage using a fetal hemoglobin monoclonal antibody by flow cytometry. Transfusion 38(8):749–756

de Laat AMM, Blaas J (1984) Flow-cytometric characterization and sorting of plant chromosomes. Theor Appl Genet 67:463–467

Dolezel J, Bartos J (2005) Plant DNA flow cytometry and estimation of nuclear genome size. Ann Bot 95(1):99–110

Doležel J, Číhalíková J, Lucretti S (1992) A high-yield procedure for isolation of metaphase chromosomes from root tips of *Vicia faba* L. Planta 188(1):93–98

Doležel J, Lucretti S, Schubert I (1994) Plant chromosome analysis and sorting by flow cytometry. Crit Rev Plant Sci 13:275–309

Dziegiel MH, Nielsen LK, Berkowicz A (2006) Detecting fetomaternal hemorrhage by flow cytometry. Curr Opin Hematol 13(6):490–495

Echelle AA, Echelle AF, DeBault LE, Durham DW (1988) Ploidy levels in silverside fishes (*Atherinidae, Menidia*) on the Texas coast: 37. Flow-cytometric analysis of the occurrence of allotriploidy. J Fish Biol 32:835–844

Escribano L, Díaz-Agustín B, Bellas C (2001) Utility of flow cytometric analysis of mast cells in the diagnosis and classification of adult mastocytosis. Leuk Res 25(7):563–570

Evenson DP, Darzynkiewicz Z, Melamed MR (1982) Simultaneous measurement by flow cytometry of sperm cell viability and mitochondrial membrane potential related to cell motility. J Histochem Cytochem 30(3):279–280

Fadok VA, Voelker DR, Campbell PA, Cohen JJ, Bratton DL, Henson PM (1992) Exposure of phosphatidylserine on the surface of apoptotic lymphocytes triggers specific recognition and removal by macrophages. J Immunol 148(7):2207–2213

Feuillet C, Eversole K (2007) Physical mapping of the wheat genome: a coordinated effort to lay the foundation for genome sequencing and develop tools for breeders. Isr J Plant Sci 55:307–313

Fugger EF (1999) Clinical experience with flow cytometric separation of human X-and Y-chromosome bearing sperm. Theriogenology 52(8):1435–1440

Fukui K (2009) Structural analyses of chromosomes and their constituent proteins. Cytogenet Genome Res 124:215–227

Gasol JM, Del Giorgio PA (2000) Using flow cytometry for counting natural planktonic bacteria and understanding the structure of planktonic bacterial communities. Sci Mar 64(2):197–224

Giaretti W, Nüsse M (1994) Light scatter of isolated cell nuclei as a parameter discriminating the cell-cycle subcompartments. Methods Cell Biol 41:389–400

Goddard KA, Dawley RM (1990) Clonal inheritance of a diploid nuclear genome by a hybrid freshwater minnows (*Phoxinus eos-neogaeus*, Pisces: Cyprinidae). Evolution 44:1052–1065

Gorczyca W, Gong J, Darzynkiewicz Z (1993) Detection of DNA strand breaks in individual apoptotic cells by the in situ terminal deoxynucleotidyl transferase and nick translation assays. Cancer Res 53(8):1945–1951

Graff RJ, Xiao H, Schnitzler MA, Ercole P, Solomon H, Pessin T, Lentine KL (2009) The role of positive flow cytometry crossmatch in late renal allograft loss. Hum Immunol 70(7):502–505

Gray JW, Caranno AV, Steinmetz LL, Van Dilla MA, Moore DH II, Mayall BH, Mendelsohn ML (1975a) Chromosome measurement and sorting by flow system. Proc Natl Acad Sci U S A 72(4):1231–1234

Gray JW, Carrano AV, Moore HH, Steinmetz LL, Minkler J, Mayall BH, Mendelsohn ML, Van Dilla MA (1975b) High-speed quantitative karyotyping by flow microfluorometry. Clin Chem 21(9):1258–1262

Hall SE, Rosse WF (1996) The use of monoclonal antibodies and flow cytometry in the diagnosis of paroxysmal nocturnal hemoglobinuria. Blood 87(12):5332–5340

Han YH, Zhang ZH, Huang SW, Jin WW (2011) An integrated molecular cytogenetic map of *Cucumis sativus* L. chromosome 2. BMC Genet 12:18–24

Hinegardner R, Rosen DE (1972) Cellular DNA content and the evolution of teleostean fishes. Am Nat 106:621–644

Jepras RI, Paul FE, Pearson SC, Wilkinson MJ (1997) Rapid assessment of antibiotic effects on *Escherichia coli* by bis-(1, 3-dibutylbarbituric acid) trimethine oxonol and flow cytometry. Antimicrob Agents Chemother 41(9):2001–2005

Jiang W, Kim BY, Rutka JT, Chan WC (2008) Nanoparticle-mediated cellular response is size-dependent. Nat Nanotechnol 3(3):145–150

Johnson LA (1995) Sex preselection by flow cytometric separation of X and Y chromosome-bearing sperm based on DNA difference: a review. Reprod Fertil Dev 7(4):893–903

Kamnasaran D, O'Brien PCM, Schuffenhauer S, Quarrell O, Lupski JR, Grammatico P, Ferguson-Smith MA, Cox DW (2001) Defining the breakpoints of proximal chromosome 14q rearrangements in nine patients using flow-sorted chromosomes. Am J Med Genet 102(2):173–182

Kejnovský E, Vrána J, Matsunaga S, Souček P, Široký J, Dolež̌e J, Vyskot B (2001) Localization of male-specifically expressed MROS genes of *Silene latifolia* by PCR on flow-sorted sex chromosomes and autosomes. Genetics 158(3):1269–1277

Koopman G, Reutelingsperger CPM, Kuijten GAM, Keehnen RMJ, Pals ST, van Oers MHJ (1994) Annexin V for flow cytometric detection of phosphatidylserine expression on B cells undergoing apoptosis. Blood 84:1415–1420

Kubaláková M, Kovářová P, Suchánková P, Číhalíková J, Bartoš J, Lucretti S, Watanabe N, Kianian SF, Doležel J (2005) Chromosome sorting in tetraploid wheat and its potential for genome analysis. Genetics 170(2):823–829

Langlois RG, Yu LC, Gray JW, Carrano AV (1982) Quantitative karyotyping of human chromosomes by dual beam flow cytometry. Proc Natl Acad Sci U S A 79:7876–7880

Lebo RV (1982) Chromosome sorting and DNA sequence localization: a review. Cytometry 3(3):145–154

Lebo RV, Gorin F, Fleterick RJ, Kao FT, Cheung MC, Bruce BD, Kan YW (1984) High-resolution chromosome sorting and DNA spot-blot analysis assign McArdle's syndrome to chromosome-11. Science 225(4657):57–59

Lee JH, Arumuganathan K, Chung YS, Kim KY, Chung WB, Bae KS, Kim DH, Chung DS, Kwon OC (2000) Flow cytometric analysis and chromosome sorting of barley (*Hordeum vulgare* L.). Mol Cell 10(6):619–625

Leitch AR, Schwarzacher T, Wang ML, Leitch IJ, Surlan-Momirovich G, Moore G, Heslop-Harrison JS (1993) Molecular cytogenetic analysis of repeated sequences in a long term wheat suspension culture. Plant Cell Tissue Organ Cult 33:287–296

López-Amorós R, Castel S, Comas-Riu J, Vives-Rego J (1997) Assessment of *E. coli* and *Salmonella* viability and starvation by confocal laser microscopy and flow cytometry using rhodamine 123, DiBAC4(3), propidium iodide, and CTC. Cytometry 29(4):298–305

Lucretti S, Doležel J (1997) Bivariate flow karyotyping in broad bean (*Vicia faba*). Cytometry 28(3):236–242

Lucretti S, Doležel J, Schubert I, Fuchs J (1993) Flow karyotyping and sorting of *Vicia faba* chromosomes. Theor Appl Genet 85(6–7):665–672

Martis MM, Zhou R, Haseneyer G et al (2013) Reticulate evolution of the rye genome. Plant Cell 25(10):3685–3698

Matsunaga S, Lebel-Hardenack S, Kejnovský E, Vyskot B, Grant SR, Kawano S (2005) An anther and petal-specific gene SlMF1 is a multicopy gene with homologous sequences on sex chromosomes. Genes Genet Syst 80(6):395–401

Mayer KFX, Martis M, Hedley PE et al (2011) Unlocking the barley genome by chromosomal and comparative genomics. Plant Cell 23(4):1249–1263

Michalak de Jimenez MK, Bassi FM, Ghavami F, Simons K, Dizon R, Seetan RI, Alnemer LM, Denton AM, Doğramacı M, Šimková H, Doležel J, Seth K, Luo MC, Dvorak J, Gu YQ, Kianian SF (2013) A radiation hybrid map of chromosome 1D reveals synteny conservation at a wheat speciation locus. Funct Integr Genomics 13(1):19–32

Michelson AD (1996) Flow cytometry: a clinical test of platelet function. Blood 87(12):4925–4936

Nakamura D, Tiersch TR, Douglass M, Chandler RW (1990) Rapid identification of sex in birds by flow cytometry. Cytogenet Cell Genet 53(4):201–205

Ng BL, Carter NP (2006) Factors affecting flow karyotype resolution. Cytometry A 69(9):1028–1036

Noguchi PD, Browne WC (1978) The use of chicken erythrocyte nuclei as a biological standard for flow microfluorometry. J Histochem Cytochem 26(9):761–763

Orfao A, Schmitz G, Brando B, Ruiz-Arguelles A, Basso G, Braylan R, Rothe G, Lacombe F, Lanza F, Papa S, Lucio P, San Miguel JF (1999) Clinically useful information provided by the flow cytometric immunophenotyping of hematological malignancies: current status and future directions. Clin Chem 45(10):1708–1717

Otto FJ, Oldiges H (1980) Flow cytogenetic studies in chromosomes and whole cells for the detection of clastogenic effects. Cytometry 1(1):13–17

Piagnerelli M, Boudjeltia KZ, Brohee D, Vereerstraeten A, Piro P, Vincent J-L, Vanhaeverbeek M (2007) Assessment of erythrocyte shape by flow cytometry techniques. J Clin Pathol 60(5):549–554

Poczai P, Varga I, Laos M, Cseh A, Bell N, Valkonen JP, Hyvönen J (2013) Advances in plant gene-targeted and functional markers: a review. Plant Methods 9(1):6–37

Požárková D, Koblížková A, Román B, Torres AM, Lucretti S, Lysák M, Doležel J, Macas J (2002) Development and characterization of microsatellite markers from chromosome 1-specific DNA libraries of *Vicia faba*. Biol Plant 45(3):337–345

Richards SJ, Rawstron AC, Hillmen P (2000) Application of flow cytometry to the diagnosis of paroxysmal nocturnal hemoglobinuria. Cytometry 42(4):223–233

Rieseberg M, Kasper C, Reardon KF, Scheper T (2001) Flow cytometry in biotechnology. Appl Microbiol Biotechnol 56:350–360

Ruf A, Patscheke H (1995) Platelet-induced neutrophil activation: platelet-expressed fibrinogen induces the oxidative burst in neutrophils by an interaction with CD11C/CD18. Br J Haematol 90(4):791–796

Salzman GC (2001) Light scatter: detection and usage. In: Robinson JP, Darzynkiewicz Z, Hoffman R, Nolan JP, Orfao A, Rabinovitch PS, Watkins S (eds) Current protocols in cytometry. Wiley, New York, pp 1.13.1–1.13.8

Schubert I, Fransz PF, Fuchs J, de Jong JH (2001) Chromosome painting in plants. Methods Cell Sci 23(1–3):57–69

Shapiro HM (1995) Practical flow cytometry. Alan R. Liss Inc, New York

Shapiro HM (2003) Practical flow cytometry. Wiley, Hoboken

Shattil SJ, Cunningham M, Hoxie JA (1987) Detection of activated platelets in whole blood using activation-dependent monoclonal antibodies and flow cytometry. Blood 70(1):307–315

Silverman GA, Schneider SS, Massa HF, Flint A, Lalande M, Leonard JC, Overhauser J, van den Engh G, Trask BJ (1995) The 18q(−) syndrome- analysis of chromosomes by bivariate flow karyotyping and the PCR reveals a successive set of deletion breakpoints within 18q21.2–q22.2. Am J Hum Genet 56(4):926–937

Sotlar K, Horny HP, Simonitsch I et al (2004) CD25 indicates the neoplastic phenotype of mast cells: a novel immunohistochemical marker for the diagnosis of systemic mastocytosis (SM) in routinely processed bone marrow biopsy specimens. Am J Surg Pathol 28(10):1319–1325

Steen HB (2004) Flow cytometer for measurement of the light scattering of viral and other submicroscopic particles. Cytometry A 57(2):94–99

Sutherland DR, Anderson L, Keeney M et al (1996) The ISHAGE guidelines for $CD34^{+}$ cell determination by flow cytometry. J Hematother 5(3):213–226

Sutton WS (1903) The Chromosomes in Heredity. Biol Bull 4:237–251

Swift H (1950) The constancy of desoxyribose nucleic acid in plant nuclei. Proc Natl Acad Sci U S A 36:643–654

Sycuro LK, Pincus Z, Gutierrez KD, Biboy J, Stern CA, Vollmer W, Salama NR (2010) Peptidoglycan crosslinking relaxation promotes *Helicobacter pylori*'s helical shape and stomach colonization. Cell 141(5):822–833

Sycuro LK, Rule CS, Petersen TW, Wyckoff TJ, Sessler T, Nagarkar DB, Khalid F, Pincus Z, Biboy J, Vollmer W, Salama NR (2013) Flow cytometry-based enrichment for cell shape mutants identifies multiple genes that influence *Helicobacter pylori* morphology. Mol Microbiol 90(4):869–883

Szinay D, Bai Y, Visser R, de Jong H (2010) FISH applications for genomics and plant breeding strategies in tomato and other solanaceous crops. Cytogenet Genome Res 129:199–210

Takata H, Uchiyama S, Nakamura N, Nakashima S, Kobayashi S, Sone T, Kimura S, Lahmers S, Granzier H, Labeit S, Matsunaga S, Fukui K (2007) A comparative proteome analysis of human metaphase chromosomes isolated from two different cell lines reveals a set of conserved chromosome-associated proteins. Genes Cell 12:269–284

Telenius H, Devos D, Blennow E, Willat LR, Ponder BAJ, Carter NP (1993) Chromatid contamination can impair the purity of flow-sorted metaphase chromosomes. Cytometry 14(1):97–101

Tiersch TR, Chandler RW (1989) Chicken erythrocytes as an internal reference for analysis of DNA content by flow cytometry in grass carp. Trans Am Fish Soc 118(6):713–717

Trask B, van den Engh G, Gray J (1989) Inheritance of chromosome heteromorphisms analysed by high-resolution bivariate flow karyotyping. Am J Hum Genet 45(5):753–760

Uchiyama S, Kobayashi S, Takata H, Ishihara T, Hori N, Higashi T, Hayashihara K, Sone T, Higo D, Nirasawa T, Takao T, Matsunaga S, Fukui K (2005) Proteome analysis of human metaphase chromosomes. J Biol Chem 280:16994–17004

Vindeløv LL, Christensen IJ, Nissen NI (1983) Standardization of high-resolution flow cytometric DNA analysis by the simultaneous use of chicken and trout red blood cells as internal reference standards. Cytometry 3(5):328–331

Vláčilová K, Ohri D, Vrána J, Číhalíková J, Kubaláková M, Kahl G, Doležel J (2002) Development of flow cytogenetics and physical genome mapping in chickpea (*Cicer arietinum* L.). Chromosom Res 10(8):695–706

Wang ML, Leitch AR, Schwarzacher T, Heslop-Harrison JS, Moore G (1992) Construction of a chromosome-enriched HpaII library from flow-sorted wheat chromosomes. Nucleic Acids Res 20:1897–1901

Zucker RM, Massaro EJ, Sanders KM, Degn LL, Boyes WK (2010) Detection of TiO_2 nanoparticles in cells by flow cytometry. Cytometry A 77(7):677–685

6 Asynapsis and Desynapsis in Plants

Aijaz Ahmad Wani and Tariq Ahmad Bhat

Abstract

The pairing of chromosomes also known as synapsis is essential for facilitating crossing over and recombination of genes during prophase-I and segregation of homologous chromosomes during anaphase-I of meiosis. Mutations in genes controlling synapsis affect normal pairing of homologues during prophase-I are give rise to synaptic mutants. The first synaptic mutants were discovered in maize and since then have been reported in large number of plant species. These synaptic mutants show complete or partial lack of chromosome pairing during meiosis. Asynapsis is the complete failure of homologous chromosomes to pair or synapse during the first meiotic division, whereas, desynapsis is a condition where homologous chromosomes pair or synapse normally at the beginning of prophase, but later fail to maintain this association in the subsequent stages of meiosis and thus separate prematurely. Both asynapsis and desynapsis have been found to play significant role in origin of popyploids via formation of 2n gametes. The meiotic disturbance due to asynapsis and desynapsis also leads to the formation of various types of aneuploids..

A.A. Wani (✉)
Department of Botany, University of Kashmir, Srinagar, Jammu and Kashmir, India
e-mail: aijazbotku@gmail.com

T.A. Bhat
Department of Education, Government of Jammu and Kashmir, Srinagar, India

T.A. Bhat, A.A. Wani (eds.), *Chromosome Structure and Aberrations*,
DOI 10.1007/978-81-322-3673-3_6

6.1 Introduction

Majority of the higher plants reproduce sexually through the union of male and female gametes ensuring restoration of parental chromosome number (2n) as well as sufficient variability in the offspring for adaptation and evolution. These male and female gametes which are haploid in nature are products of a specialized differential process called meiosis (Pankratz and Forsburg 2005). Meiosis begins with a single round of DNA replication during the S-phase of cell cycle followed by two successive nuclear divisions characterizing meiosis I and meiosis II. During zygotene stage of meiosis, homologous chromosome pairing occurs by the establishment of synaptonemal complex followed by the exchange of chromosome segments (crossing over) at pachytene stage. The homologous chromosomes are held together in the form of bivalents by various chiasmata until metaphase I of meiosis. At anaphase I, the homologous chromosomes move to opposite poles which is followed by telophase I and cytokinesis leading to the formation of two recombinant daughter nuclei with half the chromosome number than the parental cell. Each daughter cell/nucleus then further divides into two by undergoing second round of division known as meiosis II which is essentially a mitosis division characterized by chromatid segregation at anaphase II followed by telophase II and cytokinesis II (Tsubouchi and Roeder 2003). Thus after two rounds of chromosome segregation (karyokinesis) and one simultaneous or two successive cytoplasmic divisions (cytokinesis), the male meiosis generates a tetrad of haploid microspores enclosed in a callose wall. Among angiosperms, majority of the monocots are characterized by a distinct dyad stage during meiosis formed due to successive cytokinesis (i.e., one event of cytokinesis occurs after telophase I and the second one after telophase II). However, in most dicots, the cytokinesis is simultaneous (i.e., after telophase I the cell directly enters meiosis II and cytokinesis occurs only after telophase II) (Peirson et al. 1996; Cai and Xu 2007). Each meiotic event generates four haploid microspores on the male side and one egg cell on the female side. The male and female gametes later fuse with each other through a fertilization event to form the diploid offspring. The process of meiosis, on the one hand, maintains stability through faithful transmission of chromosomes and, on the other hand, induces genomic variations in the offspring through recombination during the sexual life cycles in eukaryotes (Cohen 2002; Gaut et al. 2007; Cai and Xu 2007).

The process of meiosis is controlled by a large number of genes generally dominant whose expression is stage specific, site specific, and time specific (Gottschalk and Kaul 1980a, b; Golubovskaya 1989; Pagliarini 2000; Villeneuve and Hillers 2001). Genetic, cytological, and immunological characterization of the meiotic process in model species such as *Saccharomyces cerevisiae*, *Drosophila melanogaster*, and *Arabidopsis thaliana* has led to the identification of genes conditioning distinct meiotic events, including recombination and crossing over (Dernburg et al. 1998; Grelon et al. 2001; Schwarzacher 2003; Hamant et al. 2006), spindle assembly (Musacchio and Hardwick 2002; Lew and Burke 2003), kinetochore orientation and chromosome segregation (Toth et al. 2000; Watanabe and Hauf 2004; Watanabe and Yokobayashi 2005), and cytokinesis (Nacry et al. 2000; Giansanti et al. 2004).

Among genes acting in the meiotic process, those responsible for the partitioning of the cytoplasm after nuclear division play a very important role in the formation of viable gametes (Boldrini et al. 2006).

6.2 Synaptic Mutations

Proper chromosome pairing, synapsis, and recombination are prerequisite for accurate segregation of homologous chromosomes during meiosis. Several genes are essential for encoding proteins required for pairing and synapsis (Lew and Burke 2003; Hamant et al. 2006). The mutations in genes which affect normal pairing of homologues during prophase I are called synaptic mutants. The first synaptic mutants were discovered in maize (Beadle and McClintock 1928) and since then have been reported in large number of plant species (Table 6.1). Katayama (1964) reported synaptic mutants in 20 families consisting of 50 genera and about 70 species of higher plants, whereas Koduru and Rao (1981) reported synaptic mutants in 126 species belonging to 93 genera. Majority of the synaptic mutants are found in the family Poaceae followed by Leguminosae and Liliaceae. In general synaptic mutants are found more in diploid species in higher plants (Singh 2003). These mutants may occur naturally by spontaneous mutations or may be induced by mutagenesis. Various factors such as temperature fluctuations, soil conditions, aging, structural changes in chromosomes, gene mutations, and interspecific hybridization have been reported to induce synaptic mutations in natural populations (Koduru and Rao 1981; Singh 2002; Rao and Kumar 2003; Singhal and Kumar 2010; Kumar and Singhal 2011; Singhal et al. 2012). Moreover, several mutagenic agents like X-rays, ethyl methanesulfonate (EMS), and ethylene imine (EI) have also been exploited to induce synaptic mutations (Singh and Ikehashi 1981; Naseem and Kumar 2013). In majority of the cases, synaptic mutants have been identified chiefly on the basis of cytological observations, genetic evidence, pollen and ovule abortion, etc. Various theories have been suggested by different authors for the occurrence of synaptic mutations. According to Armstrong et al. (2002), the genes responsible for the formation of synaptonemal complex (SC) proteins have highly conserved sequence. A mutation in these genes might have led to defective SC proteins, which are unable to hold the homologues together for a long time. Ji et al. (1999) proposed that recombination modifier mutation in rec gene might reduce recombination to a point where no pairing occurs. Failure of chiasma formation and recombination due to mutation in highly conserved rec gene has also been proposed by Simchen and Stamberg (1969). Complete or partial lack of chromosome pairing is expected to reduce recombination frequency in synaptic mutants. Such reduction in recombination frequency has been reported in desynaptic barley (Enns and Larter 1962) and Maize (Nel 1979). On the other hand no such reduction was found in asynaptic maize (Dempsey 1959) and tomato (Soost 1951), and even comparatively higher recombination frequency has been reported in asynaptic tomato (Moens 1969). According to Miller (1963), higher than normal rates of recombination in asynaptic plants are due to the fact that genetic markers under consideration were in distal

Table 6.1 Asynaptic and desynaptic mutants reported in plants

Species	Chromosome no.	Synaptic mutation	Origin	Reference
Avena sativa	42	Desy	I	Rines and Johnson (1988)
Avena strigosa	14	Desy	S	Dyck and Rajhathy (1965)
Brassica compestris	20	Asy	I	Stringam (1970)
Brachiaria humidicola	54	Desy	S	Calisto et al. (2008)
Capsicum annuum L.	24	Desy	S	Rao and Kumar (1983)
Carthamus tinctorius	24	Desy	I	Prasad and Prasad (1983)
Cicer arietinum L.	16	Desy	I	Kumar and Sharma (2001)
Corchorus fascicularis	14	Dcsy	S	Maity and Datta (2009)
Corchorus olitorius	14	Desy	S	Basak and Paria (1980)
Datura stramonium	24	Asy	S	Bergner et al. (1934)
Glycine max	40	Desy	S	Palmer and Kaul (1983)
		Desy	S	Bione et al. (2002)
		Desy	I	Kumar and Rai (2006)
		Desy	S	Hadley and Starnes (1964)
		Desy	S	Palmer (1974a, b)
Hordeum vulgare	14	Asy	I	Sethi et al. (1970)
		Desy	S/I	Ramage (1985)
		Desy	I	Srivastava (1974)
		Desy	I	Kumar and Singh (2002)
		Asy/Desy	I	Prasad and Tripathi (1986)
Lathyrus odoratus	14	Desy	I	Khawaja and Ellis (1987)
Lycopersicon esculentum	24	Desy	S	Soost (1951)
Oryza sativa	24	Asy	S	Kitada and omura (1984)
		Desy	I	Reddi and Rao (2000)
Pisum sativum	14	Asy	I	Gottschalk and Klein (1976)
		Desy	I	Gottschalk and Baquar (1971)
Paspalum jesuiticum	60	Asy	S	Bernardo Filho et al. (2014)
Paspalum somniferum	22	Desy	I	Naseem and Kumar (2013)
Secale cereale	14	Asy	S	Sosnikhina et al. (1992)
Sorghum vulgare	20	Asy	S	Stephens and Schertz (1965)
Solanum nigrum	48	Asy	S	Singhal et al. (2012)
Triticum durum	28	Asy	I	Martini and Bozzini (1966)
Triticum aestivum	42	Desy	S	Li et al. (1945)
Vicia faba	12	Asy	I	Sjodin (1970)
Zea mays	20	Desy	S	Bass et al. (2003)
		Asy	S	Miller (1963)

Note: part of the above table adopted from Singh (2003) with suitable modifications
Asy asynapsis, *Desy* desynapsis, *S* spontaneous, *I* induced

segments of the short arm or in segments near the centromere, where chromosomes pair more often than in the intercalary regions. Another group of workers believe that increased recombination in synaptic mutants may be due to compensation for the loss of recombination in some other parts of the genome (Sinha and Mohapatra 1969; Omara and Hayward 1978).

6.3 Asynapsis and Desynapsis

The synaptic mutations described above result in complete or partial loss of chromosome pairing during prophase and metaphase stages of meiosis. Complete failure of homologous chromosomes to pair or synapse during the first meiotic division is called **asynapsis**. On the other hand, homologous chromosomes pair or synapse normally at the beginning of prophase, but later fail to maintain this association in the subsequent stages of meiosis and thus separate prematurely. This phenomenon is called **desynapsis**. In both the cases, chromosome pairing fails due to structural and numerical reasons and genotypic or environmental anomalies may also be involved (John and Lewis 1965). The asynapsis and desynapsis are recognizable during pachytene and subsequent stages of meiosis when most or all of the chromosomes remain as univalents at diakinesis and metaphase (Visser et al. 1999). In several species where pachytene stage is not amenable due to small size of chromosomes, difficulty arises in distinguishing between the two phenomena. However, the occurrence of variable number of bivalents and univalents at metaphase I indicates desynapsis, whereas the presence of all univalents suggests asynapsis. The terminology for synaptic mutants is "as" and "ds" for asynapsis and desynapsis respectively (Srivastava 1974).

Asynapsis and desynaptic mutants have been documented in many plants species (review by Koduru and Rao 1981). Both spontaneous and induced types have been reported in different plant species such as *Zinnia haageana* (Singh and Gupta 1981), *Capsicum annum* (Kumar and Singh 2002), *Oryza sativa* (Reddi and Rao 2000), *Cicer arietinum* (Kumar and Sharma 2001), *Glycine max* (Palmer and Horner 2000; Bione et al. 2002; Kato and Palmer 2003; Kumar and Rai 2006), *Corchorus fascicularis* (Maity and Datta 2009; Naseem and Kumar 2013), Solanum *nigrum* (Singhal et al. 2012), *Zea mays* (Golubovskaya 1989; Bass et al. 2003), and *Hordeum vulgare* (Prasad and Tripathi 1986). Some of the mutants have been extensively characterized in many crop plants including wheat (Sears 1954), maize (Bass et al. 2003), rice (Kitada and Omura 1983; Nonomura et al. 2004), tomato (Moens 1969), and soybean (Kato and Palmer 2003).

According to Peirson et al. (1997), majority of the asynaptic mutants depicted irregular distribution and random dispersion of univalents in the cytoplasm at prophase I and metaphase I and they never congregate at the equatorial plate during metaphase I. On the other hand in the case of desynaptic mutant, bivalents and univalents were oriented at the equatorial plate during metaphase I. Singhal et al. (2012) performed cytological analysis of nine wild accessions of *Solanum nigrum* L. and found three interspecific cytotypes, viz., 2X (n = 12), 4X (n = 24), and 6X (n

= 36). The 4X cytotype depicted irregular chromosomal behavior during meiosis due to the presence of all 48 chromosomes as univalents. This accession was thus designated as asynaptic mutant. The meiosis in this mutant was highly irregular leading to the formation of abnormal sporads such as dyads, triads, and polyads. The microsporogenesis resulted in the formation of "n" and "2n" type of pollen grains with an overall 95% pollen sterility. Srivastava (1974) reported gamma rays induced desynaptic mutants in barley sowing varied degree of pollen and ovule sterility. The partially desynaptic mutant exhibited 60 % normal pairing and around 16–17 % pollen and ovule sterility, whereas the desynaptic mutant showed varied number of univalents and bivalents at metaphase with an overall reduced chiasma frequency per PMC and 80–100 % pollen and ovule sterility. The desynapsis was attributed to induced chromosome mutations caused by gamma radiation in one or more genes controlling chromosome pairing and chiasma formation. Prasad and Tripathi (1986) reported dES-induced asynaptic and desynaptic mutants in barley. The asynaptic mutant showed lack of complete chromosome pairing right from the pachytene to metaphase and was highly sterile (97 % pollen sterility) with very low (3–5 %) seed set. On the other hand desynaptic plant showed varying frequencies of univalents and bivalents at diplotene and diakinesis and the pollen sterility ranged from 38 to 90 %. The univalent formation at metaphase followed by unequal separation, multipolar separation, and lagging chromosomes at anaphase ultimately led to the formation of tetrads, pentads, and hexads. Naseem and Kumar (2013) reported EMS-induced desynapsis in poppy (*Papaver somniferum* L.). The desynaptic plant showed enhanced univalent frequency and loosely paired bivalents at metaphase I coupled with unequal separation of chromosomes at anaphase I. The male meiosis ended with the formation of micronuclei, abnormal tetrads, and high pollen and seed sterility. Desynapsis and abnormal cytokinesis have also been reported in one accession (HO22) of *Brachiaria humidicola* (Calisto et al. 2008).

6.4 Cytological Consequences of Asynapsis and Desynapsis

Both asynapsis and desynapsis lead to univalent formation which are usually observed at metaphase I. Univalents either get lost or are randomly transmitted to daughter cells, resulting in chromosomally unbalanced gametes and eventually aneuploids in the offspring. In addition, univalents may undergo misdivision, such as transverse division, to produce telocentric, acrocentric, acentric chromosomes and/or isochromosomes (Sears 1952; Friebe et al. 2005). Abnormal spindles and cytokinesis have been found associated with asynapsis and desynapsis in addition to abnormal chromosome behavior (Iwanaga and Peloquin 1979). All these abnormal meiotic events associated with asynapsis and desynapsis induce variations in chromosomal structure and number. A detailed cytological description of asynapsis and desynapsis is given below.

In the case of desynapsis, the bivalents are loosely associated. It has been argued that the occurrence of univalents during desynapsis may be due to small rearrangements, particularly interstitial translocations between chromosomes, with the result the chromosomes do not form rings or chains but pair loosely at metaphase I (Naseem and Kumar 2013). According to GottSchalk (1987), desynapsis (ds) genes influence microsporogenesis more strongly than megasporogenesis. Maguire et al. (1991) observed formation of synaptonemal complex and normal crossing over followed by failure of chiasma maintenance during desynapsis. Chromosomes undergo desynapsis after pachynema to diakinesis, and by metaphase I, the desynapsis is complete. The synaptonemal complex is apparently rapidly disintegrated following pachynema (Singh 2003). A reduction of chiasma frequency or complete failure of chiasma formation occurs at diplotene and diakinesis, resulting in varied frequencies of univalents and bivalents at metaphase. Due to small-size chromosomes in large number of plants, the desynapsis is usually ascertained at diakinesis and metaphase stages of meiosis. The degree of synapsis at these stages is reflected by the number of bivalents and frequency of chiasmata per cell. Chiasmata in desynaptic plants are mostly terminal at metaphase I and are rarely interstitial (Li et al. 1945). During desynapsis, the bivalents move to the equatorial plate at metaphase I, while univalents tend to be distributed at random in the cytoplasm. The bivalents may show disjunction at anaphase I, whereas univalents show an irregular behavior. Univalents that fail to move to either pole remain as laggards at the equatorial plate. At telophase I, those chromosomes that reach the poles organize dyad nuclei, while laggards often form micronuclei. The second meiotic division is essentially normal and irregularities are restricted to the first meiotic division.

The asynapsis on the other hand is characterized by complete absence of pairing and crossing over at pachytene stage and presence of univalents at diakinesis and metaphase stages of meiosis. Consequently the chromosome segregation is highly irregular with partial or complete failure of cytokinesis (Miller 1963). Meiosis I is especially irregular, whereas meiosis II is essentially normal, but the cells inherit chromosomal abnormalities resulting from 1st meiotic division. Due to irregular chromosome segregation, asynapsis produces unbalanced gametes resulting in a high level of pollen and ovule sterility. Electron microscopic studies have revealed the absence of formation of synaptonemal complex in asynaptic mutants (La Cour and Wells 1970; Golubovskaya and Mashnekov 1976). In an asynaptic mutant, Singhal et al. (2012) reported a number of meiotic anomalies such as scattered distribution of univalents at metaphase, laggards and unequal distribution of chromosomes at anaphase, and multipolar PMCs. Such irregular meiotic behaviors in asynaptic mutants have been reported by several workers (Koduru and Rao 1981; Kaul and Murthy 1985; Singh 2002; Singhal and Kumar 2010, Kumar and Singhal 2011, Kumar et al. 2011; Sharma et al. 2010, 2011).

6.5 2n Gamete Formation and Role in Polyploidy

Asynapsis and desynapsis result in abnormal microsporogenesis characterized by the presence of irregular sporads with dyads, triads, and polyads. As per rule, a dyad which has two unreduced microspores is expected to produce two 2n pollen grains, while a triad produces two "n" and one "2n" pollen grain (Singhal et al. 2012).

According to Bernardo Filho et al. (2014), the genetic constitution of 2n gametes falls into two general categories, viz., the first division restitution (FDR) or second division restitution (SDR). If the 2n gamete contains one sister chromatid of each replicated chromosome, it is considered to be an FDR 2n gamete. If on the other hand it includes both sister chromatids of one of the homologous chromosomes, it is considered to be an SDR 2n gamete.

The occurrence of "2n" pollen grains in synaptic mutants has been reported in several plants (Ortiz 1997; Calisto et al. 2008; Oselebe et al. 2010; Xue et al. 2011; Singhal et al. 2012). Thus, asynapsis and desynapsis have been considered as an important phenomenon in the formation of 2n gametes and subsequent polyploidization in higher plants (Bretagnolle and Thompson 1995). Bernardo Filho et al. (2014) from their cytological investigation in two wild accessions of *Paspalum jesuiticum* (2n = 6X = 60) reported complete asynapsis which resulted in the formation of dyads during meiosis and 2n microspores at the end of microsporogenesis. The formation of asynaptic mutant was attributed to recent natural hybridization with closely related species (*P. dilatatum*). Bernardo Filho et al. (2014) concluded that 2n gametes formed as a result of asynapsis/desynapsis may lead to auto- or allopolyploidy and that such spontaneous polyploidy is an important phenomenon both for evolution and for the development of new cultivars. Although autopolyploidy via fertilization between 2n gametes of the same species is successful right from its origin, the allopolyploidy is rarely successful depending upon the closeness between two species contributing 2n gametes. It has been argued that the presence of two different genomes in a common cytoplasm (interspecific hybrid formed by fertilization between 2n gametes) takes a long time in the evolutionary history of a new species (allopolyploid) before a fully functional interspecific hybrid operates normal meiotic behavior in nature (Bernardo Filho et al. 2014). Formation of fertile 2n gametes has been demonstrated in synaptic mutants of tomato (Ramanna 1983; Jongedijk et al. 1991), blackberry (Vorsa and Ortiz 1992), potato (Peloquin 1982), etc. The potential application of 2n gametes in the establishment of plyploid genotypes has been advocated by many researchers (Kim et al. 2009; Singhal and Kumar 2010; Dewitte et al. 2010; Silva et al. 2011; Singhal et al. 2012).

6.6 Practical Applications of Asynapsis and Desynapsis

Polyploidy is a common phenomenon among angiosperms and is regarded as an important mechanism of speciation and adaptation. Depending upon the criteria used, it has been estimated that around 40–70% angiosperms are polyploids (Grant 1981). Unlike synthetic ones, the natural polyploidy has played a very significant role in the evolution of some important crop plants. Among the cultivated plants, some of the important crops such as wheat, potato, cotton, oat, sugarcane, banana, groundnut, tobacco, and numerous horticultural crops are all polyploids. The synthetic polyploids are produced by colchicine treatment, whereas natural polyploids originate by various means of which formation of 2n gametes during meiosis followed by their fertilization is regarded as the main source. It has been well established that almost all plant species produce 2n gametes in some frequencies and that all polyploids in plants have originated through the functioning of 2n gametes. Polyploids that originate through the functioning of 2n gametes are called sexual polyploids and their usefulness for crop improvement has been demonstrated by many workers (Mendiburu and Peloquin 1971; Veronesi et al. 1996). As discussed in the preceding section, the 2n gametes originate due to disturbance in chromosome pairing during meiosis by way of asynapsis and desynapsis. A detailed review on types and formation of 2n gametes and their role in sexual polyploidization has been given by Ramanna and Jacobsen (2003). The meiotic disturbance due to asynapsis and desynapsis also leads to the formation of various types of aneuploid gametes. The fertilization event involving aneuploid gametes may lead to the formation of various types of aneuploid stock. Trisomics due to asynapsis have been obtained in different crop plants (Goodspeed and Avery 1939; Pal and Ramanujam 1940; Shah 1964; Dyck and Rajhathy 1965). Asynaptic and desynaptic mutants have been utilized to develop aneuploid stocks useful for genetic analysis. In common *Triticum aestivum* (2n = 42), a gene inhibiting asynapsis is located on the short arm of chromosome 3B (Sears 1954) and thus the nullisomic for chromosome 3B is partially asynaptic. Seventeen monosomics and 11 trisomics were isolated from the progeny of nullisomic 3B in common wheat cultivar Chinese spring (Sears 1954). Similarly in the progeny of desynaptic plants in jute (*Corchorus olitorius*, 2n = 14), all seven primary trisomics have been isolated (Basak and Paria 1980). In soybean (*Glycine max*, 2n = 40), a number of aneuploid lines with 41–43 chromosomes were isolated from different asynaptic and desynaptic plants (Palmer and Heer 1976; Palmer 1974b; Xu et al. 2000). A set of these aneuploid lines (primary trisomics, 2n = 41) have been used to assign individual genes and linage groups to specific chromosomes in soybean (Xu et al. 2000; Zou et al. 2003, 2006).

References

Anderson LK, Stack SM (2002) Meiotic recombination in plants. Curr Genomics 3:507–525

Armstrong SJ, Caryl AP, Jones GH, Franklin FCH (2002) Asy 1, a protein required for meiotic chromosome synapsis, localizes to axis-associated chromatin in *Arabidopsis* and *Brassica*. J Cell Sci 115:3644–3655

Atcheson CL, DiDomenico B, Frackman S, Esposito RE, Elder RT (1987) Isolation, DNA sequence, and regulation of meiotic specific eukaryotic recombination gene. Proc Natl Acad Sci U S A 84:8035–8039

Baker BS, Carpenter ATC, Esposito MS, Esposito RE, Sandler L (1976) The genetic control of meiosis. Annu Rev Genet 10:53–134

Basak SL, Paria P (1980) Desynapsis and spontaneous trisomy in jute (*Corchorus olitorius* L.). Theor Appl Genet 56:253–256

Bass HW, Bordoli SJ, Foss EM (2003) The desynaptic (dy) and desynaptic1 (dsy1) mutations in maize (Zea mays L.) cause distinct telomere-misplacement phenotypes during meiotic prophase. J Exp Bot 54:39–46

Beadle GW, McClintock B (1928) A genic disturbance of meiosis in Zea mays. Science 68:433

Bergner AD, Cartledge JL, Blakeslee AF (1934) Chromosome behavior due to a gene which prevents metaphase pairing in Datura. Cytologia 6:19–37

Bernardo Filho RA, Santos ACC, Souza FHD, Valls JFM, Pagliarini MS (2014) Complete asynapsis resulting in 2n pollen formation in *Paspalum jesuiticum* parody (Poaceae). Genet Mol Res 13(1):255–261

Bione NCP, Pagliarini MS, Almeida LA (2002) Anasynaptic mutation in soybean (*Glycine max* (L.) Merrill) associated with total absence of sister chromatids cohesiveness. Cytologia 67:177–183

Boldrini KR, Pagliarini MS, Valle CB (2006) Abnormal timing of cytokinesis in microsporogenesis of Brachiaria humidicola (Poaceae: Paniceae). J Genet 85:225–228

Bretagnolle F, Thompson JD (1995) Gametes with the somatic chromosome number: mechanisms of their formation and role in the evolution of autopolyploid plants. New Phytol 129:1–22

Cai X, Xu SS (2007) Meiosis driven genome variation in plants. Curr Genomics 8:151–161

Cai X, Dong F, Edelmann RE, Makaroff CA (2003) The *Arabidopsis SYN1* cohesion protein is required for sister chromatid arm cohesion and homologous chromosome pairing. J Cell Sci 116:2999–3007

Calisto V, Fuzinatto VA, Message HJ, Mendes-Bonato AB, Boldrini KR, Pagliarini MS, Do Valle CB (2008) Desynapsis and precocious cytokinesis in Brachiaria humidicola (Poaceae) compromise meiotic division. J Genet 87(1):27–31

Chaudhury AM, Lavithis M, Taylor PE, Craig S, Singh MB, Signer ER, Knox RB, Dennis ES (1994) Genetic control of male fertility in *Arabidopsis thaliana*: structural analysis of premeiotic developmental mutants. Sex Plant Reprod 7:17–28

Chelysheva L, Diallo S, Vezon D, Gendrot G, Vrielynck N, Belcram K, Rocques N, Marquez-Lema A, Bhatt AM, Horlow C, Mercier R, Mezard C, Grelon M (2005) *AtREC8* and *AtSCC* are essential to the monopolar orientation of the kinetochores during meiosis. J Cell Sci 118:4621–4632

Cohen J (1968) Sorting out chromosome errors. Science 2002(296):2164–2166. Wagenaar EB. Meiotic restitution and the origin of polyploidy. II. Prolonged duration of metaphase I as causal factor of restitution induction. Can J Genet Cytol 10:844–852

Cohen J (2002) Sorting out chromosome errors. Science 296:2164–2166

Dawson J, Wilson ZA, Aarts MGM, Braithwaite AF, Briarty LG, Mulligan BJ (1993) Microspore and pollen development in six male-sterile mutants of *Arabidopsis thaliana*. Can J Bot 71:629–638

Dempsey E (1959) Analysis of crossing over in haploid genetics of asynaptic plants. Maize Genet Coop News Lett 33:54–55

Dernburg AF, McDonald K, Moulder G, Barstead R, Dresser M, Villeneuve AM (1998) Meiotic recombination in *C. elegans* initiates by a conserved mechanism and is dispensable for homologous chromosome synapsis. Cell 94:387–398

Dewitte A, Eeckhaut T, Van Huylenbroeck J, Van Bockstaele E (2010) Meiotic aberrations during 2*n* pollen formation in *Begonia*. Heredity 104:215–223

Dyck PL, Rajhathy T (1965) A desynaptic mutant in Avena strigosa. Can J Genet Cytol 7:418–421

Enns H, Larter EN (1962) Linkage relations of ds: a gene governing chromosome behavior in barley and its effect on genetic recombination. Can J Genet Cytol 4:263–266

Friebe B, Zhang P, Linc G, Gill BS (2005) Robertsonian translocations in wheat arise by centric misdivision of univalents at anaphase I and rejoining of broken centromeres during interkinesis of meiosis II. Cytogenet Genome Res 109:293–297

Gaut BS, Wright SI, Rizzon C, Dvorak J, Anderson LK (2007) Recombination: an underappreciated factor in the evolution of plant genomes. Nat Rev Genet 8:77–84

Giansanti MG, Farkas RM, Bonaccorsi S, Lindsley DL, Wakimoto BT, Fuller MT, Gatti M (2004) Genetic dissection of meiotic cytokinesis in *Drosophila* males. Mol Biol Cell 15:2509–2522

Golubovskaya IN (1979) Genetic control of meiosis. Int Rev Cytol 58:247–290

Golubovskaya IN (1989) Meiosis in maize: *mei* genes and conception of genetic control of meiosis. In: Scandalios JG (ed) Advances in genetics, vol. 26. Academic Press, San Diego, pp 149–192

Golubovskaya IN, Mashnenkov AS (1976) Genetic control of meiosis. II. A desynaptic mutant in maize induced by N-Nitroso-N-Methylurea. Genetika 12(2):7–14

Goodspeed TH, Avery P (1939) Trisomic and other types in Nicotiana sylvestris. Genetics 38:381–458

Gottschalk W (1987) Different intensity of the action of desynaptic genes on micro- and megasporogenesis. Cytologia 52:653–656

Gottschalk W, Baquar SR (1971) Desynapsis in pisum sativum induced through gene mutation. Can J Genet Cytol 13:138–143

Gottschalk W, Kaul MLH (1974) The genetic control of microsporogenesis in higher plants. Nucleus 17:133–166

Gottschalk W, Kaul MLH (1980a) Asynapsis and desynapsis in flowering plants. I. Asynapsis. Nucleus 23:1–15

Gottschalk W, Kaul MLH (1980b) Asynapsis and desynapsis in flowering plants. II. Desynapsis. Nucleus 23:97–120

Gottschalk W, klein HD (1976) The influence of mutated genes in sporogenesis. A survey on genetic control of meiosis in *Pisum sativum*. Theor Appl Genet 48:23–34

Grant V (1981) Plant speciation, 2nd edn. Columbia University Press, New York

Grelon M, Vezon D, Gendrot G, Pelletier G (2001) *AtSPO11-1* is necessary for efficient meiotic recombination in plants. EMBO J 2001(20):589–600

Grishaeva TM, Bogdano YF (2000) Genetic control of meiosis in Drosophila. Russ J Gen 36:1089–1106

Hadley HH, Starnes WJ (1964) Sterility in soybeans caused by asynapsis. Crop Sci 4:421–424

Hamant O, Ma H, Cande WZ (2006) Genetics of meiotic prophase I in plants. Annu Rev Plant Biol 57:267–302

He C, Tirlapur U, Cresti M, Peja M, Crone DE, Mascarenhas JP (1996) An *Arabidopsis* mutant showing aberrations in male meiosis. Sex Plant Reprod 9:54–57

Hunt PA, Hassold TJ (2002) Sex matters in meiosis. Science 296:2181–2183

Iwanaga M, Peloquin S (1979) Synaptic mutant affecting only megasporogenesis in potatoes. J Hered 70:385–389

Ji YE, Stelly DM, Donato MD, Goodman MM, Williams CG (1999) A candidate recombination modifier gene for *Zea Mays* L. Genetics 151:821–830

John B, Lewis KR (1965) The meiotic system. Protoplasmatologia Bd. VI, Fr. Springer, Wein/New York

Jones GH (1974) Correlated components of chiasma variation and the control of chiasma distribution in rye. Heredity 32:375–387

Jongedijk E, Ramanna MS, Sawor Z, Hermsen JGT (1991) Formation of first division restitution (FDR) 2n megaspores through pseudohomotypic division in ds-1 (desynapsis) mutants of diploid potato: routine production of tetraploid progeny from 2xFDR x 2xFDR crosses. Theor Appl Genet 82:645–656

Katayama T (1964) Further review on the heritable asynapsis in plants. La Kromosomo 57–59:1934–1942

Kato KK, Palmer RG (2003) Molecular mapping of the male-sterile, female-sterile mutant gene (*st8*) in soybean. J Hered 94:425–428

Kaul MLH, Murthy TGK (1985) Mutant genes affecting higher plant meiosis. Theor Appl Genet 70:449–466

Khawaja HII, Ellis JR (1987) Colchicine induced desynaptic mutations in *Lathyrus odoratus* L. and *L. pretens* L. Genome 29:859–866

Kim JS, Oginuma K, Tobe H (2009) Syncyte formation in the microsporangium of *Chrysanthemum* (Asteraceae): a pathway to infraspecific polyploidy. J Plant Res 122:439–444

Kitada K, Omura T (1983) Genetic control of meiosis in rice *Oryza sativa* L. II. Cytogenetical analyses of desynaptic mutants. Jpn J Genet 58:567–577

Kitada K, Omura T (1984) Genetic control of meiosis in rice, Oryza sativa L. IV > Cytogenetical analysis of asynaptic mutants. Can J Genet Cytol 26:264–271

Koduru PRK, Rao MK (1981) Cytogenetics of synaptic mutants in higher plants. Theor Appl Genet 59:197–214
Kumar G, Rai P (2006) Induced desynaptic male sterile lines in soybean. Cytologia 71:337–343
Kumar G, Sharma V (2001) Induced desynapsis in *Cicer arietinum* L. J Cytol Genet 2:123–127
Kumar G, Singh V (2002) Desynaptic variation in barley. J Cytol Genet 3:127–131
Kumar P, Singhal VK (2011) Chromosome number, male meiosis and pollen fertility in selected angiosperms of the cold deserts of Lahaul Spiti and adjoining areas (Himachal Pradesh, India). Plant Syst Evol 297:271–297
Kumar P, Singhal VK, Kaur D (2011) Impaired male meiosis due to irregular synapsis coupled with cytomixis in a new diploid cytotype of *Dianthus angulatus* (Caryophyllaceae) from Indian cold deserts. Folia Geobot 47:59–68
La Cour LP, Wells B (1970) Meiotic prophase in anthers of asynaptic wheat. A light and electron microscopical study. Chromosoma (Berl) 29:419–427
Lew DJ, Burke DJ (2003) The spindle assembly and spindle position checkpoints. Annu Rev Genet 37:251–282
Li HW, Pao WK, Li CH (1945) Desynapsis in the common wheat. Am J Bot 32:92–101
Maguire MP, Paredes AM, Riess RW (1991) The desynaptic mutant of maize as a combined defect of synaptonemal complex and chiasma maintenance. Genome 34:879–887
Maity S, Datta AK (2009) Spontaneous desynapsis in *Corchorus fascicularis* Lamk. (Family: Tiliaceae). Indian J Sci Technol 2:34–35
Martini G, Bozzini A (1966) Radiation induced asynaptic mutations in durum wheat (Triticum durum Desf.). Chromosoma (Berl) 20:251–266
Mendiburu AO, Peloquin SJ (1971) High yielding tetraploids from 4x-2x and 2x-2x matings. Am Potato J 48:300–301
Miller OL (1963) Cytological studies in asynaptic maize. Genetics 48:1445–1466
Moens PB (1969) Genetic and cytological effects of three desynaptic genes in the tomato. Can J Genet Cytol 11:857–869
Musacchio A, Hardwick KG (2002) The spindle checkpoint: structural insights into dynamic signaling. Nat Rev Mol Cell Biol 3:731–741
Nacry P, Mayer U, Jurgens G (2000) Genetic dissection of cytokinesis. Plant Mol Biol 43:719–733
Naseem S, Kumar G (2013) Induced desynaptic variation in poppy (*Papaver somniferum* L.). Crop Breed Appl Biotechnol 13:363–366
Nel PM (1979) Effect of synaptic factor on recombination in maize. J Hered 70:401–406
Nonomura KI, Nakano M, Murata K, Miyoshi K, Eiguchi M, Miyao A, Hirochika H, Kurata N (2004) The insertional mutation of rice *PAIR2* gene, the ortholog of *Arabidopsis ASY1*, caused a defect in homologous chromosome pairing in meiosis. Mol Genet Genomics 271:121–129
Omara MK, Hayward MD (1978) Asynapsis in *Lolium perenne*. Chromosoma (Berl) 67:87–96
Ortiz R (1997) Occurrence and inheritance of 2*n* pollen in *Musa*. Ann Bot 79:449–453
Oselebe HO, Obi IU, Uguru MI (2010) Predicting hybrid performances from interploidy crosses in *Musa* species. Aust J Crop Sci 4:415–420
Page SL, Hawley RS (2003) Chromosome choreography: the meiotic ballet. Science 301:785–789
Pagliarini MS (2000) Meiotic behavior of economically important plant species, the relationship between fertility and male sterility. Genet Mol Biol 23:997–1002
Pal BP, Ramanujam S (1940) Asynapsis in chilli (Capsicum annum L.). Curr Sci 9:126–128
Palmer RG (1974a) A desynaptic mutant in the soybean. J Hered 65:280–286
Palmer RG (1974b) Aneuploids in the soybean, *Glycine max*. Can J Genet Cytol 16:441–447
Palmer RG, Heer H (1976) Aneuploids from a desynaptic mutant in soybeans (*Glycine max* (L.) Merr.). Cytologia 41:417–427
Palmer RG, Horner HT (2000) Genetics and cytology of a genic malesterile, female-sterile mutant from a transposon-containing soybean population. J Hered 91:378–383
Palmer RG, Kaul MLH (1983) Genetics, cytology and linkage studies of a desynaptic soybean mutant. J Hered 74:260–264

Pankratz DG, Forsburg SL (2005) Meiotic S-phase damage activates recombination without checkpoint arrest. Mol Biol Cell 16:1651–1660

Pawlowski WP, Golubovskaya IN, Timofejeva L, Meeley RB, Sheridan WF, Cande W (2004) Coordination of meiotic recombination, pairing, and synapsis by *PHS1*. Science 303:89–92

Peirson BN, Owen HA, Feldmann KA, Makaroff CA (1996) Characterization of three male-sterile mutants of *Arabidopsis thaliana* exhibiting alterations in meiosis. Sex Plant Reprod 9:1–16

Peirson BN, Bowling SE, Macaroff CA (1997) A defect in synapsis causes male sterility in a T-DNA-tagged *Arabidopsis thaliana* mutant. Plant J 11:659–669

Peloquin SJ (1982) Meiotic mutants in potato breeding. Stadler Genet Symp 14:99–109

Prasad G, Tripathi DK (1986) Asynaptic and desynaptic mutants in barley. Cytologia 51:11–19

Prasad G, Prasad J (1983) Induced desynapsis in safflower. Indian J Genet 43:276–278

Qu L, Vorsa N (1999) Desynapsis and spindle abnormalities leading to 2n pollen formation in *Vaccinium darrowii*. Genome 42:35–40

Ramage RT (1985) Cytogenetics. In: Rasmusson DC (ed) Barley, Agro Monograph 26. American society of Agronomy/Crop ScienceSociety of America/Soil Science Society of America, Madison, pp 127–154

Ramanna MS (1983) First division restitution gametes through fertile desynaptic mutants of potato. Euphytica 32:337–350

Ramanna MS, Jacobsen E (2003) Relevance of sexual polyploidization for crop improvement – a review. Euplytica 133:3–18

Rao KGR, Kumar OA (1983) Cytogenetics of a spontaneous desynaptic mutant in Chillies (*Capsicum annum* L). Cytologia 48:195–199

Rao SR, Kumar A (2003) Cytological investigations in a synaptic variant of *Anogeissus sericea* var. *sericea* Brandis (Combretaceae), an important hardwood tree of Rajasthan. Bot J Linn Soc 142:103–109

Reddi TVVS, Rao DRM (2000) Cytology of induced desynaptic mutant in rice. Cytologia 65:35–41

Rees H (1961) Genotypic control of chromosome form and behavior. Bot Rev 27:288–318

Rines HW, Johnson SS (1988) Synaptic mutants in hexaploid oats (*Avena sativa* L.). Genome 30:1–7

Roeder GS (1997) Meiotic chromosomes: it takes two to tango. Genes Dev 11:2600–2621

Ross JG, Sanders ME, Franzke CJ (1960) Asynapsis in sorghum. Hereditas 46:570–580

Schwarzacher T (2003) Meiosis, recombination and chromosomes: a review of gene isolation and fluorescent *in situ* hybridization data in plants. J Exp Bot 54:11–23

Sears ER (1952) Misdivision of univalents in common wheat. Chromosoma Bd. 4:535–550

Sears ER (1954) The aneuploids of common wheat, Research bulletin no. 572. University of Missouri, College of Agricultural Experiment Station, Columbia, pp 1–58

Sethi GS, Gill KS, Ghai BS (1970) Cytogenetics of induced asynapsis in barley. Indian J Genet Plant Breed 30:604–607

Shah SS (1964) Studies on a triploid, a tetrasomic triploid and a trisomic plant of Dactylis glomerata. Chromosoma 15:469–477

Sharma SK, Bisht MS, Pandit MK (2010) Synaptic mutation-driven male sterility in *Panax sikkimensis* Ban. (Araliaceae) from Eastern Himalaya, India. Plant Syst Evol 287:29–36

Sharma SK, Kumaria S, Tandon P, Rao SR (2011) Synaptic variation derived plausible cytogenetical basis of rarity and endangeredness of endemic *Mantisia spathulata* Schult. Nucleus 54:85–93

Silva N, Mendes-Bonato AB, Sales JGC, Pagliarini MS (2011) Meiotic behavior and pollen viability in *Moringa oleifera* (Moringaceae) cultivated in southern Brazil. Genet Mol Res 10:1728–1732

Simchen G, Stamberg J (1969) Fine and coarse controls of genetic recombination. Nature 222:329–332

Singh RJ (2002) Plant cytogenetics. CRC Press, London

Singh RJ (2003) Genetic control of meiosis. In: Singh RJ (ed) Plant cytogenetics, 2nd edn. CRC Press NY, Washington DC, pp 73–91

Singh S, Gupta PK (1981) Desynapsis in *Zinnia haegena* L. Cytologia 46:63–67

Singh RJ, Ikehashi H (1981) Monogenic male sterility in rice: induction, identification and inheritance. Crop Sci 21:286–289

Singhal VK, Kumar P (2010) Variable sized pollen grains due to impaired male meiosis in the cold desert plants of Northwest Himalayas (India). In: Kaiser BJ (ed) Pollen: structure, types and effects. Nova Science Publishers Inc, New York, pp 101–126

Singhal VK, Kaur M, Himshikha KP, Gupta RC (2012) High pollen sterility and 2n pollen grains in an asynaptic 4x cytotype (2n = 48) of *solanum nigrum* L. Cytologia 77(3):333–342

Sinha SK, Mohapatra BK (1969) Compensatory chiasma formation in maize. Cytologia 34:523–527

Sjodin J (1970) Induced asynaptic mutants in *Vicia faba* L. Hereditas 66:215–232

Soost RK (1951) Comparative cytology and genetics of asynaptic mutants in *Lycopersicon esculentum* Mill. Genetics 36:410–434

Sosnikhina SP et al (1992) Meiotic mutants of rye *Scale cereale* L. I. Synaptic mutant ry-1. Theor Appl Genet 84:979–985

Srivastava HM (1974) Radiation induced desynaptic mutant in barley. Cytologia 39:63–68

Stephens JC, Schertz KF (1965) Asynapsis and its inheritance in Sorghum vulgare Pers. Crop Sci 5:337–339

Stringam GR (1970) A cytogenetic analysis of three asynaptic mutants in Brassica compestris L. Can J Genet Cytol 12:743–749

Toth A, Rabitsch KP, Galova M, Schleiffer A, Buonomo SBC, Nasmyth K (2000) Functional genomics identifies monopolin: a kinetochore protein required for segregation of homologs during meiosis I. Cell 103:1155–1168

Tsubouchi H, Roeder GS (2003) The importance of genetic recombination for fidelity of chromosome pairing in meiosis. Dev Cell 5:915–925

Veronesi F, Mariani A, Bingham ET (1996) Unreduced gametes in diploid *Medicago* and their importance in alfalfa breeding. Theor Appl Genet 72:17–41

Villeneuve AM, Hillers KJ (2001) Whence meiosis? Cell 106:647–650

Visser NC, Spies JJ, Venter HJT (1999) The presence of synaptic and chromosome disjunction mutants in *Cenchrus ciliaris* (Poaceae: Paniceae). Bothalia 29:327–334

Vorsa N, Ortiz R (1992) Cytology of 2n pollen formation in a blueberry aneuploid (2n = 4x + 9 = 57). J Hered 83:346–349

Watanabe Y (2003) Monopolar attachment by Polo. Nat Cell Biol 5:379–382

Watanabe Y (2004) Modifying sister chromatid cohesion for meiosis. J Cell Sci 117:4017–4023

Watanabe Y, Hauf S (2004) Kinetochore orientation in mitosis and meiosis. Cell 119:317–327

Watanabe Y, Nurse P (1999) Cohesin Rec8 is required for reductional chromosome segregation at meiosis. Nature 400:461–464

Watanabe Y, Yokobayashi S (2005) The kinetochore protein *Moa1* enables cohesion-mediated monopolar attachment at meiosis I. Cell 123:803–817

Weaver JB (1971) An asynaptic character in cotton inherited as a double recessive. Crop Sci 11:927–928

Xu SJ, Singh RJ, Kollipora KP, Hymowitz T (2000) Primary trisomics in soybean: origin, identification, breeding behavior, and use in gene mapping. Crop Sci 40:1543–1551

Xue Z, Liu P, Liu M (2011) Cytological mechanism of 2*n* pollen formation in Chinese jujube (*Ziziphus jujuba* Mill. Linglingzao). Euphytica 182:231–238

Zickler D, Kleckner N (1999) Meiotic chromosomes: integrating structure and function. Annu Rev Genet 33:603–754

Zou JJ, Singh RJ, Lee J, Xu SJ, Cregan PB, Hymowitz T (2003) Assignment of molecular linkage groups to soybean chromosomes by primary trisomics. Theor Appl Genet 107:745–750

Zou JJ, Singh RJ, Lee J, Xu SS, Hymowitz T (2006) SSR markers exhibit trisomic segregation distortion in soybean [*Glycine max* (L.) Merr.]. Crop Sci 46:1456–1461

Cytogenetic Studies in Indian *Drimia* Jacq. (Urgineoideae: Hyacinthaceae)

7

M.M. Lekhak, P.B. Yadav, and S.R. Yadav

Abstract

Drimia (Urgineoideae: Hyacinthaceae) is a group of geophytes that bear tunicated bulbs. It is represented by more than 100 species distributed in Southern Europe, Africa and Asia. In India, *Drimia* has nine species (*Drimia indica*, *D. coromandeliana*, *D. govindappae*, *D. nagarjunae*, *D. polyantha*, *D. polyphylla*, *D. raogibikei*, *D. razii* and *D. wightii*). Species of the genus fall in two groups, namely, synanthous and hysteranthous. In the former, the leaves and flower appear simultaneously, while the latter group shows two distinct phases wherein the vegetative (leaf-bearing) and reproductive (flowering-bearing) periods are separated in time. Similarly, based on anthesis, species can be segregated into two groups, viz. night-blooming and day-blooming species. This is actually a temporal isolation that prevents gene flow amongst species. Of the nine species in India, *Drimia polyphylla* which is known to occur in the Deccan Peninsula has never been collected since its type collection. *Drimia nagarjunae*, the most robust Indian species, is probably closely related to the African *D. maritima*. *Drimia* species are well known for their large-sized chromosomes with usually subterminal constrictions. Seven species have been cytogenetically examined so far. Cytogenetical studies reveal the presence of many cytotypes at the population level. The most common diploid chromosome number for Indian *Drimia* is $2n = 20$. The species exhibiting $2n = 20$ are *Drimia indica*, *D. govindappae*, *D. nagarjunae*, *D. polyantha*, *D. razii* and *D. wightii*. In addition, diploid numbers ($2n$) of 30 (*Drimia indica*), 40 (*D. coromandeliana*, *D. indica* and *D. species*) and 60 (*D. indica*) are also on record. Geographical distribution of *Drimia* species and their cytotypes is well correlated. For instance, populations of *Drimia* species on higher altitudes where edapho-climatic conditions are harsh show

M.M. Lekhak (✉) • P.B. Yadav • S.R. Yadav
Angiosperm Taxonomy Laboratory, Department of Botany, Shivaji University, 416004, Kolhapur, Maharashtra, India
e-mail: mml_botany@unishivaji.ac.in

T.A. Bhat, A.A. Wani (eds.), *Chromosome Structure and Aberrations*,
DOI 10.1007/978-81-322-3673-3_7

presence of B chromosomes. Crossability studies in Indian *Drimia* show that *D. nagarjunae* and *D. wightii* are reproductively isolated from the rest of the species, while other species show varying degree of crossability indicating their genetic proximity.

In the present communication, we discuss morphology, cytogenetics, crossability behaviour and flowering phenology of Indian *Drimia* species.

Keywords
Urginea • Bulbous monocots • Morphology • Chromosomes • Systematics

7.1 Taxonomy

7.1.1 Present Status

Hyacinthaceae Batsch ex Borkh. comprise approximately 1000 species grouped in 35 genera (Buerki et al. 2012). The family is most speciose in Old World where it is distributed throughout Africa and the Mediterranean Basin with some species in northern Europe, the Middle East and Asia. *Oziroë* Raf. endemic to the western South America is the only genus found in the New World (Buerki et al. 2012). The genus *Drimia* was placed in subfamily Urgineoideae under Hyacinthaceae (Manning et al. 2004). As per Chase et al. (2009), the genus can be placed in the subfamily Scilloideae (Asparagaceae Juss.). In the most recent work on this group, Buerki et al. (2012) recognized *Drimia* under the family Hyacinthaceae. According to these authors, Hyacinthaceae form a monophyletic group within Asparagaceae sensu lato and should be given a familial rank.

Drimia derives its name from the Greek word *drimys* meaning acrid or pungent referring to the poisonous properties of the bulb. It is distributed in Africa, Madagascar, the Mediterranean area and Asia. It is represented by 103 species (modified after The Plant List 2013). Africa is home to 93 species. Europe harbours five species, while 12 species [*Drimia aphylla* (Forssk.) J.C. Manning & Goldblatt, *D. coromandeliana* (Roxb.) Lekhak & P.B. Yadav, *D. fugax* (Moris) Stearn, *D. govindappae* (A. Boraiah & Fathima) Lekhak & P.B. Yadav, *D. indica* (Roxb.) Jessop, *D. nagarjunae* (Hemadri & Swahari) Anand Kumar, *D. polyantha* (Blatt. & McCann) Stearn, *D. polyphylla* (Hook. f.) Ansari & Sundararagh., *D. raogibikei* (Hemadri) Hemadri, *D. razii* Ansari, *D. rupicola* (Trimen) Dassan. and *D. wightii* Lakshmin.] are reported from the Asian continent. One of the species referred in text as *Drimia* species has two cytotypes, diploid (2*x*) and tetraploid (4*x*). Since the identity of this species is not clear, we do not assign it any formal name. *Drimia indica* is a cosmopolitan species, while *D. aphylla* and *D. rupicola* are reported from temperate Asia and Sri Lanka, respectively. *Drimia fugax* is native to Africa and temperate Asia. All the remaining species except *Drimia indica* are endemic to India (Table 7.1). Amongst the Indian species, *Drimia polyphylla* has never been collected after type collection which is from Akya, a place somewhere in

Table 7.1 Indian species of *Drimia*, their distribution, gametic count (*n*) and zygotic count (2*n*)

Sr. No.	Taxa	Geographical distribution	*n*	2*n*	Authors
1.	*D. coromandeliana*	Karnataka, Maharashtra (Kagal, Kolhapur district)	–	40	Naik (1976)
			20	40	Dixit and Yadav (1989)
2.	*D. govindappae*	Karnataka (Bangalore and Mysore)	–	20	Boraiah and Fathima (1972)
			–	20	Boraiah and Khalel (1970)
			–	20 + 7B	Present communication
3.	*D. indica*	Andhra Pradesh, Bihar, Gujarat, Karnataka, Kerala, Madhya Pradesh, Maharashtra, Orissa, Rajasthan, Tamil Nadu, Uttar Pradesh. Throughout Indian plains	–	20, 30	Raghavan (1935a, b)
			–	20	Capoor (1937)
			–	20, 30	Raghavan and Venkatasubban (1940)
			–	20	Kishore (1951)
			–	20	Miege (1960)
			–	20	Mehra and Malik (1961)
			–	20 + 1–7B	Ayyangar (1966)
			–	40	Ayyangar and Sampathkumar (1968)
			–	20, 22, 30	Subramanian (1973)
			–	20 + 6–7B.	Sen (1974)
			–	20	Naik (1976)
			10, 20	20	Sen and Mukherjee (1976)
			–	10 + 1,12 + 1, 14 + 1, 16 + 1, 18 + 1, 20 + 1, 20 + B, 30, 30 + B	Subramanian (1978)
			–	20, 30	Sharma and Dash (1977)
			–	20	Zaman and Khaleque (1978)
			–	19, 20 20 + 1, 22	Subramanian (1980)
			–	30	Sheriff and Rao (1981)

(continued)

Table 7.1 (continued)

Sr. No.	Taxa	Geographical distribution	*n*	2*n*	Authors
			–	20, 30, 40	Patil and Torne (1982)
			–	20	Joshi and Ranjekar (1982)
			10, 20	20, 30, 40, 60	Jha and Sen (1983a, b)
			–	20 + 0–6B	Jha and Sen (1983a, b)
			–	20	Jha (1989)
			10	20	Dixit and Yadav (1989)
			10	20, 20 + 3B	Yadav and Dixit (1990)
4.	*D. nagarjunae*	Andhra Pradesh (Bhata village near Udayagiri Hill Fort, Nellore District), Kerala (Chinnar wildlife sanctuary, Idduki District), Tamil Nadu (Anamalai hills)	–	20	Hemadri and Swahari (1982)
			10	20	Lekhak et al. (2014)
5.	*D. polyantha*	Eastern and Western Ghats, Gujarat, Karnataka, Maharashtra, Tamil Nadu, Peninsular India.	–	20	Kambale and Ansari (1976)
			–	20	Bruyns and Vosa (1987)
			10	20	Yadav and Dixit (1990)
			–	20 + 7B	Present communication
6.	*D. polyphylla*	Deccan Peninsula	–	#20	Raghavan and Venkatasubban (1940)
7.	*D. raogibikei*	Andhra Pradesh (Nellore District)	–	–	–
8.	*D. razii*	Maharashtra (Dive Ghat, Pune District)	10	20	Yadav and Dixit (1990)
9.	*D. wightii*	Tamil Nadu (Bharathidasan University Campus, Trichy)	–	–	–
10.	*D.* species	Andhra Pradesh, Karnataka, Maharashtra, Tamil Nadu	10	20	Yadav and Dixit (1990)
			–	40	Yadav unpublished

– represents taxa where gametic count (*n*) and zygotic count (2*n*) are not known
dubious report. For narrowly distributed species exact locations are given

the Deccan Peninsula (Lekhak et al. 2014). A taxonomic revision of the genus in India is still wanting.

The genus *Drimia* is characterized by herbs with subterranean tunicated bulbs, radical, linear or lorate, green to glaucous leaves. Leaves may be erect-emergent or variously curled, touching the soil. Flowers are borne on usually long, naked scapes in racemes, usually before the appearance leaves. Flowers of the scapes may be laxly arranged or congested. They are short-lived (surviving from few hours up to a day). Pedicel is short to long and articulated at base with spurred bracts. Perianth (tepals six) is petaloid, caducous and circumscissile, modestly coloured dull cream, grey or brownish. Tepals are more or less united at the base, reflexed or non-reflexed while blooming, abscising at the base and withering as a cup on the developing capsule. Stamens are six, adnate to the tepal at base, and filaments filiform, swollen at the tip or uniformly narrowed at the apex. Anthers are linear, dehiscing longitudinally on the inner side. Gynoecium is tricarpellary, trilocular with many ovules in each cell. Style is single, thickened at the apex. Fruit is an oblong, triquetrous, loculicidal capsule. Seeds are compressed, winged and testa black, and the embryo is usually large. Figure 7.1 depicts the diagnostic morphology of *Drimia*.

7.1.2 Nomenclature

The merging of *Urginea* into *Drimia* led to many name changes as many species of the former had to be transferred to the latter. In the Indian perspective, this was done by Stearn (1978), Ansari and Sundararaghavan (1980) and Kumar (1984). Deb and Dasgupta (1987) treated *Urginea coromandeliana* and *U. govindappae* as synonymous to *Drimia indica*. Lekhak et al. (2014), however, treated them as distinct. Recently, Hemadri (2006) described a new species *Urginea raogibikei (= D. raogibikei)* from Andhra Pradesh, India. A new combination for *U. coromandeliana,* i.e. *D. coromandeliana* has been made by Yadav et al. (Yadav et al. 2016 in press). For, *U. govindappae* a new combibation has to be made so that *Urginea* ceases to exist as an accepted taxon. The new combination is as follows:

Drimia govindappae (Boraiah & Fatima) Lekhak & P.B. Yadav, **comb. nov.**
Urginea govindappae Boraiah & Fatima, Bull. Bot. Surv. India 12(1–4): 128 [Boraiah & Fatima 1970 (publ. 1972)]. Type: India, Bangalore, Lal baugh Garden, 17 May 1968, Boraiah & Fatima 601A (holotype; GKVK).

7.2 Life Cycle Pattern, Reproductive Isolation and Phenology

Drimia is a geophyte. It bears a subterranean storage organ and its life cycle includes a dormant phase. In most of the geophytes, dormant period can last for a week to most of the year (Raunkiaer 1934). Few geophytes that remain active throughout the

Fig. 7.1 Diagnostic morphology of *Drimia*. (**a**) bulbs growing in cluster, (**b**) sprouted bulbs with green shiny leaves, (**c**) a solitary bulb with thick, fleshy, emergent and glaucous leaves, (**d**) individuals with variously coiled glaucous leaves, (**e**) individuals with linear leaves, (**f**) individuals with broad, glaucous leaves, (**g**) scape with lax flowers, (**h**) scape with congested flowers, (**i**) scape without leaves (hysteranthous), (**j**) scape with leaves (synanthous), (**k**) solitary flower with non-reflexed perianth, (**l**) solitary flower with reflexed perianth, (**m**) infructescence showing capsule,(**n**) dehisced capsules with flat seeds

whole year are typical of tropics (Holtum 1955). Observations on phenology of Indian *Drimia* reveal that the dormant phase can be as short as two months as in *D. raogibikei* and as long as eight months as in case of *D. wightii* (Fig. 7.2). Geophytes exhibit two life cycle patterns, viz. synanthous and hysteranthous (Dafni et al. 1981). Leaves and flowers are simultaneous and the course of events is growth, storage, flowering and dormancy in synanthous pattern (Dafni et al. 1981). On the other hand, in a hysteranthous pattern, leaves and flowers appear in different seasons and the course of events is growth, storage, dormancy and flowering (Dafni et al. 1981). Amongst Indian *Drimia*, *D. wightii* and *D. polyphylla* are the only synanthous species.

Hybridization and gene flow between or among species are regulated by isolating mechanisms. Hybridization breaks adaptively valuable gene combinations of a species which is manifested in hybrid breakdown (Grant 1971). Reproductive isolating mechanisms aid in protecting the adaptive gene combinations from disintegration (Grant 1971). Grant (1971) referred to them as "blocks to gene exchange between populations which stem from genotypically controlled differences in their reproductive organs, reproductive habits, or fertility relationships." These isolating mechanisms mainly operate either before pollination or postpollination. The former kind has been referred as external, while the latter internal isolating mechanisms (Grant 1971). External mechanisms include mechanical, ethological and seasonal isolation, while internal is due to incompatibility, hybrid inviability, hybrid sterility and hybrid breakdown. Levin (1971) gave a detailed classification of reproductive isolating mechanisms in flowering plants. He recognized three categories, viz. premating, postmating-prezygotic and postmating-postzygotic isolating barriers.

In *Drimia*, premating isolating barriers are more pronounced and can be easily observed. However, one cannot rule out postmating barriers as our knowledge on interspecific hybrids is far from sufficient. Dixit and Yadav (1989) made some interspecific hybrids but do not mention about their behaviour, i.e. whether the hybrids obtained were stable, sterile or hydrid breakdown was noticed. Hence, no data are there on postmating-postzygotic isolating barriers. As already mentioned above, crosses between *Drimia coromandeliana* and *D.* species (as *Urginea congesta*), *D. coromandeliana* and *D. polyantha* and *D. razii* and *D. polyantha* failed. This indicates the presence of pollen-style incompatibility, a postmating-prezygotic barrier. It is well known that if pollen of one species reaches the stigma of the other, the stigma and style usually will not allow the foreign pollen to germinate.

Premating barriers operating in *Drimia* can be easily noticed. Patil and Yadav (1992–1993) based on their studies on floral phenology and palynology segregated *Drimia* into two groups, one with a small pollen (65–73 × 62–69 μm) and fine reticulate walls and another with large (80–83 × 78–79 μm) and coarse reticulate walls. The former includes day-blooming species (*Drimia polyantha*, *D. razii* and *D.* species) and the latter night-blooming species (*D. coromandeliana*, *D. govindappae*

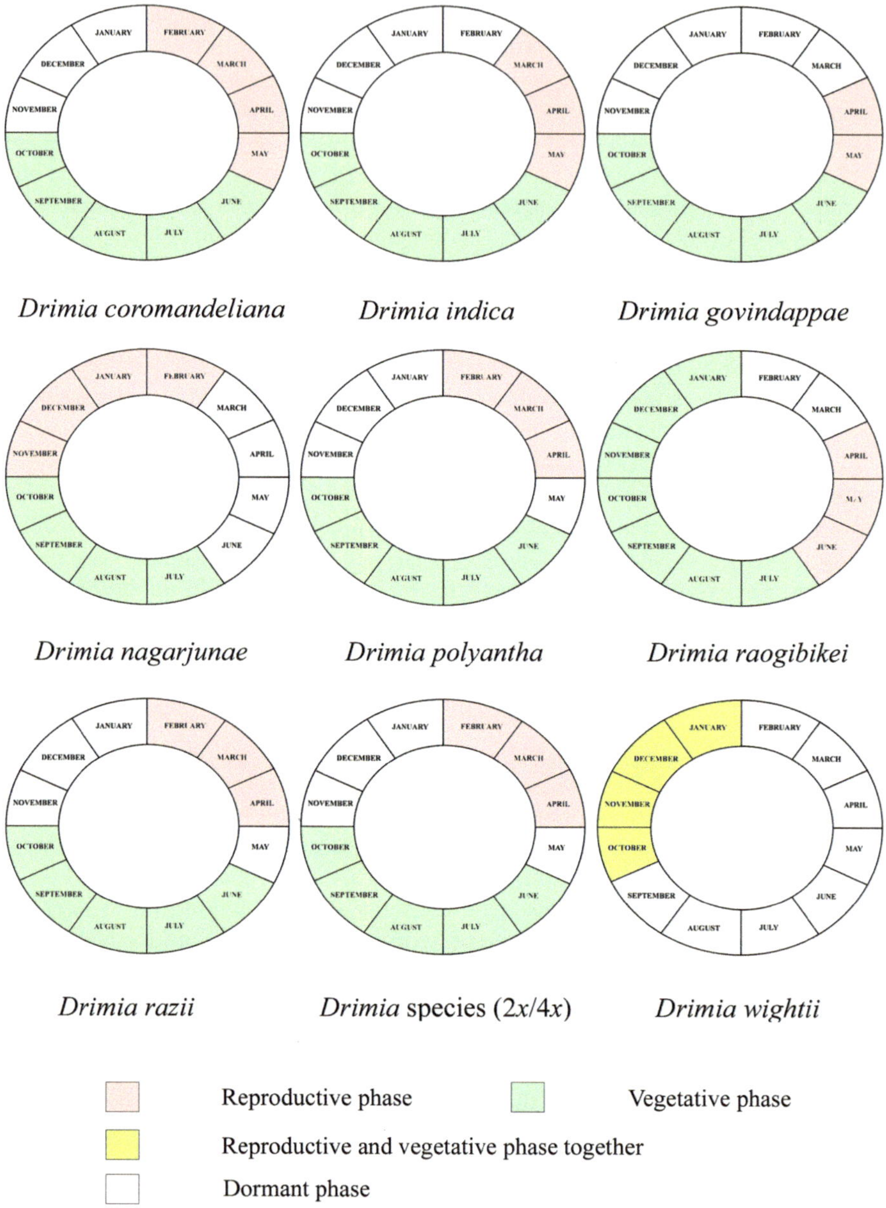

Fig. 7.2 Vegetative and reproductive phenology of *Drimia* species

and *D. indica*). *Drimia nagarjunae* and *D. wightii* are also day-blooming species, but their palynological details have not been worked out so far. The significance of different exine ornamentations from the point of view of pollination or pollinators in day- and night-blooming species, is yet to be assessed. But the difference in the

Table 7.2 Timing of floral events and life cycle pattern of *Drimia* species

Sr. No.	Species	Flowering		Anther dehiscence	Life cycle pattern
		Opening	Closing		
1.	*D. coromandeliana*	5 pm–6 pm	6 pm–7 pm	2 pm–4 pm	Hysteranthous
2.	*D. govindappae*	5 pm–6 pm	6 pm–7 pm	2 pm–4 pm	Hysteranthous
3.	*D. indica*	5 pm–6 pm	6 pm–7 pm	2 pm–4 pm	Hysteranthous
4.	*D. nagarjunae*	7 am–8 am	3 pm–4 pm	8 am–9 am	Hysteranthous
5.	*D. polyantha*	7 am–9 am	4 pm–5 pm	10 am–12 noon a day before anthesis	Hysteranthous
6.	*D. polyphylla*	–	–	–	Synanthous
7.	*D. raogibikei*	9 am–11 am	4 pm–5 pm	10 am–12 noon	Hysteranthous
8.	*D. razii*	9 am–11 am	4 pm–5 pm	10 am–12 noon	Hysteranthous
9.	*D. wightii*	9 am–11 am	3 pm–4 pm	12.30 pm–01 pm	Synanthous
10.	*D.* species (2*x*/4*x*)	12 noon–01 pm	4 pm–5 pm	10 am–12 noon	Hysteranthous

Modified after Patil and Yadav (1992–1993)

blooming time of *Drimia* species serves as a premating temporal isolation (Table 7.2). This is actually a diurnal isolation (flowering at different times in a day) which ceases the gene flow in day and night-blooming species. Natural hybridization within (as species of the group show overlap in the timing of their floral events; see Table 7.2) and between night-blooming and day-blooming species is not possible as their geographical distribution rarely overlaps. *Drimia nagarjunae* is reproductively isolated from all the remaining species on account of seasonal isolation. It flowers in winter, while for other species, summer is the flowering season.

7.3 Hybridization and Breeding System

Hybridization forms an integral part of biosystematic studies. It assesses the presence or absence of sterility barriers amongst the taxa under study. Hybridization experiments are also a very important parameter for delimiting species boundaries for the followers of the *Biological Species* concept. However, botanists are not stringent followers of this concept as in plants, morphological and genetic limits do not coincide (Stace 1989). Many plant species which are morphologically distinct can interbreed, while morphologically similar species may not. Hence, there is no point in uniting many species of genera such as *Dactylorrhiza* Neck., *Euphrasia* L. or *Salix* L. (Stace 1989). The ability to interbreed or not is not an absolute character for defining species limits as many species do not hybridize in wild and also for every species examining interbreeding capacity is impractical (Stace 1989). The ability to

Table 7.3 Interspecific crosses between *Drimia* species

Sr. No.	Crosses	Flowers crossed	Fruits/seeds obtained
1.	*Drimia* species (2*x*) x *D. nagarjunae*	06	–
2.	*D. polyantha* x *D. nagarjunae*	31	–
3.	*D. nagarjunae* x *D. polyantha*	47	–
4.	*D.* species (4*x*) x *D. nagarjunae*	28	–
5.	*D. nagarjunae* x *D.* species (4*x*)	09	–
6.	*D. polyantha* x *D.* species (4*x*)	04	–
7.	*D.* species (4*x*) x *D. polyantha*	08	–

hybridize is a measure of close relationship and proves taxonomically significant when it can be used differentially, i.e. if hybridization is uncommon in a genus, species pairs hybridizing are more closely related as long as they are not phenetically less similar than other species pairs or groups (Stace 1989).

Drimia taxonomy in India has always been confused (Lekhak et al. 2014). An exhaustive taxonomic revision of this polymorphic taxon is awaited. However, morphological knowledge alone may not prove handy for species circumscription and understanding species interrelationships. Interspecific hybridization is one additional criteria to understand affinities amongst taxa. Dixit and Yadav (1989) carried out interspecific hybridization in six species of *Drimia*, viz. *D. coromandeliana*, *D. govindappae*, *D. indica*, *D.* species (as *Urginea congesta*), *D. polyantha* and *D. razii*. They found that reciprocal crosses between *Drimia coromandeliana* and *D. indica*, *D. indica* and *D. govindappae*, *D. indica* and *D. polyantha*, *D. coromandeliana* and *D. polyantha* and *D. coromandeliana* and *D. govindappae* were successful. Crosses between *Drimia* species and *D. indica*, *D. coromandeliana* and *D.* species (as *Urginea congesta*), *D. coromandeliana* and *D. polyantha* and *D. razii* and *D. polyantha* failed. Two one-sided crosses (*D. coromandeliana* x *D. razii* and *D. indica* x *D. razii*) were successful with hot water and gibberellic acid (GA_3) treatment. Dixit and Yadav (1989) concluded that the genomes of *D. coromandeliana*, *D. govindappae*, *D. indica* and *D. polyantha* are closely related, and these species form a complex. *Drimia nagarjunae* was recently collected by Lekhak et al. (2014) from Chinnar Wildlife Sanctuary, Idukki District, Kerala. In 2014 seven more crosses were made (Table 7.3) of which five involved *Drimia nagarjunae* as one of the parents. None of these crosses was successful indicating that *Drimia nagarjunae* occupies an isolated position amongst Indian *Drimia*. In terms of both morphology and cytology, *Drimia nagarjunae* is closer to *D. indica* than to any other species (Lekhak et al. 2014). However, more interspecific crosses would give a better understanding of its affinities to other species. Jehan et al. (2014) based on their studies on genetic diversity (using RAPD and SRAP markers) on Indian *Drimia* concluded that *D. razii* occupies a distinct clade while *D.* species (as *D. wightii*) is the most distinct species. Most of these crosses are unsuccessful. All the crosses made amongst the Indian species have been depicted in Fig. 7.3.

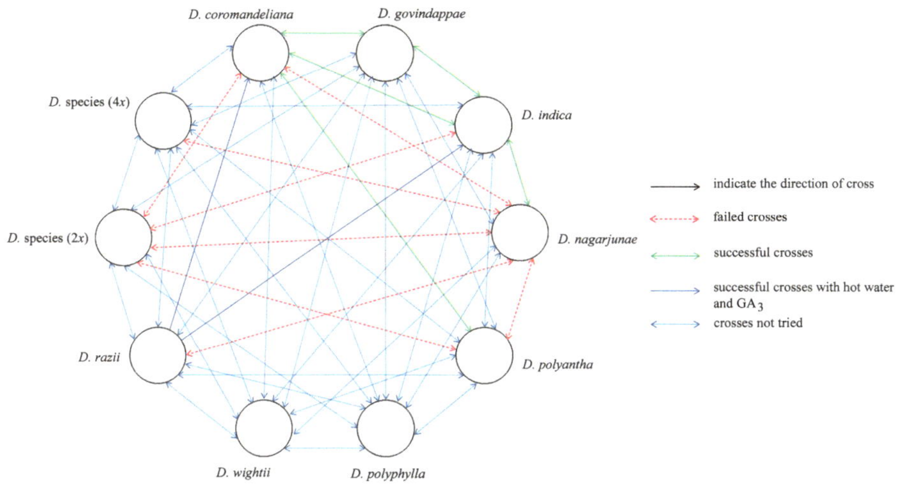

Fig. 7.3 Crossability polygon of *Drimia* species (Modified after Dixit and Yadav 1989)

Apart from assessing interrelationships, interspecific hybridization is a means to produce polyploids/hybrids which can prove to be elite sources of bufadienolides (cardioactive C-24 steroids) (Lekhak et al. 2014). Hence, hybridization studies in *Drimia* hold great promise for further studies with regard to exploiting its medicinal principles (Lekhak et al. 2014).

Breeding system of a plant is the mode, pattern and extent to which it interbreeds with other plants of the same or of different taxa (Stace 1989). It includes mechanisms of pollination and pollen transfer (Shivanna 2003). Pollination systems in Hyacinthaceae are specific and highly diverse even within genera; however, these mechanisms have only occasionally been studied thoroughly (Speta 1998). In southern African plants, melittophily (*Lachenalia*, *Litanthus*, *Massonia*, *Polyxena*, *Rhadamanthus* and *Urginea*), psychophily (*Dipcadi* and *Massonia*) and ornithophily (*Lachenalia*, *Massonia*, *Veltheimia* and *Daubenya*) have been reported (Speta 1998). Genera like *Litanthus*, *Rhadamanthus* and *Urginea* are now synonymous with *Drimia* meaning that bee pollination is quite common in *Drimia*. As far as Indian species are concerned, there are no reports on pollination biology. All the Indian species are outbreeders. *Drimia* species have a very short span of flowering, and sometimes the flowers would open merely for an hour or two. This might be the reason for lack of reports on the pollination biology or pollinators of Indian *Drimia*.

7.4 Cytogenetics

Cytological data offer significant insights into the understanding of species concepts in various taxonomic groups. Sharma (1964) anticipated the importance of cytological data in taxonomy. There are ample instances where cytology has bailed out taxonomists from situations where morphology had been misleading or not of much help in defining species, genera or family limits. Lavania and Lavania (2007) have cited many examples on the utility of chromosome data in resolving taxonomic misfits or confusions. Chromosome number is one of the widely known cytotaxonomical parameters for a very large number of plant taxa (Guerra 2000). Data from chromosome morphology, nuclear DNA content, interphase nuclear structure, banding pattern or in situ hybridization are some less common methods employed in karyosystematic studies (Guerra 2000). These approaches can prove very useful in refining initial cytotaxonomic analysis based on chromosome number and reveal the structural basis of the mechanism involved in evolutionary change (Guerra 2000). Since such approaches demand more laborious and specialized methods, they have been applied only in certain cases; however, chromosome counting has been routine work in many botanical laboratories (Guerra 2000). The use of chromosome number in monocots is advantageous due to high variability found in many families (Guerra 2000). The lowest gametic number (n) is known for four monocots, viz. *Colpodium versicolor* (Steven) Schmalh. (Poaceae), *Ornithogalum gussonei* Ten. (= *Ornithogalum tenuifolium* Redouté) (Hyacinthaceae), *Rhynchospora tenuis* Link (Cyperaceae) and *Zingeria biebersteiniana* (Claus) P.A. Smirn. (Poaceae) (Vanzela et al. 1996; Roberto 2005). In Orchidaceae, the lowest ($2n = 12$) and the highest chromosome number ($2n =$ c. 240) are reported for *Erycina pusilla* (L.) N.H. Williams & M.W. Chase [≡ *Psygmorchis pusilla* (L.) Dodson & Dressler] and *Epidendrum cinnabarinum* Salzm. ex Lindl., respectively (Guerra 2000). The highest known chromosome number ($2n =$ c. 298) in the monocots is reported for a palm *Voanioala gerardii* J. Dransf. (Johnson et al. 1989).

Drimia, owing to its large chromosomes, has been a classical material for observing chromosomes by botanists across the globe. In India alone there is a bulk of data on cytology of *Drimia* species (Table 7.1). However, most of this information is on mitosis or somatic chromosome number. This is because of the ephemeral flowering of *Drimia* that makes harvesting of the buds in appropriate stages of division difficult. We discuss cytogenetics of Indian *Drimia* with respect to its base number, cytotypes, karyotype evolution and B chromosomes.

7.4.1 Base Number

Guerra (2000) defined base number "as one of the haploid numbers observed in a taxon that most parsimoniously explains the chromosomal variability of that group and shows a clear relationship with the base numbers of its closest related groups." Base number is denoted by "x." According to Raven (1975), the original basic chromosome number in angiosperms has been $x = 7$, which is characteristic of all major groups of both dicots and monocots except Caryophyllidae, with $x = 9$. Stebbins (1971) suggested that original basic numbers of angiosperms (both woody and herbaceous) were $x = 6$ and $x = 7$. Goldblatt (1980) and Sharma (1984) proposed base numbers between 6 and 10 for many monocot families. Soltis and Soltis (1990) supported the original base numbers suggested by Stebbins (1971); Raven (1975) and Grant (1981). They also noted that the basic chromosome numbers near $x = 11$ or 12 may be of polyploid origin.

The basic chromosome numbers $x = 5$ (Darlington and Wylie 1955) and $x = 7$ (Jones and Smith 1967) are known for *Drimia*. These basic chromosome numbers were proposed when a broad concept for *Drimia* was not even dreamt of or in other words it can be said that many taxa (genera or species) which were earlier included under *Drimia* are now not a part of it and vice versa. For instance, Jones and Smith (1967) proposed base number of $x = 7$ based on $2n = 14$ for *Urginea volubilis*, an endemic species of Madagascar, which is now treated as *Albuca volubilis* (H. Perrier) J.C. Manning & Goldblatt (Manning et al. 2009). The most common chromosome number amongst Indian *Drimia* species is $2n = 20$ (Table 7.1). *D. indica*, *D. govindappae*, *D. nagarjunae*, *D. polyantha*, *D. razii*, *D. wightii* and *D.* species ($2x$) possess $2n = 20$. The only species for which a diploid count of $2n = 40$ is reported is *Drimia coromandeliana* (Naik 1976; Dixit and Yadav 1989). *Drimia indica* which occurs throughout the country has been shown to have diploid numbers of 30 (Patil and Torne 1982; Jha and Sen 1983a, b), 40 (Patil and Torne 1982; Jha and Sen 1983a, b) and 60 (Jha and Sen 1983a, b). Moorthy and Sampathkumar (1968) reported $2n = 10$, 30, 40 and 80 in somatic cells of root tips of *Drimia indica*. *Drimia* species ($4x$) exhibits $2n = 40$. (Yadav unpublished). In addition to the counts mentioned above, some unusual diploid chromosome numbers are also mentioned in the literature. Subramanian (1978) recorded nine cytotypes for *Urginea indica* (*Drimia indica*). Of these, somatic chromosome counts of four cytotypes deviated from the normal $2n = 20$. These cytotypes exhibited $2n = 11$ (10 + 1), 15 (14 + 1), 19 (18 + 1) and 22 (20 + 2). Subramanian (1978) also mentioned the percentage of cells with deviating chromosome numbers in the nine cytotypes. Some of the cells he observed had $2n = 12 + 1$, $16 + 1$. These cytotypes have never been confirmed by any other worker till date. Nevertheless, the observations of Moorthy and Sampathkumar (1968) and Subramanian (1976) reveal that the phenomenon of polysomaty might be operating in some population of *Drimia indica*. These changes or deviation in the chromosome numbers may be attributed to polyploidy, aneuploidy or dysploidy. Except for *Drimia coromandeliana*, *D. indica* and *D.* species($4x$), all other Indian species are at diploid level. *Drimia raogibikei* has not been cytologically screened yet while the occurrence of *D. polyphylla* is doubtful. The report of the diploid number $2n = 20$ for *D. polyphylla* by Raghavan and Venkatasubban (1940) is questionable and requires validation as the species has never been collected since its type collection. Considering the case of Indian species of *Drimia*, it is quite evident that the most frequent haploid number or number that can be linked to the most extensive polyploid series is ten. Therefore, $x = 10$ is the basic chromosome number for Indian *Drimia*. Goldblatt et al. (2012) also hypothesized that $x = 10$ is the basic, ancestral chromosome number in Urgineoideae.

7.4.2 Cytotypes

Cytological differences are of common occurrence in *Drimia* species. Variations can either be distributed across populations (interpopulation) or within/between individuals of a single population (intrapopulation). These cytologically different populations/individuals are referred as cytological races or cytotypes. Species with wider geographical distribution show well-developed cytological races on account of their broad ecological amplitude. Consequently, these species have adapted to different ecological niches. On the other hand, species with narrow distribution such as endemics are quite habitat specific and do not exhibit cytological differences. The former category includes *Drimia indica*, while the rest of the species can be grouped under the latter. Many cytotypes have been reported for *Drimia indica* (Table 7.1). These cytotypes are either polyploids or aneuploids/dysploids. Polyploid cytotypes in *Drimia indica* have been reported by Raghavan and Venkatasubban (1940), Ayyangar and Sampathkumar (1968), Sharma and Dash (1977), Patil and Torne (1982) and Jha and Sen (1983a, b). However, except for triploids, nothing is known about the morphology, pollen viability and sexuality of tetraploids and hexaploids. Triploids show less than 40 % pollen fertility (Patil and Yadav 1992–1993.) They are sterile, vegetatively more robust and have higher growth rate of floral scape than their diploid counterparts (Raghavan and Venkatasubban 1940). Triploid meiosis has been detailed by Raghavan and Venkatasubban (1940). The authors report many meiotic anomalies such as elimination of chromosomes, micronuclei formation, ameiosis, formation of restitution nucleus, etc. Such detailed analysis is lacking for tetraploids and hexaploids. Jha and Sen (1983a, b) studied meiosis in tetraploids and observed the formation of quadrivalents. They also noticed laggards and bridges and concluded that the tetraploids have originated through autotetraploidy followed or associated with structural changes as chromosomes cannot be represented in groups of four. The exact mode of origin of polyploid races of *Drimia indica* is yet to be traced. In the present investigation, it has been found that the triploid cytotypes of *Drimia indica* occur throughout the east and west coast. These plants grow in clumps as they reproduce vegetatively (Fig. 7.1a, b). The cytological configuration of each individual in the clump may differ, and hence individual bulbs need to be screened. We have noticed that bulbs brought from Bhatye, a coastal region in Ratnagiri District, Maharashtra, represent both diploid and triploid races. In addition to polyploids, many aneuploids/dysploids have also been recorded for *Drimia indica* by Moorthy and Sampathkumar (1968) and Subramanian (1973, 1978, 1980). The variation in chromosome number is seen even in the same root tip cells. This presence of different chromosome counts within the same root tip cells have been referred as polysomaty. The frequency of these cells is very low as compared to the normal cells. Polysomaty can arise due to non-disjunction and endomitotic reduplication of all or a few chromosomes of the complement (Sen 1973). Since many monocots show vegetative reproduction, the occurrence of polysomaty is the main feature of their chromosome behaviour (Sen 1973). Sen (1973) observed polysomaty in Liliaceae members (*Disporum* Salisb., *Ornithogalum*, *Polygonatum* Mill. and *Smilacina* Desf.) collected from extreme alpine conditions in the Himalayas.

Here, polysomaty is responsible for the origin of new genotypes (with different chromosomal configuration) through vegetative propagation (Sen 1973). Such genotypes may exploit a particular habitat better than the normal genotypes. Polysomaty has provided the species with wide adaptability under varying environmental conditions (Sen 1973). However, in polysomatic tissues balance of nucleo-cytoplasmic ratio which is crucial for cell division is upset, nevertheless the cell divides. According to Sharma (1956) nucleo-cytoplasmic balance (in plants exhibiting polysomaty) is maintained at the tissue level rather than the individual cell.

Drimia coromandeliana, *D. govindappae*, *D. nagarjunae*, *D. razii*, *D. polyantha* and *D. wightii* do not exhibit any variation in their chromosome number (Fig. 7.4). *Drimia* species has two cytotypes, viz. diploid and tetraploid (Table 7.1). All these species have restricted geographical distribution reflecting their inability to colonize new ecological niches on account of little or no chromosomal variation.

B chromosomes are also reported in *Drimia* (Fig. 7.4h, i). Their number varies from one to seven in case of *Drimia indica* (Table 7.1). In the present investigation, we report seven B chromosomes in *Drimia govindappae* and *D. polyantha* (Fig. 7.4h, i, respectively).

The presence of cytotypes in *Drimia* necessitates further investigation regarding their origin and establishment. Also, thorough meiotic studies in polyploid cytotypes will reveal their auto or alloploid nature. Pollen viability studies will also throw light on the sexuality of cytotypes. At the same time, their morphology needs immediate attention by taxonomists. Differences in morphology may provide a means to circumscribe their limits and recognize them taxonomically. This has been done in Nigeria by Oyewole and Mustapha (2000) who recognized three subspecies of *Urginea indica* (*Drimia indica*) based on variation in their cytology, morphology and ecological niches. Additionally, instances where morphology of cytotypes is indistinguishable may represent the case of cryptic or sibling species which are similar in morphology but do not interbreed and hence warrant the status of a distinct taxon. Chemical examination of cytological races of *Drimia* may also prove rewarding as cytotypes differ in their chemistry (presence of active principles such as bufadienolides) and are commercially very important (Jha and Sen 1981). Furthermore, cytotypes are tools to understand the origin of polyploidy and its role in speciation and diversification of *Drimia*.

7.4.3 B Chromosomes

B chromosomes also known as supernumerary or accessory chromosomes were first described by E. B. Wilson in the leaf-footed bug *Metapodius* (now known as *Acanthocephala*) (Wilson 1907a, b). In plants they were first reported in rye (*Secale cereale*) by Gotoh (1924) and then in maize (*Zea mays*) by Kuwada (1925) and Longley (1927). Randolph (1928) used the term B chromosomes, to distinguish them from the normal set of chromosomes (referred as A chromosomes or As). B chromosomes do not obey Mendelian law of inheritance and have been reported in numerous animals, plants and fungi species (Houben et al. 2013). The

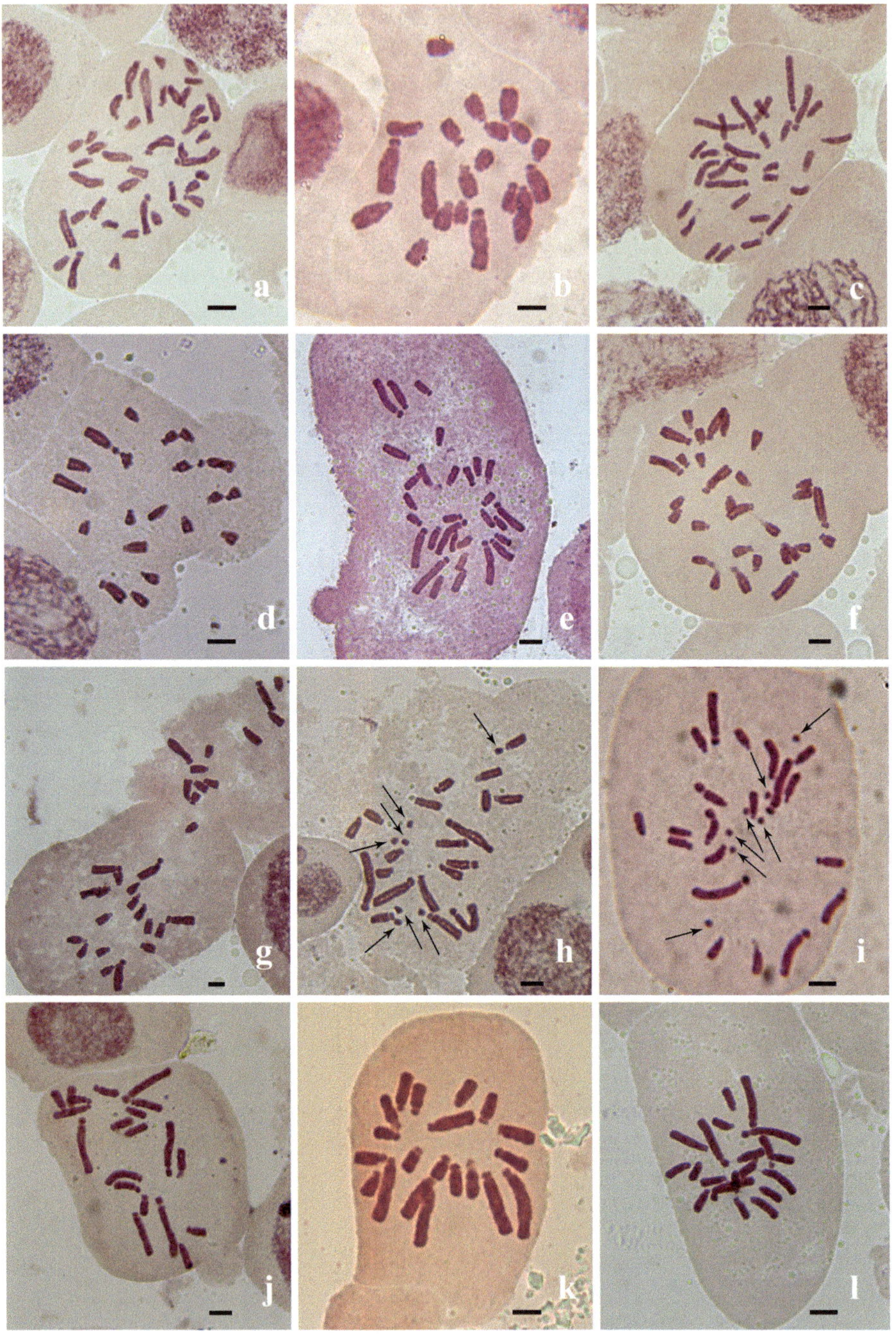

Fig. 7.4 Mitotic metaphase chromosomes of *Drimia* species. Arrows indicate B chromosomes (**a**) *Drimia coromandeliana* ($2n = 4x = 40$), (**b**) *D. indica* Form 1 ($2n = 2x = 20$), (**c**) *D. indica* Form 2 ($2n = 3x = 30$), (**d**) *D. indica* Form 3 ($2n = 2x = 20$), (**e**) *D. indica* Form 4 ($2n = 3x = 30$), (**f**) *D. indica* Form 5 ($2n = 3x = 30$), (**g**) *D. indica* Form 6 ($2n = 3x = 30$), (**h**) *D. govindappae* ($2n = 2x = 20 + 7B$), (**i**) *D. polyantha* ($2n = 2x = 20 + 7B$), (**j**) *D. razii* ($2n = 2x = 20$), (**k**) *D. wightii* ($2n = 2x = 20$), (**l**) *D.* species ($2n = 2x = 20$) *Scale bar = 5µm*

distribution or presence of B chromosomes across angiosperms is not random but heterogeneous (Jones et al. 2007). B chromosomes have been reported in many species of Asteraceae, Campanulaceae, Fabaceae, Liliaceae, Orchidaceae, Poaceae and Ranunculaceae (Levin et al. 2005). They are recorded in more than 1500 species of plants (Jones et al. 2007). Amongst angiosperms, some 8 % monocots and 3 % eudicots possess B chromosomes (Levin et al. 2005). Commelinales and Liliales are the ordinal hot spots for B chromosomes (Levin et al. 2005). Commelinales in monocots and Santalales in eudicots have the highest frequency of species with B chromosomes (Levin et al. 2005). B chromosomes are usually found in low numbers (0–4, 5) but high numbers are recorded for *Allium schoenoprasum* L. (0–18), *Brachyscome lineariloba* (DC.) Druce (0–22) and *Silene maritima* (0–15) (Mehra and Pandita 1979; Cobon and Murray 1983; Jones et al. 2007). In *Zea mays* ($2n$ = 20 As + 0–34B) B chromosomes exceed the number of As (Houben et al. 2013).

Drimia is placed in the order Asparagales (APG III 2009). Hyacinthaceae exhibit the highest percentage (of species) with B chromosomes in the order (Levin et al. 2005). It has been suggested that higher frequency of B chromosomes is found in families with larger genome (Trivers et al. 2004), and Hyacinthaceae being one such family has numerous species possessing B chromosomes. As far as Indian *Drimia* is concerned, B chromosomes have been recorded for only three species, viz. *Drimia govindappae*, *D. indica* and *D. polyantha*. Of these, *Drimia indica* shows the highest frequency of occurrence of B chromosomes. For *Drimia govindappae* and *D. polyantha*, B chromosomes are reported here for the first time. The maximum number reported so far is seven. Nature, behaviour and significance of these accessory chromosomes have not been studied thoroughly and seek attention from cytogeneticists. Sen (1974) suggested the origin of B chromosomes in *Drimia* through elimination of euchromatin segments from triploid or aneuploids. However, whether this mode of origin stand up to the scrutiny of time or not needs to investigated. Origin of B chromosomes in interspecific hybrids of *Coix aquatica* and *C. gigantea* has been proposed by Sapre and Deshpande (1987). For *Drimia indica* where many cytotypes exist and the frequency of B chromosomes is higher than the other species, interspecific hybridization may have some role in the origin of B chromosomes. Presence of accessory chromosomes, in some cases such as *Centaurea scabiosa* L., *Festuca pratensis* Huds. and *Lilium medeoloides* A. Gray has been related to their restriction to a geographic range which is associated with specific factors of environment or population structure (Stebbins 1971). In the present investigation, *Drimia polyantha* seems to exhibit the aforesaid correlation. The B chromosome containing population of *Drimia polyantha* is restricted to a high altitude lateritic plateau (Kas plateau) of southwestern Maharashtra. These plateau ecosystems are unique in exhibiting both extremes of heavy monsoon and a long period of dry season. Similarly, present collection of *Drimia govindappae* is from rocky habitats in Mysore district.

B chromosomes are unessential components of the genome and usually do not show phenotypic effects. This holds true in case of Indian *Drimia* also as the B chromosome containing morphs do not differ morphologically from non B morphs. However, they may have physiological effects as in case of *Allium schoenoprasum* where seeds with B chromosomes have advantage in germination over seeds without them. Effects of B chromosomes are cumulative, i.e. depending on their number not the presence or absence (Jones 1995; Bougourd and Jones 1997). They reduce hybrid vigour and fertility (Bougourd and Jones 1997; Jones and Houben 2003; Camacho 2005). Investigations concerning fertility of B chromosome containing *Drimia* individuals have not been undertaken yet. What remains an enigma in case of B chromosomes of *Drimia* is their origin, behaviour during cell division and genes they possess.

7.4.4 Karyotype Diversity and Evolution

Karyotype is generally defined as the phenotypic appearance of a somatic chromosome complement. The data [such as chromosome number, total chromatin length (TCL), centromeric position of chromosomes, symmetry of chromosome complement, etc.] obtained through karyotype analysis has been used and is still used, though not on a very large scale, as an additional criterion for understanding species limits and interrelationships. Karyotype data have been explored to a great extent in families where chromosomes are fairly large sized. Monocot families like Amaryllidaceae, Asparagaceae, Colchicaceae, Hyacinthaceae, Liliaceae, etc. have been a classical source of karyotype mining. Stebbins (1971) cited example of Poaceae where karyomorphology in conjunction with anatomy, histology, geographic distribution, etc. has clarified taxonomic relationships. However, karyotype alone must not be given overriding importance (Stebbins 1971).

Drimia has continued to be a cynosure of cytogeneticists and cytotaxonomists for its large and amenable chromosomes. Bulk of karyotype information exists for Indian *Drimia.* However, the lack of taxonomic understanding of the group and attempts to correlate knowledge of chromosomes with the morphology, ecology and other aspects contributed to the confused systematics. Consequently, the genus enjoys the status quo which existed five decades ago with no refinement in its systematics. Here, we analyse the karyotypes of 12 accessions (across six species) in the light of the available information (Fig. 7.5). The measures do not take into account the secondary constriction and B chromosomes. Karyological parameters are depicted in Table 7.2. Data for *Drimia nagarjunae* have been taken from Lekhak et al. (2014). The most common diploid chromosome number for Indian *Drimia* is $2n = 20$. *Drimia indica, D. govindappae, D. nagarjunae, D. polyantha, D. razii, D. wightii* and *D.* species exhibit this number. A diploid number of 30 ($2n = 3x = 30$) and 60 ($2n = 6x = 60$) is reported so far for *Drimia indica* only. Tetraploids ($4x$) with $2n = 4x = 40$ are represented by *Drimia coromandeliana, D. indica* and *D.* species. The chromosome length ranged from 12.94 ± 2.28 μm to 3.07 ± 0.14 μm. The longest pair was observed in *Drimia govindappae* and the lowest in *D. coromandeliana*. Amongst the Indian species, *Drimia govindappae* has the maximum mean chromosome length measuring 6.84 ± 2.92 μm. In contrast *Drimia indica* (Bhatye population) had the shortest mean chromosome length, i.e. 4.49 ± 1.57 μm. *Drimia coromandeliana* has the highest TCL value (99.19 ± 1.82 μm), while the lowest was

Fig. 7.5 Karyograms of *Drimia* species. (**a**) *Drimia coromandeliana*, (**b**) *D. indica* Form 1, (**c**) *D. indica* Form 2, (**d**) *D. indica* Form 3, (**e**) *D. indica* Form 4, (**f**) *D. indica* Form 5, (**g**) *D. indica* Form 6, (**h**) *D. govindappae*, (**i**) *D. polyantha*, (**j**) *D. razii*, (**k**) *D. wightii*, (**l**) *D.* species *Scale bar = 5μm*

recorded for *Drimia indica* collected from Bhatye, a diploid coastal population in Ratnagiri District, Maharashtra. The longest to shortest chromosome ratio (R) ranged from 2.43 to 2.99 indicating similarity in the position of centromere which is usually subterminal. *Drimia coromandeliana* is a natural tetraploid, and the highest TCL value can be attributed to its large chromosome complement. Similarly, polyploids such as hexaploids in case of *Drimia indica* may have even greater values for TCL. A_1 and A_2 indices which refer to intrachromosomal and interchromosomal asymmetry, respectively, did not show marked variation in their values. The highest value of A_1, i.e. 0.85, was observed in *Drimia nagarjunae*. For the rest of the species, the value ranged from 0.73 to 0.80. As a whole intrachromosomal asymmetry, which is due to centromere position, is high in all the species. The values obtained for interchromosomal asymmetry index (A_2) was more or less alike for all the species. It ranged from 0.35 to 0.40 indicating that there is little variation in chromosome size in *Drimia* species. All the analysed species fall in 1b category of Stebbins asymmetry class which reveals that the species of the genus exhibit gross homogeneity in their karyotypes. This is further corroborated by the karyotypic formulae observed across the species. There is sheer predominance of subterminal primary constrictions (Table 7.4). Some species do show terminal constrictions but such instances are very few (Table 7.4). *Drimia indica* (Bhatye population) differs from all other accession in having a chromosome pair with submedian region centromere. This can be expected in a polymorphic species like *Drimia indica* which has a very wide range of geographical distribution within India. Another important fact worth mentioning is that of all the *Drimia indica* populations screened presently, the Bhatye population is the only one where diploid and triploids are occurring together, i.e. they are sympatric. This has not been observed anywhere else in the coastal populations. Coastal populations being triploid are usually sterile, and then how these diploids arise need to be investigated. Another interesting implication of the co-occurrence of diploids and triploids is that if these individual exchange genes, progenies with different chromosome number/ploidy can be expected. Whether they cross or not is again a matter of investigation. Similarly, pollen viability/fertility of coastal diploids demands thorough investigation.

The gross chromosome morphology of *Drimia* species is very similar as has been mentioned earlier (Fig. 7.5). Therefore, conventional chromosome markers (primary constriction, arm ratio, etc.) cannot differentiate between morphologically similar chromosomes. This led Nath et al. (2014) to undertake cytological characterization of *Drimia* based on fluorochrome banding and genome size variation. These authors concluded that reproductive isolation of polymorphic species has led to the diversification of *Drimia* in India, probably from a common ancestor similar to *Drimia indica*. This study did not include *Drimia nagarjunae* which was recently collected by Lekhak et al. (2014) from Chinnar Wildlife Sanctuary, Idukki District, Kerala. *Drimia nagarjunae* is the most robust species of the genus in India which shows affinities with African *Drimia maritima*. It could be the common ancestor Nath et al. (2014) referred to. *Drimia indica* shows tremendous intraspecific variation, and the prevalence of its many cytotypes is a testimony to this fact. For that reason *Drimia indica* should be regarded in a state of evolutionary flux and evolutionarily young. Also, it also shows higher amount of heterochromatin (Nath et al. 2014), and the general tenet of karyotype evolution says that ancient species had less heterochromatin. *Drimia wightii* and *D.* species with congested inflorescence

Table 7.4 Comparative karyotypes of *Drimia* species and their cytotypes

Sr. No.	Taxa	2*n*	TCL ± SD (μm)	Haploid karyotype formulae	Asymmetry (A)		L ± SD (μm)	r ± SD (μm)	R (μm)	St	Collection Locality
					A_1	A_2					
1	`*D. coromandeliana*	40	99.19 ± 1.82	20st	0.75	0.37	4.96 ± 1.82	4.36 ± 1.23	2.99	1b	Kagal, Kolhapur district, Maharashtra
2	*D. govindappae*	20 + 7B	68.41 ± 2.92	8st + 2 t + 7B	0.80	0.43	6.84 ± 2.92	5.22 ± 1.34	2.97	1b	Jodabetti, Mysore district, Karnataka
3	*D. indica* (Form 1)	20	52.72 ± 2.04	9st + 1 t	0.76	0.39	5.27 ± 2.04	4.42 ± 1.41	2.77	1b	Delhi
4	*D. indica* (Form 2)	30	90.86 ± 2.27	15st	0.80	0.37	6.06 ± 2.27	5.15 ± 0.96	2.78	1b	Hyderabad, Andhra Pradesh
5	*D. indica* (Form 3)	20	44.89 ± 1.57	1sm + 9st	0.73	0.35	4.49 ± 1.57	3.80 ± 0.82	2.43	1b	Bhatye, Ratnagiri district, Maharashtra
6	*D. indica* (Form 4)	30	84.51 ± 2.01	15st	0.77	0.36	5.63 ± 2.01	4.51 ± 0.79	2.70	1b	Bhatye, Ratnagiri district, Maharashtra
7	*D. indica* (Form 5)	30	76.70 ± 1.88	15st	0.76	0.37	5.11 ± 1.88	4.37 ± 0.80	2.76	1b	Shirgaon, Ratnagiri district, Maharashtra
8	*D. indica* (Form 6)	30	79.40 ± 1.99	15st	0.77	0.38	5.29 ± 1.99	4.53 ± 0.96	2.83	1b	Shiroda, Sindhudurg district, Maharashtra
9	*D. nagarjunae*	20	60.87 ± 2.40	7st + 3 t	0.85	0.39	6.09 ± 2.40	6.69 ± 0.63	2.74	1b	Chinnar wild life sanctuary, Idduki district, Kerala
10	*D. polyantha*	20 + 7B	57.25 ± 2.20	9st + 1 t + 7B	0.77	0.38	5.73 ± 2.20	4.62 ± 1.36	2.80	1b	Thoseghar, Satara district, Maharashtra
11	*D. razii*	20	66.84 ± 2.70	10st	0.79	0.40	6.68 ± 2.70	4.95 ± 1.25	2.93	1b	Dive ghat, Pune district, Maharashtra
12	*D.wightii*	20	64.75 ± 2.42	8st + 2 t	0.79	0.37	6.47 ± 2.42	5.34 ± 2.00	2.69	1b	Tiruchirapalli, Tamil Nadu
13	*D.* species	20	60.30 ± 2.26	8st + 2 t	0.80	0.38	6.03 ± 2.26	5.38 ± 1.56	2.65	1b	Badami, Bagalkot district, Karnataka

TCL haploid chromosome length, A_1 and A_2 mean asymmetry indices (intrachromosomal and interchromosomal, respectively), *L*, mean chromosome length, *r* mean arm ratio, *R* ratio between the largest and the smallest chromosome of the complement, *St* type of asymmetry (an apomorphic character) and the presence of cytological races (2*x* and 4*x*) in the latter rule out their basal position. *Drimia govindappae*, *D. polyantha* and *D. razii* with their restricted distribution seem to be recent in origin. Eventually, it can be said that there is no pattern of variation in *Drimia* karyotype. Hence, tracing the path of karyotype evolution requires further studies. Nath et al. (2014) recommend the application of finer tools like GISH, FISH and phylogenetic analysis. Phylogenetic analysis involving samples from throughout India will shed light on the interrelationships within the genus. Coupled with this, the data on morphology, cytology and crossability will definitely reveal the direction of evolution in *Drimia*.

References

Ansari MY, Sundararaghavan R (1980) Nomenclature of some bulbous liliaceae of India. J Bombay Nat Hist Soc 77:172–174

APG III (2009) An update of the Angiosperm Phylogeny Group classification for the orders and families of flowering plants: APG III. Bot J Linn Soc 161:105–121

Ayyangar KR (1966) On the occurrence of B chromosomes in *Urginea indica* Kunth. Proc 53rd Indian Sci Congr Part 3:273–274

Ayyangar KR, Sampathkumar R (1968) A note on the artificial induction of polyploidy in *Urginea indica* Kunth. Proc 55th Indian Sci Congr Part 3:376

Boraiah A, Fathima T (1972) Cytotaxonomy of a new species of *Urginea*. Proc 59th Indian Sci Congr Part 3:359–360

Boraiah G, Khaleel TF 1970(1972) Cytotaxonomy of *Urginea govindappae* sp. nov. Bull Bot Surv India 12:128–131

Bougourd SM, Jones RN (1997) B chromosomes: a physiological enigma. New Phytol 137(1):43–54

Bruyns PV, Vosa CG (1987) Taxonomic and cytological notes on *Bowiea* Hook. f. and allied genera (Liliaceae). Caryologia 40:287–297

Buerki S, Jose S, Yadav SR, Goldblatt P, Manning JC, Forest F (2012) Contrasting biogeographic and diversification patterns in two Mediterranean-type ecosystems. PLoSONE 7(6):e39377. doi:10.1371/journal.pone.0039377

Camacho JPM (2005) B chromosomes. In: Gregory TR (ed) The evolution of the genome. Elsevier, San Diego, pp 223–228

Capoor SP (1937) Contribution to the morphology of some Indian Liliaceae. II. The gametophytes of *Urginea indica* Kunth. Beih Bot Centralbl 56(Abt.A):156–170

Chase MW, Reveal JL, Fay MF (2009) A subfamilial classification for the expanded Asparagalean families Amaryllidaceae, Asparagaceae and Xanthorrhoeaceae. Bot J Linn Soc 161:132–136

Cobon AM, Murray BG (1983) Unstable B chromosomes in *Silene maritima* with. (Caryophyllaceae). Bot J Linn Soc 87(3):273–283

Dafni A, Cohen D, Noy-Meir I (1981) Life-cycle variation in geophytes. Ann Mo Bot Gard 68:652–660

Darlington CD, Wylie AP (1955) Chromosome atlas of flowering plants. George Allen & Unwin, London

Deb DB, Dasgupta S (1987) On the identity of three new species of *Urginea* (Liliaceae). J Bombay Nat Hist Soc 84:409–412

Dixit GB, Yadav SR (1989) Cytotaxonomical and genetical studies in *Urginea* Steinh. species from India. Cytologia 54:715–721

Goldblatt P (1980) Polyploidy in angiosperms: monocotyledons. In: Lewis WH (ed) Polyploidy: biological relevance. Plenum Press, New York, pp 219–239

Goldblatt P, Manning JC, Forest F (2012) A review of chromosome cytology in Hyacinthaceae subfamilies Urgineoideae and Hyacinthoideae (tribes Hyacintheae, Massonieae, Pseudoprospereae) in Sub-Saharan Africa. S Afr J Bot 83:134–144

Gotoh K (1924) U¨ ber die Chromosomenzahl von *Secale cereale* L. Bot Mag Tokyo 38:135–152

Grant V (1971) Plant Speciation. Columbia University Press, New York

Grant V (1981) Plant speciation, 2nd edn. Columbia University Press, New York

Guerra M (2000) Chromosome number variation and evolution in monocots. In: Wilson KL, Morrison DA (eds) Monocots: systematics and evolution. CSIRO, Melbourne, pp 127–136

Hemadri K (2006) Medicinal Flora of Pragati resorts and surrounding (Ranga Reddy district, Andhra Pradesh, India). Pragati Green Meadows and Resorts Ltd, Hyderabad, pp 385–388

Hemadri K, Swahari S (1982) *Urginea nagarjunae* Hemadri et Swahari a new species of Liliaceae from India. Anc Sci Life 2:105–110

Holtum RE (1955) Growth-habits of monocotyledons, variations on a theme. Phytomorphology 5:399–413

Houben A, Banaei-Moghaddam AM, Klemme S, Timmis JN (2013) Evolution and biology of supernumerary B chromosomes. Cell Mol Life Sci 71:467–478

Jehan T, Vashishtha A, Yadav SR, Lakhanpaul S (2014) Genetic diversity and genetic relationships in Hyacinthaceae in India using RAPD and SRAP markers. Physiol Mol Biol Plants 20(1):103–114

Jha S (1989) Karyology of callus induction in inflorescence and bulb scale explant of diploid Indian squill. Proc Indian Sci Congr Assoc 76(3–VI):207

Jha S, Sen S (1981) Bufadienolides in different chromosomal races of Indian squill. Photochemistry 20(3):524–526

Jha S, Sen S (1983a) Quantification of principle Bufadienolides in different cytotypes of *Urginea indica*. Planta Med 47:43–45

Jha S, Sen S (1983b) Chromosome study of diploid Indian squill, *Urginea indica* Kunth. Cytologia 48:79–86

Johnson MAT, Kenton AY, Bennett MD, Brandham PE (1989) *Voanioala gerardii* has the highest known chromosome number in the monocotyledons. Genome 32:328–333

Jones RN (1995) B chromosomes in plants. New Phytol 131:411–434

Jones RN, Houben A (2003) B chromosomes in plants: escapees from the A chromosome genome? Trends in Plant Sci 8:417–423

Jones K, Smith JB (1967) The chromosomes of the Liliaceace: I. The karyotypes of twenty-five tropical species. Kew Bull 21:31–38

Jones RN, Viegas W, Houben A (2007) A century of B chromosomes in plants: so what? Ann Bot 101:767–775

Joshi CP, Ranjekar PK (1982) Visualization and distribution of heterochromatin in interphase nuclei of several plant species as revealed by a new Giemsa banding technique. Cytologia 47:471–480

Kambale S, Ansari MY (1976) A note on the somatic chromosomes of *Urginea polyantha* Blatt. Bull Bot Surv India 18:1–4

Kishore H (1951) A note on the chromosome numbers of some plants. Indian J Genet Plant Breed 11:217

Kumar A (1984) A new combination in *Drimia* Jacq. ex Wild. (Liliaceae). J Econ Taxon Bot 5(4):962

Kuwada Y (1925) On the number of chromosomes in maize. Bot Mag Tokyo 39:227–234

Lavania S, Lavania UC (2007) In: Chauhan SVS et al (eds) Plant reproduction biology and biotechnology. Avishkar publishers, distributors, Jaipur, pp 126–144

Lekhak MM, Adsul AA, Yadav SR (2014) Cytotaxonomical studies on *Drimia nagarjunae* and a review on the taxonomy of Indian species of *Drimia*. Nucleus 57(2):99–103

Levin DA (1971) The origin of reproductive isolating mechanisms in flowering plants. Taxon 20:91–113

Levin DA, Palestis BG, Jones RN, Trivers R (2005) Phyletic hot spots for B chromosomes in angiosperms. Evolution 59:962–969

Longley AE (1927) Supernumerary chromosomes in Zea mays. J Agric Res 35:769–784

Manning JC, Goldblatt P, Fay MF (2004) A revised generic synopsis of Hyacinthaceae in subSaharan Africa, based on molecular evidence, including new combinations and the new tribe Pseudoprospereae. Edinb J Bot 60(3):533–568

Manning JC, Forest F, Devey DS, Fay MF, Goldblatt P (2009) A molecular phylogeny and a revised classification of Ornithogaloideae (Hyacinthaceae) based on an analysis of four placid DNA regions. Taxon 58:77–107

Mehra PN, Malik CP (1961) Cytology of Indian medicinal plants. III. Some Liliaceous drugs. Proc 48th Indian Sci Congr Part 3:293–294

Mehra PN, Pandita TK (1979) In IOPB chromosome number reports LXIV. Taxon 28:405

Miege J (1960) Nombreschromosimiques de plantesd'Afrique Occidentale. Rev Cytol Biol Veg 21:373–384

Moorthy VD, Sampathkumar R (1968) Chromosomal variability in the Indian squill (*Urginea indica* Kunth). Curr Sci 21(5):620–621

Naik VN (1976) Cytotaxonomic studies in two species of *Urginea* Steinh. From India. J Indian Bot Soc 55:60–64

Nath S, Jha TB, Mallick SK, Jha S (2014) Karyological relationships in Indian species of *Drimia* based on fluorescent chromosome banding and nuclear DNA amount. Protoplasma 252(1):283–299

Oyewole SO, Mustapha OT (2000) Cytotaxonomy of the genus *Urginea* Stein. V. Taxonomic treatment of *Urginea indica* (Roxb.) Kunth complex. Biosci Res Commun 12(3):259–264

Patil MM, Torne SG (1982) Different cytotypes of *U. indica* Kunth. Proc Indian Sci Congr Assoc 69(3–VI):233

Patil BR, Yadav SR (1992–1993) Studies in the genus *Urginea* Steinh. (Liliaceae). I Phenology and Palynology. J Shivaji Univ (Science) 29:69–78

Raghavan TS (1935a) Observations on somatic chromosomes of *Urginea indica* Kunth. J Indian Bot Soc 14:151–158

Raghavan TS (1935b) Observations on somatic chromosomes of *Urginea indica* Kunth. Proc 22nd Indian Sci Congr 3:268–269

Raghavan TS, Venkatasubban KR (1940) Studies in the Indian Scilleae III. The cytology of diploid Urginea indica Kunth. Cytologia 11:55–70

Randolph LF (1928) Types of supernumerary chromosomes in maize. Anat Rec 41:102

Raunkiaer C (1934) The life forms of plants and statistical plant geography. Clarendon Press, Oxford, p 631

Raven PH (1975) The basis of angiosperms phylogeny: cytology. Ann Mo Bot Gard 62:724–764

Roberto C (2005) Low chromosome number angiosperms. Caryologia 58(4):403–409

Roxburgh W (1832) Flora Indica; or descriptions of Indian Plants. Thacker & Co., Serampore, Calcutta and Allen & Co., Parbury, London 2:147 pp

Sapre AB, Deshpande DS (1987) Origin of B chromosomes in *Coix* L. through spontaneous interspecific hybridization. J Hered 78:191–196

Sen S (1973) Polysomaty and its significance in Liliales. Cytologia 38:737–751

Sen S (1974) Nature and behaviour of B chromosomes in *Allium stracheyi* Baker and *Urginea indica* (Roxb.) Kunth. Cytologia 39:245–251

Sen S, Mukherjee S (1976) Analysis of chromosomes and chemical constituents of *Urginea indica* and *Scilla indica*. Res Bull Univ Calcutta Cytogenet Lab 3:58

Sharma AK (1956) A new concept of a means of speciation in plants. Caryologia 9:93–103

Sharma AK (1964) Cytology as an aid in taxonomy. Bull Bot Soc Bengal 18:1–4

Sharma AK (1984) Chromosome evolution in the monocotyledons- an overview. In: Sharma A, Sharma AK (eds) Chromosomes in evolution of eukaryotic groups, vol II. CRC Press, Boca Raton, pp 169–187

Sharma CBSR, Dash SK (1977) Cytopolymorphism in *Urginea indica* (Roxb.) Kunth. Proc 46th Indian Sci Congr 3:158

Sheriff A, Rao UG (1981) Cytogeographical studies on *Scilla indica* in India-triploids. Cytologia 46:69–74

Shivanna KR (2003) Pollen biology and biotechnology. Oxford & IBH Publishing Co Pvt Ltd, New Delhi, p 104

Soltis DE, Soltis PS (1990) Isozyme evidence for ancient polyploidy in primitive angiosperms. Syst Bot 15:328–337

Speta F (1998) Hyacinthaceae. In: Kubitzki K (ed) The families and genera of vascular plants, vol 3. Springer Verlag, Berlin, pp 261–285

Stace CA (1989) Plant taxonomy and biosystematics, 2nd edn. Edward Arnold, London

Stearn WT (1978) Mediterranean and Indian species of *Drimia*, (Liliaceae) a nomenclatural survey with special reference to the medicinal squill *D. maritima* (syn. *Urginea maritima*). Ann Mus Goulandris 4:199–210

Stebbins GL (1971) Chromosomal evolution in higher plants. Edward Arnold, London

Subramanian D (1973) Meiotic studies in *Urginea indica* Kunth with special reference to the formation of pollen grains. Proc 63rd Indian Sci Congr Part 3:323

Subramanian D (1976) Karyomorphological analyses of the different cytotypes in Urginea indica (Roxb.) Kunth. Proc 63rd Indian Sci Congr Part 3:124

Subramanian D (1978) Cyto-genetical studies in *Urginea indica* (Roxb) Kunth. J Indian Bot Soc 57:211–218

Subramanian D (1980) Cytomorphological studies in *Urginea indica* (Roxb) Kunth. J Indian Bot Soc 3rd All India Botanical Conference 59(Suppl):103–104

The Plant List (2013) Version 1.1. http://www.theplantlist.org. Accessed 01 Jan 2015

Trivers R, Burt A, Palestis BG (2004) B chromsomes and genome size in flowering plants. Genome 47:1–8

Vanzela ALL, Guerra M, Luceño M (1996) *Rhynchospora tenuis* link (Cyperaceae), a species with the lowest number of holocentric chromosomes ($n = 2$). Cytobios 88:213–228

Wilson EB (1907a) The supernumerary chromosomes of *Hemiptera*. Science 26:870–871

Wilson EB (1907b) Note on the chromosome groups of *Metapodius* and *Banas*. Biol Bull XII:303–313

Yadav SR, Dixit GB (1990) Cytotaxonomical studies in Indian Squill Steinhill species. Cytologia 55:293–300

Yadav BR, Lekhak, MM, Yadav SR (in press) Typifications in *Drimia* with notes on the identity of *D. wightii* (Asparagaceae, Scilloideae). Phytotaxa.

Zaman MA, Khaleque Y (1978) The cytology of *Urginea indica* Kunth from Bangladesh. Caryologia 31:135–145

Banding Techniques in Chromosome Analysis

8

Peerzada Arshid Shabir, Aijaz Ahmad Wani, and Irshad A. Nawchoo

Abstract

Chromosome identification has been traditionally based on morphological features of individual chromosomes such as chromosome length, arm ratio and primary and secondary constriction collectively called as karyotype. A number of stains such as acetocarmine, Feulgen and aceto-orcein all of them being whole chromosome stains have been used in these studies. Although classical staining helps in studying chromosome morphology, structural and numerical variations, however, morphologically similar chromosomes cannot be distinguished. The utilization of fluorescent and other dyes together with various modifications in pretreatment of cytological material in the late 1960s led to the discovery of various banding techniques which proved to be additional tool for identification of individual chromosomes. New and reliable staining procedures were introduced; each was capable of revealing a unique banding pattern of the chromosomes of a given species. The advantage with banding techniques is that they can resolve morphologically similar as well as different chromosomes and help in understanding the chromosome organization. Chromosome banding is a lengthwise variation in staining properties along a chromosome based on the GC- or AT-rich regions or constitutive heterochromatin. A single dye or fluorochrome can often be used to produce a banding pattern on a chromosome. A band is a part of chromosome which is clearly distinguishable from its adjacent segment by appearing darker or lighter with various banding methods. The Paris Conference – 1971 – classified banding techniques as Q–banding (fluorescence based), C–banding (constitutive heterochromatin (AT- or GC-rich DNA)), G–banding (whole length

P.A. Shabir (✉) • I.A. Nawchoo
Economic Botany and Reproductive Biology Laboratory, Department of Botany, University of Kashmir, 190006 Srinagar, Jammu and Kashmir, India
e-mail: peerzadarshid@gmail.com

A.A. Wani
Department of Botany, University of Kashmir, Srinagar, Jammu and Kashmir, India

T.A. Bhat, A.A. Wani (eds.), *Chromosome Structure and Aberrations*,
DOI 10.1007/978-81-322-3673-3_8

banding (Giemsa staining)), R–banding (reverse of G–banding) and Ag-NOR stain (nucleolar organizing regions). All these banding techniques have led to a more precise cytogenetic and phylogenetic analysis of various eukaryotes. The major applications of banding techniques have been the mapping of genes on chromosomes and identification of chromosome alterations such as deletions, duplications, translocations and aneuploidy. They have also played an important role in measuring the amount of heterochromatin among individuals.

8.1 Introduction

Cytogenetical studies have played an important role to obtain basic information indispensable to plant genetics and breeding. The field of plant cytogenetics was heavily influenced by Barbara McClintock's pioneering work on maize (*Zea mays*). Her scrupulous work in developing a methodology for unequivocal identification of individual chromosomes permitted major discoveries regarding the structure and dynamic behaviour of the maize genome (Creighton and McClintock 1931; McClintock 1932, 1938, 1941, 1984). While performing karyotype analysis, using carmine-based chromatin staining procedures, McClintock showed that all of the individual chromosomes could be uniquely identified from its specific relative length and arm ratio (McClintock 1929). Karyotype analysis based on relative chromosome lengths and arm ratio has described plant chromosomes for over 30 years, and this approach has proved useful while developing cytogenetic maps for many plant species, including rice (*Oryza sativa*) (Shastry et al. 1960; Misra and Shastry 1967), sorghum (*Sorghum propinquum*) (Magoon and Shambulinguppa 1961) and tomato (*Lycopersicon esculentum*) (Barton 1950; Ramanna and Parkken 1967). However, in almost all species the usefulness of standard staining procedures has been limited. Although they have facilitated the ascertainment of chromosome numbers and gross morphological features of chromosomes, they have not permitted an accurate and unequivocal identification of all the chromosomes. An exhaustive analysis of the karyotype requires the use of staining procedures that can reveal each chromosome as a specific, unique and constant pattern of alternating dark and light banding regions, topologically equivalent to the bands in the polytene chromosomes in salivary gland cells of *D. melanogaster.* Darlington and La Cour (1940) demonstrated that with cold treatment of somatic cells of *Trillium erectum*, some regions of chromosomes revealed unique patterns by appearing thinner and less intensely stained than the rest of the chromosomes. Caspersson et al. (1968), while using fluorescence microscopy, discovered one of the first specific chromosome banding techniques (Q–banding), which involved staining chromosomes with a fluorochrome, such as quinacrine mustard or quinacrine dihydrochloride. This technique, however, was less than optimal for routine studies, as the fluorescent staining quickly quenched. Hence, several other banding techniques were developed, for example, G–, R–, C– and NOR–banding, each having its own specific properties and applications (Rooney 2001). The utilization of fluorescent and other dyes together with various modifications in pretreatment of cytological material in the late 1960s heralded a new era of cytogenetics. New and reliable staining procedures were introduced, each of which was capable of revealing a unique banding pattern of the chromosomes of a given species. By 1972 the application of one or another of five major banding techniques (Q–, G–, R–, C– and N–) for the purpose of karyotypic analysis came into vogue. These have led to a more precise cytogenetic and phylogenetic analysis of various eukaryotes.

8.2 Chromosome Banding Techniques and Mechanisms

8.2.1 *Q-Banding*

Caspersson and colleagues at the Karolinska Institute at Harvard Medical School set out to test or design fluorescent molecules that would preferentially bind to specific nucleotide pairs in DNA, which they hoped to be able to detect spectrophotometrically. One molecule tested was quinacrine mustard dihydrochloride (QM) on *Vicia faba* and *Trillium erectum*. The spectrophotometric analysis revealed brightly fluorescent bands that distinguished the individual plant chromosomes. By applying QM to human chromosomes, Caspersson et al. (1970) discovered that the end of the long arm of the Y-chromosome was brightly fluorescent and could be easily detected in interphase as well as in metaphase cells. With refinements, QM produced banding patterns that were specific for each individual human chromosome. Several investigators showed that the bright fluorescent bands obtained with quinacrine mustard correspond to AT-rich regions of DNA and the differences in intensity of fluorescence reflected a difference in quenching of the QM molecule (Weisblum and de Haseth 1972; Comings et al. 1975). It has been observed that the amino group at position 2 of the guanine bases of the DNA quenches the fluorescence of quinacrine thus causing the AT-rich regions of the chromosomes to fluoresce more brightly than the GC-rich regions. Moreover, Weisblum and Haenssler (1974) demonstrated that the fluorescence of quinacrine can also be efficiently quenched even in AT-rich regions of the chromosome if guanine resides are present and spaced with periodicity. Thus the AT-richness alone is not the sole determinant of the intensity of Q–banding, but it is the periodicity of interspersed GC, within short and highly repetitive AT-rich sequences, as well as the presence of specific nucleoproteins appears to play a significant role in quenching the fluorescence of quinacrine (Michelson et al. 1972).

8.2.1.1 Protocol: Barch et al. (1991)

8.2.1.2 Preparations

Quinicine dihydrochloride: Dissolve 0.25 g in 50 ml distilled water and store away from light.

McIlvaine's buffer: Make a solution of 0.1 M anhydrous citric acid (19.2 g/L) (solution A) and a solution of 0.4 M anhydrous sodium phosphate dibasic (56.8 g/L) (solution B). The buffer solution consists of 92 mL of solution A and 50 mL of solution B; adjust pH to 5.6 if required.

8.2.1.3 Procedure

Soak fixed slides in methanol for 5 min. Pour methanol off and add the stain.
Stain for 5 min with QM.
Rinse three times in distilled water.
Expose in McIlvaine's buffer for 30 s.
Mount with a cover slip using McIlvaine's buffer diluted one part to four parts distilled water.
Examine by fluorescent microscope (390–490 excitations; suppression: 515 nm).
Scan cells and photograph.

Q–banding permits an identification of all the chromosomes and their homologues in most species. For example, in man all 23 pairs of homologous chromosomes can be distinguished on the basis of their Q–banding patterns (Caspersson et al. 1970). In *Scilla siberica* all eight chromosome pairs can be identified (Caspersson et al. 1969). Q–banding does not require any pretreatment of the chromosomes and is the simplest of all the banding methods. Compared to other banding techniques, it has several disadvantages; the fluorescent bands are not permanent; the technique requires the use of ultraviolet light and does not stain the ends of chromosomes. As a consequence Q–banding has been used to a limited extent, and since the late 1970s (Pinkel et al. 1986) has been largely replaced by other banding methods. In plants, Q–banding studies have been limited to a few species belonging to *Trillium*, *Scilla*, *Allium*, *Crepis*, *Lilium*, *Secale* and *Vicia* (Caspersson et al. 1969; Vosa and Marchi 1972; Kongsuwan and Smyth 1977; Schweizer 1979; Rowland 1981).

8.2.2 *G-Banding*

Soon after the discovery of Q–banding, Sumner et al. (1971) introduced the Giemsa (G-) banding that utilized the common Giemsa stain following various chemical and enzymatic treatments of the chromosome preparations. In comparison to Q–banding, G–banding is nonfluorescent and is advantageous in the aspect of stability and resolution of the bands produced (Liu 2006). This method therefore overcame two significant problems of Q–banding (stability and cost) as it offered the advantage of producing permanent slides that can be studied under a standard light microscope and thus became the more widely used banding technique in cytogenetic studies.

Giemsa is a visible light dye that binds to DNA through intercalation. It is a mixture of cationic thiazine dyes, most importantly azure B, and anionic eosin dyes such as eosin Y (Fig. 8.1) (Horobin 2011).

While performing G–banding, staining of the chromosomes involves the formation of a thiazine-eosin precipitate in a 2:1 molar ratio (Sumner 1982; Horobin 2011). During G–banding preparation, small thiazine dye molecules diffuse fast and first intercalate between the base pairs of the DNA in a configuration that favours the binding of the large, slow diffusing eosin molecule. The chromosomes stain blue as a result of thiazine binding. The eosin molecule then forms a precipitate with the

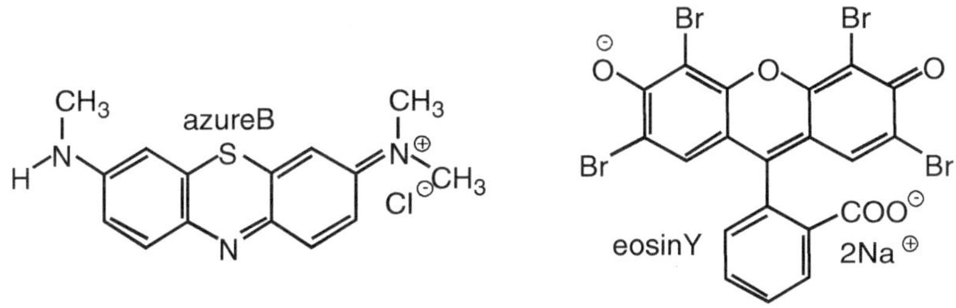

Fig. 8.1 Structure of azure b and eosin Y

thiazine molecules, thus causing the chromosomes to stain purple. The formation of this precipitate is favoured on a hydrophobic environment. The type of interaction between the thiazine and eosin molecules is still a matter of debate (Horobin 2011). A study by Zanker (1981) suggests the formation of a charge transfer complex between the thiazine and eosin molecules with the thiazine as the receptor and the eosin as the donor. Wittekind and Gehring (1985) suggest the formation of H–bonds between thiazine and eosin molecules and thus facilitate the electron transfer between the molecules. Furthermore, Wittekind and Gehring (1985) are of the opinion that the association between thiazine and eosin molecules involves weak hydrophobic interactions.

For G–banding, the chromosomes must first undergo a pretreatment process before staining with Giemsa (Sumner 1982), and the most common pretreatment is the digestion of the chromosomes with a protease such as trypsin. Other pretreatments like incubation of the chromosomes in hot saline citrate or subjecting the chromosomes into a detergent or urea solution are also variously recommended. These pretreatment methods are known to extract a characteristic subset of proteins from the chromosomes (Burkholder and Weaver 1977; Sumner 1982). This differential extraction of the proteins throughout the length of a chromosome is responsible for the banding and is a reflection of the difference in the structure of the various chromosomal regions. The positive G–bands, which are the dark bands, correspond to the hydrophobic regions of the chromosome that favour the formation of the thiazine-eosin precipitate (Sumner 1982). These regions are identified as the late replicating heterochromatin, which are characterized as condensed and rich in protein disulphide cross-links. Due to the condensed structure and richness in disulphide cross-links of these regions, the hydrophobic proteins needed for the formation of the Giemsa complex are retained in position or appropriate conformation during the pretreatment process. Since the heterochromatic regions are AT rich, the positive G–bands can be revealed by fluorochromes that are specific for AT-rich regions of the chromosomes. Meanwhile, the negative G–bands, which are the light bands, correspond to the less hydrophobic regions of the chromosomes that do not favour the formation of the thiazine-eosin precipitate (Sumner 1982). These regions are identified as the early replicating euchromatin that have relatively loose structure and have their protein sulphur predominantly as sulphhydryls. Hence, the hydrophobic proteins in these regions are easily removed during the pretreatment.

8.2.2.1 Protocol: Barch et al. (1991)

8.2.2.2 Preparations

pH 6.8 phosphate buffer
Working Giemsa stain: 4 mL Giemsa; 96 mL buffer with pH 6.8
Leishman's stain: Leishman's stain diluted 1:4 with pH 6.8 buffer

8.2.2.3 Procedure

Place fixed, dry slides on slide rack in 95 °C oven and bake for 20 min. Then allow them to cool.
Immerse slide in 0.025 % trypsin for 10–120 s.
Remove slide from trypsin and immerse immediately in buffer to stop trypsin action.
Now stain for 2 min with solution of one part Leishman's stain and three parts of Giemsa working solution.
Rinse slides thoroughly with distilled water, allow slides to drain, and then place on 60 °C slide warming tray until completely dry.

For routine analysis, however, the G–banding technique using trypsin and Giemsa became the most accepted methods worldwide (Seabright 1971). The generated banding pattern is specific for the organisms and allows the identification of the individual chromosomes or their parts. On a typical metaphase preparation 300–550 bands or so can be distinguished which is appropriate for karyotyping. In prometaphase cells the chromosomes are much longer; the bands are divided in sub-bands. The analysis of prometaphase chromosomes is used only for specific purposes (high-resolution karyotyping). The banding pattern enabled the detection of various structural aberrations like translocations, inversions, deletions and duplications in addition to the well-known numerical aberrations. However, although the technique has been attempted in many plant species, G–bands have been generated in the chromosomes of only a few species: *Tulipa gesneriana* (Filion and Blakey 1979), *Pinus resinosa* (Drewry 1982) and *Vicia hajsatana* (Wang and Kao 1988). The failure to produce G–bands in the chromosomes of most plant species, including those in the *Triticeae*, has attributed to the increased condensation of the plant chromosomes (Greilhuber 1977; Drewry 1982). Anderson et al. (1982), however, failed to show consistent differences in the degree of compaction, based on measurements of lengths and volumes of chromosomes from several plant and animal species. Wang and Kao (1988) demonstrated that improper pretreatment of plant chromosomes alters the organization of their chemical constituents and renders them unresponsive to the *G*–banding procedure.

8.2.3 *C-Banding*

Pardue and Gall (1970) and Jones (1970) independently demonstrated a procedure which with stringent treatment of chromosomes prior to fixation and staining with Giemsa, stained only the regions of constitutive heterochromatin in chromosomes of mouse (*Mus musculus*). The regions, now referred to as C–bands, were observed to be proximal to the centromeres of all the chromosomes. Constitutive heterochromatin usually appears as satellite DNA when nuclear chromosomal DNA is fragmented and centrifuged (Kit 1961) and consists of short, highly repeated base pair sequences in tandem (Southern 1970; Corneo et al. 1970; Brutlag et al. 1977).

Arrighi and Hsu (1971) developed a modified technique in which they applied Giemsa staining to preparations that were first denatured with 0.07 M NaOH and then incubated in two times standard saline concentration (2 × SSC) for several hours. Arrighi and Hsu (1971) showed that C–bands are located next to the centromeres of each chromosome, next to the secondary constrictions of chromosomes 1, 9 and 16 as well as the satellites of acrocentric chromosomes in man. Sumner (1972) substituted barium hydroxide for sodium hydroxide, producing the same C–banding pattern but with less distortion of the chromosome morphology. Both procedures result in intense staining of the heterochromatin around the centromeres, whereas the rest of the chromosome stains pale blue. Arrighi and Hsu (1971) initially postulated that this differential staining was due to faster reannealing of repetitive DNA in heterochromatin than in the less repetitive DNA sequence elsewhere. McKenzie and Lubs (1973) produced C–banding by simply treating chromosomes with HCl and prolonged incubation in 2 × SSC. Studies by Comings et al. (1975) demonstrated considerable extraction of nucleoprotein and DNA from non-heterochromatic regions by various C–banding treatments, while heterochromatic regions were resistant to such extraction. Furthermore, they demonstrated that hybridization of repetitive sequences in solution was not required for enhancement of staining, but in fact those regions reassociated instantaneously when they were removed from the NaOH solution. Subsequent incubation in 2 × SSC extracted additional non-heterochromatic DNA.

8.2.3.1 Protocol: Friebe and Gill (1994)

8.2.3.2 Preparations

1. 3:1 fixative: three parts absolute methanol with one part glacial acetic acid.
2. 0.2 N HCL: add 4.15 ml concentrated hydrochloric acid to 200 ml of distilled water, mix and add water to 250 ml.
3. 2 × SSC: dissolve 17.53 g sodium chloride crystal and 8.82 g sodium citrate crystal in 1 L of distilled water.
4. 2.5 % $Ba(OH)_2$: dissolve 2.5 g of barium hydroxide crystal in 100 ml distilled water.

8.2.3.3 Procedure

1. Remove oil from G–banded slide thoroughly with at least two rinses of fresh xylene substitute.
2. Destain by dipping in 3:1 fixative; wipe the bottom of slide, place on 40–60 °C slide warming tray until beads of solution are formed, and then blot gently with bibulous paper. These four steps should be repeated until beads are clear.

3. Place dry, destained slide in 0.2 N HCL for 1 hour. After 1/2 hour turn on preset water bath and start to filter $Ba(OH)_2$ through #1 Whatman filter paper into Coplin jar.
4. Rinse slide (treated in 0.2 N HCL) in Coplin jar filled with distilled water.
5. Place rinsed slide in freshly filtered $Ba(OH)_2$ solution for 2 minutes.
6. Rinse with distilled water in squirt bottle (some force is required to remove $Ba(OH)_2$ crystals).
7. Place rinsed slide in Coplin jar (in water bath) filled with 2 × SSC at approximately 62.5 °C for 1 h.
8. Remove slide slowly and rinse gently in Coplin jar filled with freshly distilled water.
9. After drying, stain in 2 % Giemsa for 10–20 min.
10. Rinse in xylene.
11. Rinse in tap water. Air-dry.

C–banding is of limited use in the clinical laboratory and is primarily of value in the identification of the gene coding potential of various segments of the genome, especially when small marker chromosomes of unknown origin are present, and for the study of chromosomal polymorphisms in the population. The short arms and satellites of acrocentric chromosomes, pericentric heterochromatin and much of the long arm of the Y-chromosome are all C–band positive, contain no active genes and show variations in size in normal individuals.

8.2.4 *R-Banding*

Utrillaux and Lejeune (1971) introduced a new chromosomal banding technique, involving treatment of chromosomes in saline at high temperature (87 °C) that resulted in a reverse pattern of G– or Q–band. They called this as "reverse (R) banding". Mild denaturation by heat and subsequent staining of chromosomes with Giemsa or a fluorochrome dye revealed a banding pattern that is the reverse of the patterns produced by the G- and Q–banding methods (Bobrow et al. 1972; Comings 1975). Specifically, if the chromosomes are stained with Giemsa, the dark bands (R–bands) produced with this technique are equivalent to the light bands produced by the G–banding technique and vice versa (Dutrillaux and Lejeune 1971). If a fluorochrome dye such as acridine orange or olivomycin is used, fluorescent R–banding is the reverse of Q–banding in that the R–bands fluoresce bright green and the non-R–bands show a faint red colour (Schweizer 1976; Lin et al. 1980; Schmid and Guttenback 1988). Since the euchromatic regions are GC rich, the negative G–bands or R–bands can be produced by GC-specific fluorochromes (Schweizer 1976; Holmquist et al. 1982). A study by Burkholder and Weaver (1977) shows the role of the nonhistone proteins in banding which are known for uniform distribution and binding throughout the chromosomes and, hence, will not be extracted differentially during the pretreatment process. On the other hand, the nonhistone proteins are differentially bound throughout the chromosomes and may thus be responsible for the banding process. These nonhistones are more tightly bound in the condensed chromatin regions than in the less condensed regions and, thus, will be extracted differentially during the pretreatment process.

8.2.4.1 Protocol: Verma and Babu (1989)

8.2.4.2 Preparations

1. Buffer: 10.0 mL of Earle's balanced salt solution (EBSS), 0.1 mL 7.5 % sodium bicarbonate and 89.9 mL distilled water. Place buffer in water bath, and heat to 88 °C to 89 °C.
2. 2 % Giemsa in distilled water.

8.2.4.3 Procedure

1. Incubate slides in hot EBSS for 10–20 min.
2. Allow slides to cool quickly in tap water.
3. Stain in 2 % Giemsa for 10–20 min.
4. Rinse in xylene.
5. Rinse in tap water. Air-dry.

R–bands have the theoretical advantage of staining the gene-rich chromatin, thus enhancing the ability to visualize small structural rearrangements in the parts of the genome that are most likely to result in phenotypic abnormalities. R–banding is particularly useful in the detection of structural rearrangements involving ends of chromosomes in that it stains telomeres as T–bands (Dutrillaux and Lejeune 1975). R–bands have been detected in only a few plant species, e.g. *Vicia faba* and *Allium* spp. (Deumling and Greilhuber 1982; Loidl 1983). Moreover, since the R–bands in these species are few in number and faint in expression, they have not been used for karyotyping and cytogenetic studies.

8.2.5 *N-Banding*

In 1973 Matsui and Sasaki developed a technique they called N–banding, which selectively stained NORs in the chromosomes of mammalian species. Funaki et al. (1975) improved this procedure and demonstrated that N–bands were confined to the NORs of the chromosomes of 27 eukaryotic species that they studied. Faust and Vogel (1974) and Pimpinelli et al. (1976) observed that the bands obtained with this procedure are not NOR specific in *D. melanogaster* and the mammalian species studied. Nevertheless, these non-NOR bands are and continue to be referred to as N–bands. Using the method of Funaki et al. (1975), with slight modifications, Gerlach (1977) and Jewell (1979), working with *Triticum* and *Aegilops* species, respectively, clearly demonstrated that N–bands do not necessarily correspond to NORs. Gerlach (1977), and subsequently others, noted that many of the N–bands occupy the same positions as C–bands, implying that the N–banding technique, like the C–banding one, identifies constitutive heterochromatin and that at least two classes of heterochromatin occur in wheat, rye and other species. This was confirmed by Schlegel and Gill (1984) and Endo and Gill (1984). Some heterochromatic regions in each chromosome stain positively using both C– and N–banding procedures. These regions are referred to as C + N bands. Other such regions stain positively only with C–banding techniques. Schlegel and Gill (1984) have shown that, among these C + N bands, only N–bands possess multiple copies of the (GAA) n (GAG)n sequence DNA. The common base pair sequences in C + N bands have not been identified.

8.2.5.1 Protocol: Gill et al. (1991)

8.2.5.2 Preparations

1. 50 % silver-nitrate solution: dissolve 5 g silver nitrate in 10 ml distilled water and pour into clean and dry amber bottle. Label and store at 4 °C.

8.2.5.3 Procedure

1. Place unstained slide in Coplin jar with distilled water. Place in 37 °C water bath for 2 h. Remove slide and allow to air-dry.
2. Prepare moist chamber by putting two 12.5 cm circles of Whatman #2 filter in bottom of glass petri dish. Saturate paper with distilled water. Place plastic lids or rings on wet filter paper to support each slide at both ends.
3. Using 18 gauge needle with 1 cc syringe, draw 1 cc of silver-nitrate solution into syringe.
4. Remove needle and replace with Acrodisc filter.
5. Lay 5–7 drops of filtered solution near labelled end of slide.
6. Transfer treated slide to moist chamber and support on lids above wet filter paper. Cover dish.
7. Carefully place moist chamber in 37 °C incubator and incubate for 7 h.
8. Carefully remove slides from moist chamber and rinse slide in distilled water.
9. After air-drying, check treated slide under 10X phase to judge effectiveness of first treatment.
10. If stained NORs are unapparent, repeat steps 3–9 for 1 h.
11. Repeat steps 10 and 11.
12. Counterstaining is not necessary but may be desired. Counterstaining with quinacrine is ideal; however counterstaining with Wright's stain (see G–banding procedure) may work satisfactorily. If Wright's stain is too dark, it is difficult to distinguish between satellite and silver staining on NORs.

Nucleolus organizer regions (NORs) are the sites of rRNA genes in the chromosomes of animal (Ritossa and Spiegelman 1965; Wallace and Bimstie 1966; Henderson et al. 1974) and plant species (Phillips et al. 1971; Flavell and O'Dell 1976; Hutchinson and Miller 1982). These regions are located in the satellite stalks of acrocentric chromosomes and house genes for ribosomal RNA. Methods have been developed for the selective staining of these chromosomal regions. Howell et al. (1975) showed that NORs on chromosomes could be stained with silver nitrate and called their techniques “Ag-SAT”. The silver staining method selectively stains those sites on chromosomes which correspond exactly to regions that can be detected by in situ hybridization with rDNA probes (Howell et al. 1975; Miller et al. 1976). It seems that this procedure stains only the NORs that are functionally active during the preceding interphase (Howell et al. 1975). There is evidence to suggest acidic or nonhistone proteins associated with the rDNA regions are responsible for the selective staining of NORs (Howell et al. 1975; Howell 1985). Nonetheless, N–banding is useful in clinical practice to study certain chromosome polymorphisms, such as double satellites. This method is also helpful to identify satellite stalks that are occasionally seen on non-acrocentric chromosomes.

References

Anderson LK, Stack SM, Mitchell JB (1982) An investigation of the basis of current hypothesis for lack of G-banding in plant chromosomes. Exp Celt Res 138:433–436

Arrighi FE, Hsu TC (1971) Location of heterochromatin in human chromosomes. Cytogenetics 10:81–86

Barch MJ, Lawce HJ, Arsham MS (1991) Peripheral blood culture. In: Barch MJ (ed) The ACT cytogenetics laboratory manual, 2nd edn. Raven, New York, pp 17–30

Barton DW (1950) Pachytene morphology of the tomato chromosome complement. Am J Bot 37:639–643

Bobrow M, Madan K, Pearson PL (1972) Staining of some specific regions of human chromosomes, particularly the secondary constriction of no. 9. Nat New Biol 238:122–124

Brutlag D, Appels R, Dennis EG (1977) Highly repeated DNA and *Drosophila melanogaster*. J Mol Biol 112:31–47

Burkholder GD, Weaver MG (1977) DNA-protein interactions and chromosome banding. Exp Cell Res 110(2):251–262

Caspersson T, Farber S, Foley GE, Kudynowski J, Modest EJ, Simonsson E, Wagh U, Zech L (1968) Chemical differentiation along metaphase chromosomes. Exp Cell Res 49:219–223

Caspersson T, Zech L, Modest EJ, Foley GE, Wagh U, Simonsson E (1969) DNA-binding fluorochromes for the study of the organization of the metaphase nucleus. Exp Cell Res 58:141–152

Caspersson T, Zech L, Johansson C (1970) Analysis of human metaphase chromosome set by aid of DNA-binding fluorescent agents. Exp Cell Res 62:490–492

Comings DE (1975) Mechanisms of chromosome banding. IV. Optical properties of the Giemsa dyes. Chromosoma 50:89–110

Comings DE, Kovacs BW, Avelino E, Harris DC (1975) Mechanism of chromosome banding. V. Quinacrine banding. Chromosoma 50:111–145

Corneo G, Ginelli E, Polli E (1970) Repeated sequences in human DNA. J Mol Biol 48:319–327

Creighton HB, McClintock B (1931) A correlation of cytological and genetical crossing-over in Zea mays. Proc Natl Acad Sci U S A 17:492–497

Darlington CD, La Cour LF (1940) Nucleic acid starvation of chromosomes in *Trillium*. J Genet 40:185–213

Deumling B, Greilhuber J (1982) Characterization of heterochromatin in different species of the *Scilla Siberia* group (Liliaceae) by *in situ* hybridization of satellite DNA and fluorochrome banding. Chromosoma 84:535–555

Drewry A (1982) G-banded chromosomes in *Pinus resinosa*. J Hered 73:305–306

Dutrillaux B, Lejeune J (1971) Sur une nouvelle technique d’analyse du caryotype humain. CR Acad Sci Paris 272:2638–2640

Dutrillaux B, Lejeune J (1975) New techniques in the study of human chromosomes; methods and applications. Adv Hum Genet 5:119–156

Endo TR, Gill BS (1984) The heterochromatin distribution and genome evolution in diploid species of *Elymus* and *Agropyron*. Can J Genet Cytol 26:669–678
Faust J, Vogel W (1974) Are N-bands selective staining of specific heterochromatin? Nature 249:352–353
Filion WG, Blakey DH (1979) Differential Giemsa staining in plants VI. Centromeric banding. Can J Genet Cytol 21:373–378
Flavell RB, O'Dell M (1976) Ribosomal RNA genes on homoeologous chromosomes of group 5 and 6 in hexaploid wheat. Heredity 37(3):377–385
Friebe B, Gill BS (1994) C-band polymorphism and structural rearrangements detected in common wheat (*Triticum aestivum*). Euphytica 78:1–5
Funaki K, Matsui S, Sasaki M (1975) Localization of nucleolar organizers in animal and plant chromosomes by means of an improved N-banding technique. Chromosoma 49:357–370
Gerlach WL (1977) N-banded karyotypes of wheat species. Chromosoma 62:49–56
Gill BS, Friebe B, Endo TR (1991) Standard karyotype and nomenclature system for description of chromosome bands and structural aberrations in wheat (*Triticum aestivum*). Genome 34(5):830–839
Greilhuber J (1977) Why plant chromosomes do not show G-bands. Theor Appl Genet 50:121–124
Henderson SA, Warburton D, Atwood KC (1974) Localization of rDNA in the chromosome complement of Rhesus (*Macaca mulatta*). Chromosoma 44:367–370
Holmquist G, Gray M, Poster T, Jordan J (1982) Characterization of Giemsa dark and light banded DNA. Cell 31:121–129
Horobin RW (2011) How Romanowsky stains work and why they remain valuable – including a proposed universal Romanowsky staining mechanism and a rational troubleshooting scheme. Biotech Histochem 86(1):36–51
Howell WM (1985) Selective staining of nucleolus organizer regions (NORs). In: Busch H, Rothblum L (eds) The cell nucleus vol. XI. rDNA part B. Academic Press Inc, New York, pp 89–142
Howell WM, Denton TE, Diamond JR (1975) Differential staining of the satellite regions of human acrocentric chromosomes. Experentia 31:260–262
Hutchinson J, Miller TE (1982) The nucleolus organizers of tetraploid and hexaploid wheats revealed by *in situ* hybridization. Theor Appl Genet 61:285–288
Jewell DC (1979) Chromosome banding of *Triticum aestivum* cv. Chinese spring and aegilops variables. Chromosoma 71:129–134
Jones KW (1970) Chromosomal and nuclear location of mouse satellite DNA in individual cells. Nature 225:912–915
Kit S (1961) Equilibrium sedimentation in density gradients of DNA preparations from animals tissues. J Mol Biol 3:711–716
Kongsuwan K, Smyth DR (1977) Q-band in *Lilium* and their relationship to C banded heterochromatin. Chromosoma 60:169–178
Lin CC, Jorgenson KF, van de Sande JH (1980) Specific fluorescent bands on chromosomes produced by acridine orange after prestaining with base specific non-fluorescent DNA ligands. Chromosoma 79: 271–276
Liu JY (2006) Detection of human chromosomal abnormalities using a new technique combining 4',6-diamidino-2-phenyl-indole staining and image analysis. Clin Genet 69(1):65–71
Loidl J (1983) Some features of heterochromatin of wild *Allium* species. Plant Syst Evol 143:117–131
Magoon ML, Shambulinguppa KG (1961) Karyomorphology of *Sorghum propinquum* and its bearing on origin of 40- chromosome sorghum. Chromosoma 12:460–465
Matsui S, Sasaki M (1973) Differential staining of nucleolus organizers in mammalian chromosomes. Nature 246:148–150
McClintock B (1929) Chromosome morphology in Zea mays. Science 69:629

McClintock B (1932) The stability of broken ends of chromosomes in *Zea mays*. Genetics 26:234–282

McClintock B (1938) The production of homozygous deficient tissues with mutant characteristics by means of the aberrant mitotic behavior of ring-shaped chromosomes. Genetics 23:315–376

McClintock B (1941) The stability of broken ends of chromosomes in *Zea mays*. Genetics 26:234–282

McClintock B (1984) The significance of responses of the genome to challenge. Science 226:792–801

McKenzie WH, Lubs HA (1973) An analysis of the technical variables in the production of C-bands. Chromosoma 41:175–182

Michelson AM, Monny C, Kovoor A (1972) Action of quinacrine mustard on polynucleotide. Biochimie 54:1129–1136

Miller DA, Dev VG, Tantravahi R, Miller DJ (1976) Suppression of human nucleolus organizer activity in mouse human somatic hybrid cells. Exp Cell Res 101:235–243

Misra RN, Shastry SVS (1967) Pachytene analysis in *Oryza*. VIII. Chromosome morphology and karyotypic variation in O. sativa. Indian J Genet & Plant Breed 27:349–368

Pardue ML, Gall JG (1970) Chromosomal localization of mouse satellite DNA. Science 108:1356–1358

Phillips RL, Kleese RA, Wang SS (1971) The nucleolus organizer region of maize (*Zea mays* L.): chromosome site of DNA complementary to ribosomal RNA. Chromosoma 36:79–88

Pimpinelli S, Sanuni G, Gatti M (1976) Characterization of *Drosophila* heterochromatin. II. C-and N-banding. Chromosoma 57:377–386

Pinkel D, Straumem T, Gray JW (1986) Cytogenetic analysis using quantitative, high sensitivity, fluorescence hybridization. Proc Natl Acad Sci U S A 83:2934–2938

Ramanna MS, Parkken P (1967) Structure of and homology between pachytene and somatic metaphase chromosomes of tomato. Genetica 38:115–133

Ritossa FM, Spiegelman S (1965) Localization of DNA complementary to ribosomal RNA in the nucleolus organizer region of *Drosophila melanogaster*. Proc Natl Acad Sci U S A 53:737–745

Rooney DE (2001) Human cytogenetics: constitutional analysis. Oxford University Press, New York

Rowland RE (1981) Chromosome banding and heterochromatin in *Vicia faba*. Theor Appl Genet 60:275–280

Schlegel R, Gill BS (1984) N-banding analysis of rye chromosomes and the relationship between N-banded and C-banded heterochromatin. Can J Genet Cytol 26:765–769

Schmid M, Guttenback M (1988) Evolutionary diversity of reverse (R) fluorescent chromosome bands in vertebrates. Chromosoma 97:101–114

Schweizer D (1976) Reverse fluorescent chromosome banding with chromomycin and DAPI. Chromosoma 58:307–324

Schweizer D (1979) Fluorescent chromosome banding in plants: applications, mechanisms, and implications for chromosome structure. In: Davies DR, Hopwood DA (eds) Proceedings of the Fourth John Innes Symposium. Norwich, Crowe

Seabright M (1971) A rapid banding technique for human chromosomes. Lancet 2:971–972

Shastry SVS, RangaRao DR, Misra RN (1960) Pachytene analysis in *Oryza*. I. Chromosome morphology in Oryza sativa. Indian J Genet & Plant Breed 20:15–21

Southern EM (1970) Base sequence and evolution of guinea a-satellite DNA. Nature 227:794–798

Sumner AT (1972) A simple technique for demonstrating centromeric heterochromatin. Exp Cell Res 75:304–306

Sumner AT (1982) The nature and mechanisms of chromosome-banding. Cancer Genet Cytogenet 6(1):59–87

Sumner AT, Evans HJ, Buckland RA (1971) New technique for distinguishing between human chromosomes. Nat New Biol 232:31–32

Utrillaux B, Lejeune J (1971) Cytogenetique humaine. Sur une nouvelle technique d' analyse du caryotype humain. C R Acad Sci 272:2638–2640

Verma RS, Babu A (1989) Human chromosomes. Manual of basic techniques. Pergamon press, New York, p 240

Vosa GG, Marchi P (1972) Quinacrine fluorescence and Giemsa staining in plants. Nat New Biol 237:191–192

Wallace H, Bimstie ML (1966) Ribosomal cistrons and the nucleolus organizer. Biochem Biophys Acta 144:295–310

Wang HC, Kao KN (1988) G-banding in plant chromosomes. Genome 30:48–51

Weisblum B, de Haseth PL (1972) Quinacrine, a chromosome stain specific for deoxyadenylate-deoxythymidylate rich regions in DNA. Proc Natl Acad Sci U S A 63:629–632

Weisblum B, Haenssler E (1974) Fluorometric properties of bibenzimidazole derivative hoechst 33258, a fluorescent-probe specific for AT concentration in chromosomal DNA. Chromosoma 3:255–260

Wittekind DH, Gehring T (1985) On the nature of Romanowsky-Giemsa staining and the Romanowsky-Giemsa effect. 1. Model experiments on the specificity of azure B-eosin Y stain as compared with other thiazine dye-eosin Y combinations. Histochem J 17(3):263–272

Zanker V (1981) Fundamentals of pigment-substrate relationships in histochemistry. Acta Histochem 3:151–168

Karyotype Evolution: Concepts and Applications

9

Kátia Ferreira Marques de Resende

Abstract

The karyotype is the characterization of number, size, and morphology of the set of chromosomes of a species, as seen under the microscope. Studying the karyotypic evolution of a group of organisms is necessary to obtain data such as number of chromosomes, the position of the centromeres, and the number and positions of secondary constrictions and banding patterns. The karyotype study is important to help identify karyotypic polymorphism and interspecific and intraspecific cytogenetic variation and evolution. It also helps us in understanding the relationship between chromosome number, amount of DNA, and complexity of the organisms. Thus, the main application of studies of evolution karyotype is in the (cyto)taxonomy/systematics and phylogeny which will be discussed in the present chapter with examples in plants of the genus *Carex* (Cyperaceae), which exhibits karyological peculiarities with holocentric chromosomes and continual variation in chromosome number; of the genus *Allium*, which has polymorphic species with karyotype analysis and size genome described; of the genus *Crepis*, which has been a model of cytological studies with karyotype evolution; of the *Crocus* series (Iridaceae), which is characterized by high intra- and interspecific variability of karyotypes; of the genera *Cassia*, *Chamaecrista*, and *Senna*; and, finally, in epiphytic cacti (*Lepismium*) and species of the *Rhipsalis* and *Delphinium*, in species of scorpions (*Androctonus*), and in the genus *Oligoryzomys* (Rodentia).

Keywords

Chromosome number • Cytogenetics • Cytotaxonomy • DNA content • Karyotype • Phylogeny

---

K.F.M. de Resende (✉)
Lab of Plant Cytogenetics, Department of Biology/DBI, Federal University of Lavras/UFLA, 3037, 37.200-000 Lavras, MG, Brazil
e-mail: katia.ufla@gmail.com

T.A. Bhat, A.A. Wani (eds.), *Chromosome Structure and Aberrations*,
DOI 10.1007/978-81-322-3673-3_9

Abbreviations

CMA	A3 chromomycin
DAPI	4′,6-diamidino-2- phenylindole
RON	Nucleolus organizer regions
FISH	Fluorescent in situ hybridization
GISH	Genomic in situ hybridization
m	Metacentric
sm	Submetacentric
a	Acrocentric
t	Telocentric
AR	Ratio of arms
CI	Centromeric index
Tli	Total length of chromosome i
TLLH	Total length lot haploid
RL	Relative length of each chromosome
FN	Fundamental number
A1	Intrachromosomal asymmetry index
A2	Interchromosomal asymmetry index
SA	Short arm
LA	Long arm
OMO	*Oligoryzomys moojeni*
pg	Picograms

9.1 Introduction

The karyotype is the characterization of number, size, and morphology of the set of chromosomes of a species, as seen under the microscope. Studying karyotype evolution of a group of organisms means initially being concerned with the information provided by karyotypic data, which are chromosome number, position of centromeres, and number and position of secondary constrictions and banding patterns. The characterization of karyotype is interesting for identification of wild and cultivated crop species (Dutta et al. 2015), resolving dubious chromosome counting (Correia-da-Silva et al. 2014), and identifying the ploidy level between related species (Wang et al. 2015) and in explant culture (Sacristan 1971; Gomes et al. 2014). It also helps in understanding the hybridization and chromosome duplication events (Mavrodiev et al. 2015), in testing the genetic stability of the clones (Rao et al. 1992), and, in humans, for identifying clinical syndromes due to aneuploidy.

The karyotype study is important to help identify polymorphisms caused by numerical and structural chromosomal abnormalities and to compare cytogenetically different species (interspecific variation) and individuals of the same species (intraspecific variation), besides deriving evolutionary relationships. Thus, comparative analysis of karyotypes in related species allows us to distinguish the characteristics that are exclusive to each species (autapomorphies[1]) and those that are common to all species (synapomorphies[2]), enabling the study of the diversification of the same. Since chromosomes are composed of the genetic material, therefore, changes in them are always significant to the evolutionary course of species, assuming the original idea that the current organisms are derived from primitive ancestors. Thus, in evolutionary history of species, we can distinguish some major modes of evolution (Futuyma 2009) such as:

- Anagenesis – gradual transformation of one species into another by the fixing of new characters, with the replacement of ancestral species by newly derived one.
- Cladogenesis – one species is divided into two or more through reproductive isolation mechanisms, which ensure that each species keep their exclusive genetic characteristics.

After a sufficiently long period of time, the populations tend to differ for small cumulative changes, due to adaptive processes of species to biotic and abiotic factors, to instability of their habitat, and to genetic variability of the species, generating therefore intra- or interspecific divergence. Regarding the utility of karyotype study, this can be used in:

- The taxonomy/systematics for understanding evolutionary processes
- Plant breeding to understand barriers to the introduction of genes and to identify structural changes and association with sterility
- The construction of genetic and chromosomal maps
- The planning of transfer of genes/chromosomes
- Genomic analysis, aimed toward monitoring of crossings
- The detection of syndromes
- The study of environmental stress and effect of chemicals on chromosomes

Thus, all these aspects associated with the karyotype analysis contribute to karyotype evolution. The term cytotaxanomy refers to the area where in all aspects of karyotype are considered for identification and classification of species and has been widely used to understand evolutionary processes.

9.2 Evolutionary Aspects of Chromosome Polymorphisms

Chromosomal variation can occur between different populations of the same species, between individuals of the same population, and even between different cells of the same tissue of the individual. The unprogrammed variations are chromosomal alterations, which, compared to normal karyotype, may be advantageous, disadvantageous, or neutral. When alterations are neutral or advantageous, they are transmitted to the progeny, contributing to the natural karyotype variation of the species. Thus, for a characteristic observed among members of a population that possess structures or arrangements alternative of a chromosome, i.e., the existence of two or more different shapes for the same chromosome is called chromosomal polymorphism (Klug et al. 2014).

There are two main types of karyotypic changes, viz., (1) changes in chromosome number which include euploidy and aneuploidy and (2) changes in chromosome structure which include deletions, duplications, inversions, translocations, and transpositions (translocation intrachromosomal). The numerical variations are comparatively easy to detect, even in species with small and numerous chromosomes, and cause significant effect in the individual carrying them and consequently for the evolution of a species. Structural changes are recognized when they affect the chromosome morphology or change the position of certain chromosomal markers as the centromere, the secondary constriction, the banding pattern, etc. Consequently, the eukaryotic chromosomes may differ in size, shape, and composition of DNA, RNA, and proteins, as well as their number and quantity (Sharma and Sen 2002). These characteristics are subject to evolutionary changes that vary between individuals of the same species and within each individual. Thus, the set of chromosome differs distinctly between organisms, which can define phylogenetic relationships (Eichler and Sankoff 2003).

In some species, polymorphism occurs by small intra- or interspecific chromosomal variations. In such cases, the species are considered karyotypically stable so that the evolution has mainly processed by gene alterations or changes in gene frequencies. In other cases, the species of a given group are morphologically identical but show a strong intra- and interspecific polymorphism. For example, different populations of *Anopheles* species which are morphologically similar, however, the polytene chromosomes of both contain inversions (Coluzzi et al. 2002), and the genus *Carex* (Cyperaceae), that exhibits karyological peculiarities with holocentric chromosomes and continual variation in chromosome number (Lipnerová et al. 2013). In *Carex*, the authors study the genome size and GC content in 157 taxa, using flow cytometry, and the chromosome numbers for 96 samples and from the available literature for other taxa. A strong negative correlation between genome size and chromosome number was documented in non-polyploid taxa. Genomes of *Carex* are relatively small and very GC-poor compared with other angiosperms. The authors conclude that the evolution of genome and karyotype in *Carex* is promoted by frequent chromosomal fissions/fusions, rare polyploidy, and common repetitive DNA proliferation/removal. Thus, probably there is no direct relationship between chromosomal and morphological variability. Each type of polymorphism (enzyme, chromosomal, morphological, etc.) appears to act independently in the evolutionary process. Probably, the more polymorphic for a characteristic, the greater its role is in the evolution of species.

The chromosome polymorphisms can be identified by analysis of chromosome number and karyotype morphology, variation in the pattern of bands (C, G, Q, etc.), location of the nucleolus organizer regions (RON) and of secondary constrictions, amount and location of repetitive sequences by fluorescent in situ hybridization (FISH), and allo- and autopolyploidy by genomic in situ hybridization (GISH). For example, karyotype analysis was made in *Allium roseum* L. which has been treated as a polymorphic species, comprising 12 different taxa, using fluorescent in situ hybridization of rDNA sites. Guetat et al. (2015) studied more than 80 individuals in the *Allium roseum* complex from 20 populations at different localities in Tunisia. Three cytotypes, diploid ($2n = 2x = 16$), triploid ($2n = 3x = 24$), and tetraploid ($2n = 3x = 32$), were found. The chromosome set of studied populations was found nonuniform at the species level and within cytotypes. The diploid showed 6 rDNA sites; the triploid and tetraploid, respectively, showed 9 and 12 sites. However, 5S rDNA showed two sites for the diploid genome and four loci for both triploid and tetraploid. The results obtained by the authors regarding the 45S rDNA showed intra- and interpopulational numerical variations. However, the data obtained with the 5S rDNA probe revealed highly conserved sequences in the chromosomes of individuals from each population, as well as among the analyzed populations.

However, correlate chromosomal variation with adaptive characteristics of individuals is rather complex, though it is possible to observe that certain chromosomal arrangements are much more frequent in certain environmental conditions suggesting that different chromosomal rearrangements have different adaptive values. One example is the polymorphism due to centric fusion-fission and inversions in the marine snail *Nucella lapillus* L. In this case, individuals who are located in regions affected by the high tide have karyotype with $2n = 36$ acentric chromosomes, while those individuals located in low-tide regions have $2n = 26$ chromosomes, with 16 acentric and 10 metacentric. And individuals with chromosome numbers of $2n = 26–36$ occur in intermediate regions. Thus, the mechanism of centric fusion-fission and inversions acts as mediator of the correlation between karyotype and environment (Hoxmark 1970; Bantock and Cockayne 1975). Another example is the relationship between the environment and chromosomal inversions. *Drosophila willistoni* Sturtervant populations have different chromosomal polymorphism rates in different environments. In appropriate climatic conditions with a great variety of food sources, the chromosomal polymorphism is more intense than in unfavorable conditions for the species. This increase of polymorphism in favorable conditions is attributed to the fact that in these conditions, the species can explore different microhabitats, making use of different chromosomal rearrangements (Da Cunha and Dobzhansky 1954). So, the higher the environmental diversity is effectively exploited, the larger the chromosomal polymorphism, i.e., greater adaptation. Thus, chromosomal changes can have an adaptive effect of the organism on the environment, even when they don't cause apparent phenotypic changes.

As a consequence of chromosomal polymorphism, we have (1) change in the expressed gene pool of an organism, (2) change in the frequency of genetic recombination, and (3) reduction, in the free gene flow within a population. A chromosomal variation modifies the gene expression due to physical alteration of the gene (deletion, duplication, transposon activation, etc.) or the position effect. The position effect is a change in the capacity of expression of a gene due to the presence or absence of a block of heterochromatin in their neighborhood. In this case, rearrangements affecting the sequence of segments of a chromosome, especially inversion and translocation, can suppress a gene, for bringing it closer to heterochromatin, or allow its activation, pushing of the heterochromatin (Griffiths et al. 2006). In *Drosophila melanogaster* Meigen, there is a correlation between the number of repetitions of the segment 16A region on the X chromosome and the number of facets of eye. However, the number of facets also depends on how these repeated segments are arranged, showing the influence of the distribution of the genes in the chromosomes upon the phenotype (Sturtevant 1925).

The gene recombination frequencies can be changed due to the formation of new linkage groups by reciprocal translocations, which can reduce or block the occurrence of proximal chiasma. Similarly, the simple translocations, centric fusions and fissions, and fusions in tandem can also form new linkage groups and alter the recombination frequency. Among the changes recognized as significant in the process of restriction of gene flow between the members of a population with greater potential for reproductive isolation is polyploidy, which can lead to complete reproductive isolation between the ancestral species and the polyploid derivative. An example is an instantaneous speciation in chemotypes of *Lippia alba* (Mill.) N.E. Brown (Verbenaceae), having chromosome numbers of $2n = 30$ for the chemotype La1-citral, $2n = 60$ for La2-carvone, and $2n = 12–60$ in La3-linalool, for which there is a reproductive isolation (Pierre et al. 2011).

In many groups of organisms, karyotypic alterations are due mainly to a particular type of chromosome change, i.e., structural or numerical change, such as paracentric inversion, centric fissions, etc., so that the karyotype evolution did not proceed in the same way in all evolutionary lines (Lawlor 1974; Schubert 2007). In this way, there are four evolutionary alternatives: (1) conservation of the amount of DNA with stable karyotype, (2) conservation of the quantity of DNA with reorganization of karyotype, (3) increased amount of DNA, and (4) reduction in the amount of DNA (Schubert 2007). Most groups with stable karyotype have proved to be polymorphic when analyzed with more technical details. Some groups comprise of a small number of species with small and numerous chromosomes. One example is the family Camelidae (camels and llamas), consisting of only four to six species, all with $2n = 74$ chromosomes (Taylor et al. 1968). Among the groups that have a relatively constant amount of DNA in the mammalian exemplify the alternatives (1) and (2) above. On the other hand, with respect to the chromosome number, some orders of mammals are very constant, while others are extremely diversified. For example, in order Cetacea (whales), all species have $2n = 44$ (Árnason 1974). Other orders, such as Rodentia (rodents) (Lawlor 1974), Artiodactyla (ox, deer) (Balmus et al. 2007; Rubes et al. 2011), and Perissodactyla (horse, tapir) (Trifonov et al. 2008), vary greatly in the number and chromosomal form. The extreme case is found in the genus *Muntiacus* (deer), where the chromosome number ranges from $2n = 6$ to $2n = 46$ (Tanomtong et al. 2005; Huang et al. 2006). Thus, there are different lines of thinking when it comes to evolution in mammalian karyotypes. Some researchers support the hypothesis that the change in chromosome numbers was unidirectional in the sense that many chromosomes have evolved to a few, based on reducing the number of acrocentric chromosomes (telocentrics) and increased meta- and submetacentric chromosomes coming from centric "fusion," being that it has several cytological evidence that this process is relatively common. But the question remains, how large chromosome numbers evolved at an early stage? There are also numerous indirect evidences supporting the vision of centric "fission," by which acrocentric chromosomes were derived from the centromeric division of metacentric chromosomes or submetacentrics, also being equally important in chromosomal evolution of mammals. Therefore, trends of chromosomal evolution in mammals are also open to different interpretations and should be evaluated further within their evolutionary lineage (Lawlor 1974).

Depending on the type of change involved, the evolution of the group may occur with or without significant variation in the amount of DNA. In general, plant amplitude variation in the amount of DNA is much more intense than animals, which are quantitatively more stable for not representing polyploidy. For example, variation in the amount of DNA of haploid nuclear genome is referred to as the C value and is measured in picograms (pg). This C value ranges from $C = 0.2$ pg (wherein 1 pg = 0.965 × 109 nucleotide pairs) in species such as *Arabidopsis thaliana* (L.) Heynh. (Brassicaceae) to $C = 127.4$ pg in species such as *Fritillaria assyriaca* Baker (Liliaceae) (Leitch and Bennett 1997; Levin 2002). This variation was presented by Leitch et al. (1998), which showed the distribution of the genome size by more than 2,800 plant species. Large-scale surveys of genome size evolution in angiosperms show that the ancestral genome was most likely small, with a tendency toward an increase in DNA content during evolution (Enke et al. 2011). Due to polyploidization and self-replicating DNA elements, angiosperm genomes were considered to have a "one-way ticket to obesity" (Bennetzen and Kellogg 1997). However, the plant karyotypic evolution has taken place with respect to both increase as well as decrease in the amount of DNA (genome size). In genus *Allium*, Gurushidze et al. (2012) observed that 2C genome size in 160 accessions of 70 species of its subgenus *Melanocrommyum* had high variation (26.26–78.73 pg). Estimation of ancestral genome sizes using generalized least squares revealed lineages with increasing as well as decreasing DNA content. Thus, the authors concluded that genome size variation reflects incipient speciation or diversification in *Allium* subgenus *Melanocrommyum*.

The genus *Crepis* has been a model of cytological studies (Hollingshead 1930; Tobgy 1943; Sherman 1946; Babcock 1947a, b) with regard to karyotype evolution. Babcock studies the karyotypes of over 100 *Crepis* species and proposed several hypotheses. Chromosome number decreases during evolution so that species with small chromosome numbers are derived. Short-lived annuals have undergone a reduction in the quantity of nuclear DNA. Thus chromosomal rearrangements are the main means of evolution in the genus. Currently, with the advances of molecular phylogeny new works have re-interpreted these hypotheses of Babcock. Enke et al. (2011) to investigate variation in genome size in the genus Crepis and its relations to speciation and phylogeny. According to these authors within 30 diploid Crepis species there is a striking trend towards genome contraction. Genome size was found significantly higher in perennials than in annuals. Within sampled species, very small genomes are only present in Mediterranean or European species, whereas their Central and East Asian relatives have larger 1C values. Enke et al. (2015) continued studies with on the Mediterranean Crepis sect. Neglectoides, which includes species with some of the smallest genomes within the whole genus. The occurrence of genome size contraction in *Crepis* is believed mostly due to elimination of dispersed repetitive elements, whereas chromosomal reorganization was involved in the karyotype formation. According to Schubert (2007), chromosomal rearrangements were responsible for the progressive reduction of an ancestral chromosome complement with eight pairs of chromosomes for the current karyotype *Arabidopsis thaliana* (L.) Heynh. ($n = 5$). The comparison of the genetic maps of *Arabidopsis thaliana* and *Arabidopsis lyrata* L. ($n = 8$) has shown that collinear regions of the genomes suggest reduction in the number of eight to five chromosomes by involving three pairs of chromosomes into two reciprocal translocations and at least three inversion events.

Thus, *there is no relationship between the amount of DNA and chromosome number and biological complexity of the organism.* The "C-value paradox" (Thomas 1971) refers to the phenomenon where the amount of DNA is not reflected in the complexity of an organism. More complex organisms do not necessarily have more DNA; in fact, the basic set of genetic information required for normal development is similar in most plants (Flavell 1980). Studies have shown that lower taxa tend to have small genomes, with a smaller amount of DNA (Leitch et al. 1998). However, in some ancestral species, smaller genomic complexity is associated with a greater amount of DNA. One example is the *Latimeria*, the only living representative of Crossopterygii order, considered as a living fossil. The DNA content of *Latimeria* is about three times larger than the average of finfish found today (Makapedua et al. 2011).

9.3 Description and Procedure for Obtaining the Karyotype

As we could see, to perform chromosome characterization of a particular species, it is necessary to work out the number, size, and morphology of chromosomes.

Regarding the chromosome number, there is an enormous interspecific variation reported in the literature, with extremes ranging from species with $2n = 2$ chromosomes (*Parascaris univalens*) (Goday and Pimpinelli 1986) to $2n = 1260$ (*Ophioglossum reticulatum* L.) (Abraham and Ninan 1954). When it comes to chromosome numbers, it is worth to emphasize the definition of *somatic number (2n)* = number of chromosomes in somatic cells, *gametic number (n)* = number of chromosomes contained in a gamete, and *basic number (x)* = number of chromosome sets that an individual has in their cells. Remembering that base number indicates the individual ploidy, e.g., diploid ($2x$), triploid ($3x$), tetraploid ($4x$), etc., and is widely used in evolutionary studies (Guerra 2000). Not only interspecific variation occurs in chromosome number but also intraspecific variation within a certain group of individuals of the same species, and, most recently, there are reports of variation in chromosome number among cells from the same meristem of the species such as in *Senna rugosa* (Fig. 9.1) (Resende et al. 2014). There are techniques of direct and indirect analysis. As techniques for direct analysis, we have size of the metaphase plate, which is proportional to the number of chromosomes, number of prochromosomes, and counting the number of heterochromatic segments conspicuous in interphase nuclei and the number of nucleoli/nucleus which depends directly on the number of genomes. The measurement of the metaphase plate is preferred in case of a high number of small chromosomes and when it is not possible to detect aneuploidy, it is a preferable technique in relation to the measuring interphase nuclei of

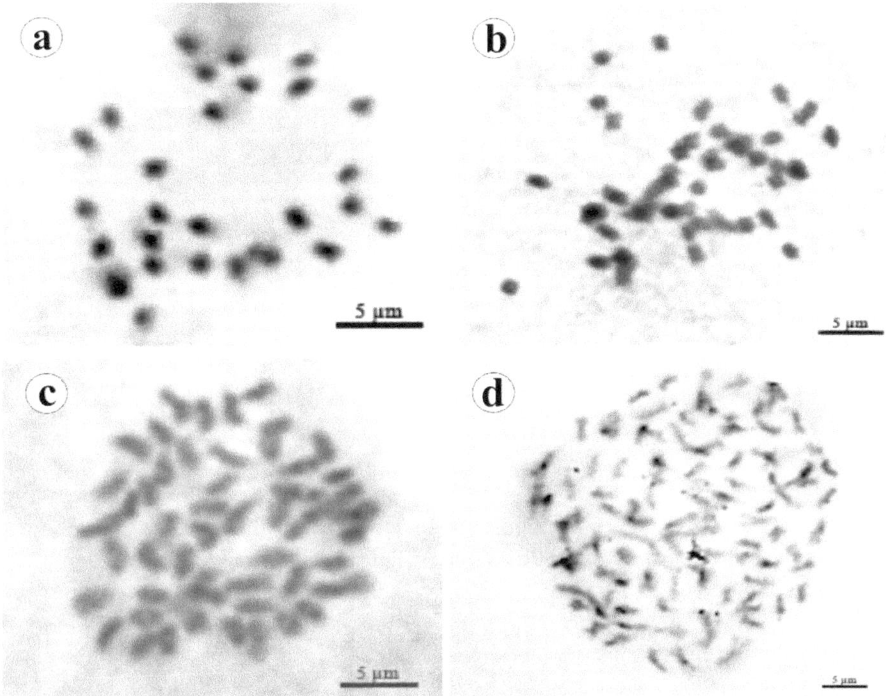

Fig. 9.1 Mitotic metaphases of accessions from *Senna rugosa* (G. Don) H.S. Irwin & Barneby. Metaphases with (**a**) $2n = 28$, (**b**) $2n = 42$, (**c**) $2n = 56$, and (**d**) $2n = 112$ chromosomes (Resende et al. 2014)

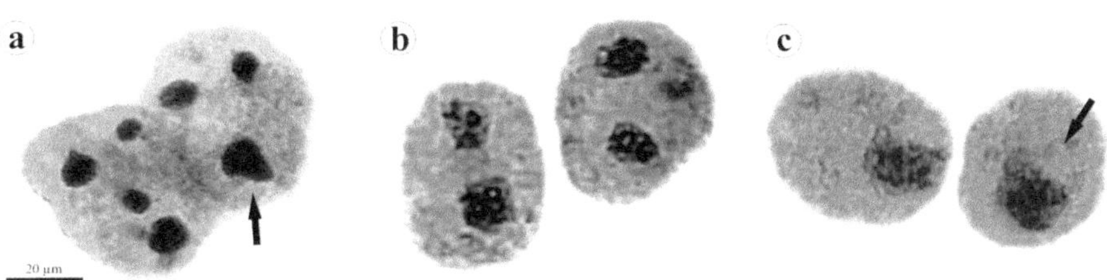

Fig. 9.2 Different number of nucleoli in interphasic nuclei of *Senna occidentalis* (L.) Link stained with silver nitrate. (**a**) A cell with four and another with three nucleoli, (**b**) two cells with two nucleoli, and (**c**) two cells with one nucleolus. *Arrows* indicate the phenomenon of nuclear fusion (Ferreira et al. 2010).

the same tissues. However, there are some unclear issues for this type of direct analysis such as the fact of the instability of heterochromatic segments in interphase nuclei and the phenomenon of nucleolar fusion, which can be shown in species of *Senna occidentalis* (L.) (Fig. 9.2) (Ferreira et al. 2010). Such as analysis indirect techniques there is the correlation between the number of chromosomes and nucleus size and determining the amount of DNA/nucleus, estimated by absorption of visible light nuclei stained with Feugen or, which is currently most widely used, the technique of 6 flow cytometry. In relation to chromosome size, there is also variation of 1.6 μm in case of *Lepidothamnus intermedius*, to 20 μm, in case of *Manoao colensoi*, both species of gymnosperms in New Zealand. However, in majority of the cytogenetically studied organisms, the average chromosome size measures 5–6 μm with exceptions like microchromosome, which measure around 0.5 μm, and giant chromosomes with a size range of 10 μm (King et al. 1981).

In relation of the existing groups of cytogenetically studied organisms, there are reports that the lungfish have the largest chromosomes, and lepidopterans have the smaller chromosomes. Among plants, monocotyledons have larger chromosomes than dicotyledons (except Cyperaceae, the monocot with small chromosomes) and hardwood plants having smaller chromosomes than their herbaceous relatives. As such, the species with large chromosomes are generally considered primitive than those with smaller derivatives (Stebbins 1971; Stace 1980; Sharma and Sharma 2014). Within a particular karyotype, the chromosome size variation can be gradual, as in human beings and *Pinus* (Hizume et al. 2002), resulting in a symmetrical type of karyotype or asymmetrical type as found in *Agave*, birds, and reptiles. As suggested by Stebbins (1971), there is a correlation between the asymmetry of the karyotype and evolutionary specialization of a taxon in habit and morphology. For example, primitive genus *Helleborus* of family Ranunculaceae has symmetric karyotype, and its advanced genera such as *Aconitum* and *Delphinium* have asymmetric karyotypes (Stace 1980; Jabbour and Renner 2012; Sharma and Sharma 2014). Kolar et al. (2012) compared the karyotype of the *Delphinium malabaricum* var. *malabaricum* (Huth) Munz. and *Delphinium malabaricum* var. *ghaticum* Billore. Both taxa were found to possess eight pairs of homologous chromosomes (one metacentric, one submetacentric, and six subtelocentric). The length of metaphase chromosomes ranged from 1.42 to 5.35 μm in *D. malabaricum* var. *malabaricum* and from 1.77 to 5.30 μm in *D. malabaricum* var. *ghaticum* with an average length of 2.88 μm and 2.92 μm, respectively. According to these authors, the prevalence of chromosomes with subtelocentric centromeres indicated a tendency toward asymmetric karyotype; hence, both the taxa belong to Stebbins's 2B type karyotypic symmetry class. The results of this work revealed that *D. malabaricum* var. *malabaricum* and *D. malabaricum* var. *ghaticum* have a similar karyotype pattern, despite differing morphologically. Thus, the authors concluded that the external morphological variation occurred independently of the chromosomal variation, which may have played a prominent role in the intraspecific differentiation of *D. malabaricum*. Another interesting example of this type of asymmetric karyotype (bimodal) is described for the three kinds of Italian sturgeon. For *Huso huso* species, 34 pairs of meta- and submetacentrics chromosomes, 3 pairs of acrocentric (1 large and 2 small), and approximately 38 microchromosomes have been described. However, for the species *Acipenser sturio*, 35 pairs of meta- and submetacentrics chromosomes, only 2 pairs of small acrocentric, and approximately 40 microchromosomes were described, while for *Acipenser naccarii*, 75 pairs of meta- and submetacentrics chromosomes, 6 pairs of acrocentric, and approximately 76 microchromosomes have been described (Fontana and Colombo 1974). In the case of bimodal karyotype, it is possible to observe two types of chromosome sets within the same cell, one with large chromosomes and other with small chromosomes, well illustrated in the example quoted above.

Thus, for karyotype description, chromosome measurements are carried out, which often result in the construction of karyograms and idiogram. So the karyotypes can be represented by a karyogram which is constructed by photomicrography, or an idiogram, which is a graphical representation of the karyotype, using average values of the position of the centromere and size of each chromosome of the haploid set (Fig. 9.3). During measurements, some chromosomal characteristics must be considered, and in preparation of the karyogram, the chromosomes are arranged in pairs (homologous chromosomes) obeying to some order of criterion, such as chromosome size (largest to smallest), relative measurements of the short arm and the long arm, position of the centromere or primary constriction, the presence of secondary constrictions, satellite or nucleolus organizer regions (RON), and bands, among others.

The chromosome morphology is related to the position of the centromere in metaphase chromosomes, which can be located at any position between the middle and the end of the chromosome. In majority of cases, the chromosome is monocentric (one centromere), but sometimes it can be acentric (without centromere); dicentric (with two centromeres), occasionally formed due to irregularities; and holocentric where the centromere has no morphological differentiation with the fibers of the spindle connecting themselves along the length of chromatid and not to a specific region of the chromosome. In this way, with respect to morphology, we may have metacentric (m), submetacentric (sm), acrocentric (a), and telocentric (t) types of chromosomes (Levan et al. 1964; Guerra 1986; 1988). The position of the centromere can be defined numerically by the arm ratio (AR = long arm/short arm) and by the centromeric index (CI = short arm × 100/long arm + short arm). So the chromosome morphology is determined from these values of the AR and CI (Levan et al. 1964; Guerra 1986). There are also other important parameters such as total length of chromosome *i* (Tl*i* = long arm + short arm); total length of the haploid lot, in other words, $\sum$ total length of all chromosomes of the genome ($\text{TLLH} = \sum \text{Tl}i / 2$); relative length of each chromosome ($\text{RL} = \text{Tl}i \times 100 / \text{TLLH}$); and, finally, fundamental number (FN), which corresponds to the number of arms of the diploid complement (Guerra 1986; Guerra 1988).

The indices obtained above can also be used to numerically determine the symmetry of the karyotype, according to the methods proposed by Stebbins (1958) and Zarco (1986). The latter being estimated by intrachromosomal asymmetry index (A_1), given by centromere position, and interchromosomal ($A2$), given by the difference in size of chromosomes, by the equations

$$A_1 = 1 - \frac{\sum_{i=1}^{n} \frac{b_i}{B_i}}{n} \qquad A_2 = \frac{s}{\bar{x}}$$

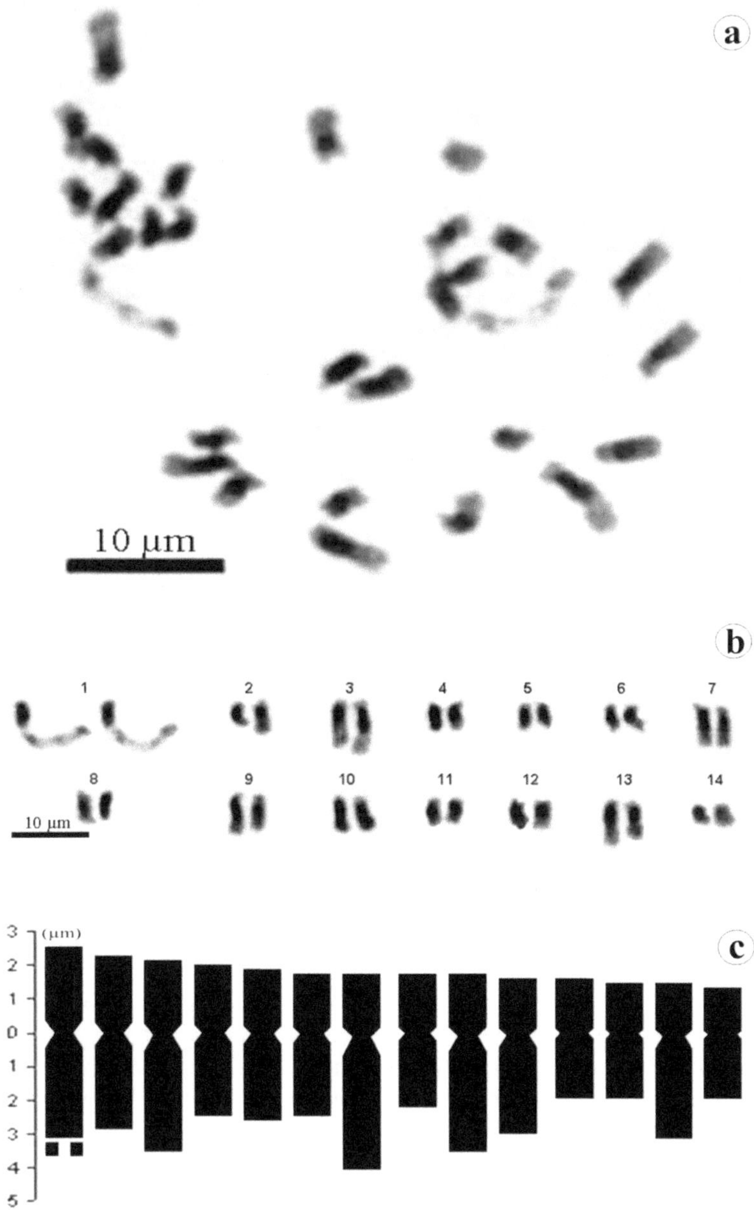

Fig. 9.3 Karyotype representations of *Senna occidentalis* (L.) Link. (**a**) Mitotic metaphase with $2n = 28$ chromosomes, (**b**) karyogram, and (**c**) idiogram ($n = 14$) (Ferreira et al. 2010)

where A_1 = intrachromosomal asymmetry index, varying between 0 and 1 (not dependent on the number and size of the chromosome), n = number of pairs of homologous chromosomes, bi = average length of short arms in each pair of homologous chromosomes, Bi = average length of the long arms of each pair of homologous chromosomes, A_2 = interchromosomal asymmetry index (does not depend on chromosome number and has no unit of measure), s = standard deviation, and x = average length of chromosomes.

Therefore, for studying karyotypes and perform the chromosome morphology is needed obtain well-defined metaphase chromosomes, high stained, with good morphology and no overlapping. The metaphase preparations are then scanned by microcamera and processing images, to obtain the required measurements (SA, LA, AR, CI, TL, TLLH, RL, and FN), as seen above through specific programs for this purpose, and finally classify the chromosomes according to the position of the centromere (Levan et al. 1964; Guerra 1986). For species where chromosomes are of similar size and morphology, it is difficult to identify pairs of homologous chromosomes. In such cases, chromosome banding techniques, which allow to identify a particular chromosome pair within the full complement, can be used for karyotype preparation and analysis.

9.4 Applications of Karyotype Analysis

Many species show no distinct morphological characters that allow the delimitation of the same, thereby causing contrapositions between many authors. Thus, the study of chromosomes by means of karyotypic data has been recognized as an auxiliary tool for the taxonomists, i.e., for the separation and classification of species in an area called cytotaxonomy. The cytotaxonomy in turn has been widely used in the reformulation of phylogenetic trees in various groups contributing to the evolutionary studies (Guerra 2008). This way, many cytogenetic techniques may be used for evolutionary studies using karyotype analysis. These include from studies that used only the classical or conventional cytogenetic techniques and to those where cytomolecular techniques have been used, and furthermore, those in which has been developed the integration of the cytogenetic data and phylogenetic.

Studies on karyotype evolution help in generating data on comparison of karyotypes among species and related genera that allow us to infer characteristics of higher taxa, among others, for example, the cytotaxonomic studies in plants of the subtribe Cassiinae, the tribe Cassieae, and the subfamily Caesalpinioideae, which consists of three genera: *Cassia* L. sensu stricto, *Chamaecrista* Moench, and *Senna* Mill (Lewis 2005). However, the phylogenetic relationships of this subtribe are still conflicting regarding its monophyletic character and about the relationship between genera. In genus *Senna*, there are reports in literature of karyotype variability with potential for cytotaxonomic studies. Therefore, the study of karyotype evolution started with 11 species of eight series and three sections of genus *Senna*, of the Minas Gerais State, Brazil, to subsidize the taxonomy of *Senna* taxa. Thus, until the moment, interspecific variability was observed with respect to some characters,

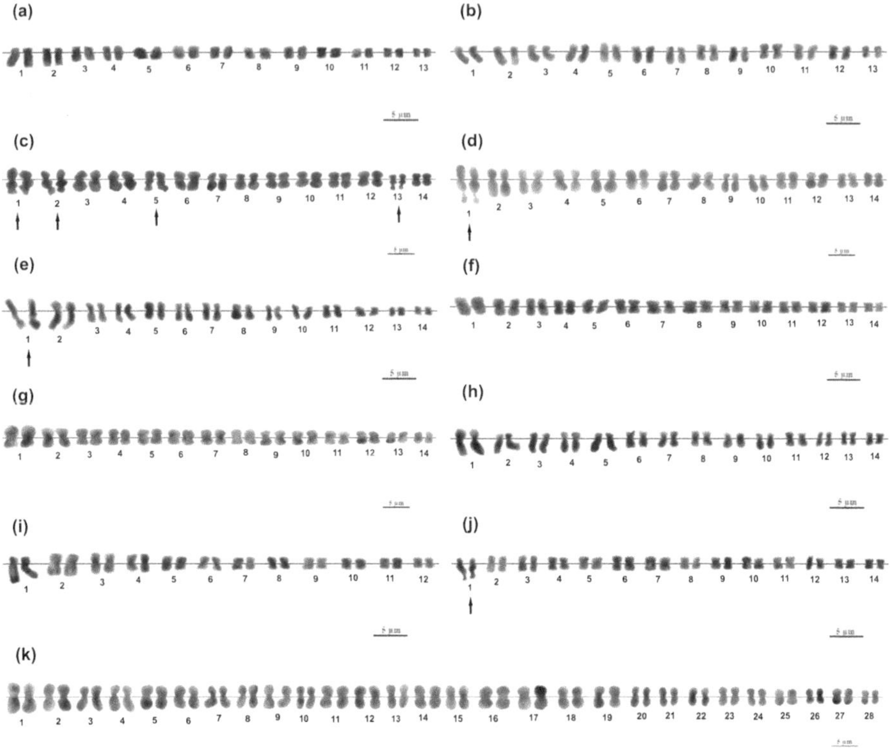

Fig. 9.4 Karyograms of species of *Senna* Mill. (**a**) *S. macranthera macranthera*, (**b**) *S. splendida splendida*, (**c**) *S. cernua*, (**d**) *S. corymbosa*, (**e**) *S. pendula*, (**f**) *S. siamea*, (**g**) *S. silvestres bifaria*, (**h**) *S. spectabilis*, (**i**) *S. multijuga*, (**j**) *S. alata*, and (**k**) *S. rugosa*. *Arrows* indicate the pairs of chromosomes with secondary constriction. Scale in μm

such as number and chromosome morphology, emphasizing the event of polyploidy and diploidy as main characters in this evolutionary scenario (Fig. 9.4). Similarly, Harpke et al. (2015) evaluated the phylogeny, karyotype evolution, and taxonomy of *Crocus* series *Verni* (Iridaceae), which is characterized by high intra- and interspecific variability of karyotypes (2*n* = 8–23). The authors combined morphological (12 characters), molecular (chloroplast DNA (trnL-*trn*F, *ndh*F); nuclear DNA (ITS, pCOSAt103)), and karyological features, in samples of different populations of *C. etruscus*, *C. ilvensis*, *C. kosaninii*, *C. tommasinianus*, *C. vernus* sensu lato, and *C. longiflorus* (series Longiflori). The comparison of total haploid genome lengths suggested that in the evolution of the group, polyploidization only played a role within the *C. vernus* species complex, where two hybridization events have been detected. In all other taxa, chromosome evolution is probably characterized by chromosome fusions and fissions, sometimes affecting the entire haploid chromosome set. According to the authors, comparative cytogenetics of the group indicates that series *Verni* is subject to a peculiar type of unequal change in chromosome size, i.e., both chromosome arms did not gain or lose equally in DNA content.

Regarding the analysis derived from data obtained by chromosome banding techniques and molecular cytogenetics, it is possible to identify a particular chromosome pair in the full complement and distinguish unique characteristics of each species (autapomorphies) and those common to all or most (synapomorphies), determining basal characteristics which may indicate a possible ancestral karyotype, which has diversified in the karyotype of the current species of a genus. Sadílek et al. (2015) analyzed the karyotypes of four geographically distant species of scorpions of the genus *Androctonus* Ehrenberg (*Androctonus australis*, *Androctonus bourdoni*, *Androctonus crassicauda*, and *Androctonus maelfaiti*) (Scorpiones: Buthidae) using both classic and molecular cytogenetic methods. According to the authors, this family has an extraordinary karyotype variation and is unique among scorpions that have holokinetic chromosomes (hypothesis of the tendency of holokinetic chromosomes to fuse and fission). However, these species studied seem to be the exception proving the rule. All species presented $2n = 24$ chromosomes, with the largest chromosome pair in all species (as examined by the fluorescence in situ hybridization – FISH), with a single interstitial rDNA locus of the 18S ribosomal RNA gene. The 18S rDNA (using FISH) thus did not provide any evidence of karyotype variation in the genus *Androctonus* and confirmed the remarkable stability of its conserved genome architecture despite the high levels of genetic divergence reported recently from northwestern Africa (Coelho et al. 2014). The findings of this study thus support the evolutionary stasis of the *Androctonus* karyotypes, which provides a unique opportunity for contrasting current hypotheses on the chromosome evolution both within and beyond the family Buthidae (Sadílek et al. 2015). The molecular cytogenetic techniques of chromosome painting have also been used for studying evolution in the genus *Oligoryzomys* (Rodentia, Sigmodontinae), which was earlier studied only by classical cytogenetic approaches (Di-Nizo et al. 2015). Diploid numbers ranging from $2n = 44$ in *Oligoryzomys* sp. to $2n = 72$ in *Oligoryzomys utiaritensis* have been reported, whereas phylogenetic relationships are not well defined. The authors developed chromosome probes from a female of *Oligoryzomys moojeni* (OMO) with $2n = 70$ and hybridized to other five *Oligoryzomys* species. Some probes showed a syntenic association in three species (*O. fornesi*, *O. microtis*, and *O.nigripes*). According to the authors, extensive chromosomal rearrangements, which could not be detected by classical cytogenetic techniques, such as pericentric inversions or repositioning of centromeres, Robertsonian rearrangements, and tandem fusions/fissions, as well as gain/activation or loss/inactivation of centromeres and telomeric sequences, can now be discussed with advances in molecular cytogenetics and probably have driven the huge genome reshuffling in these closely related species. According to the phylogeny, the authors could infer the occurrence of many tandem fusion events, since single OMO probes (the most derived species of the genus) painted two or more regions of different chromosomes of *Oligoryzomys microtis* (that have diverged earlier in the phylogeny). And these results indicate that all those closely related species have experienced recent autosomal rearrangement. However, the phylogenetic relationships of the species *Oligoryzomys* remained unclear, and this was the first phylogeny in which *O. moojeni* belonged to the most derived clade. The integration of cytogenetic and molecular data has thus been considered important to integrate in order to understand the karyotype evolution of such a complex genus as *Oligoryzomys* (Di-Nizo et al. 2015). Similarly Moreno et al. (2015) used integration of cytogenetic and molecular data in Neotropical epiphytic cacti of the genus *Lepismium* spp. and some species relative of the *Rhipsalis* spp., using rDNA probes. These authors generated a time-calibrated phylogenetic tree to place the karyological information and biogeographical history in an explicit evolutionary context. All species in this study exhibited $2n = 22$ with symmetrical karyotypes comprising only metacentric chromosomes of similar sizes. The heterochromatin bands ($CMA^+/DAPI^-$) were always associated with chromosome satellites coinciding with the location and number of the 18S-5.8S-26S rDNA loci. The 5S rDNA loci had more heterogeneous profiles with one or two loci per haploid genome. Phylogenetic analysis suggested an ancient duplication event of the 5S rDNA loci and more recent post-speciation translocation and deletion events (Moreno et al. 2015). The authors concluded that these genome restructurings occurred approximately 13.98 Mya in the Middle Miocene, after *Lepismium* and *Rhipsalis* diverged. The ancestor of *Lepismium* may have had a similar karyotype to *Lepismium lumbricoides* and the *Rhipsalis* spp. Both genera hypothetically originated in the northeastern Argentina and southern Bolivia (Yungas forests), but diversification of the *Lepismium* crown group probably originated from populations with duplicated 5S loci in the Parana forests of southeastern Brazil (8.70 Mya in the late Miocene). The authors suggested two migration events between the Yungas and Parana forests to explain the extant distribution of *Lepismium* spp. These results make *Lepismium* a model system for the study of the complex chromosomal evolution in plants (Moreno et al. 2015).

Therefore, the karyotype evolution studies nowadays tend to be related to this multidisciplinary issue of cytogenetic data integration, whether they are classic, conventional, and molecular, with other data (morphological, molecular, biogeographic) in a phylogenetic/cladistic approach, assisting cytotaxonomy and understanding evolutionary processes of various groups of organisms. And, finally, in relation of the future prospects, we must be concerned not only with the space/time location of important DNA sequences in the karyotypes of various organisms, which can accurately direct the research of chromosomal evolution, as in cytotaxonomy and as in the phylogeny/cladistic, but also with possible relationships of these sequences located and their karyotype evolution with their phenotypic expressions.

Glossary

Autapomorphies Derived character that is present exclusively in a single terminal taxon of a given cladogram

Chromosome painting Technique related to the differential staining of all chromosomes of the species, using a pool of differently labeled specific fluorescent probes

Chromosome polytene Formed by several overlapping chromatids, derived from repeated duplication of DNA strands without cell division, found in the salivary glands of *Drosophila melanogaster*

Flow cytometry Technique for counting, examining, and classifying microscopic particles suspended in a liquid medium through an opto-electronic detection apparatus, for example, determining the amount of DNA/nucleus

Fluorescent in situ hybridization Cytogenetic technique used to detect and localize specific DNA sequences in chromosomes using fluorescent probes

Genomic in situ hybridization Cytogenetic technique like FISH to locate genomic probes, i.e., an entire genome, in chromosomal set of a given species

Haploid set Has only one chromosome of each pair of homologous chromosome

Nucleolar fusion The cells may have several nucleoli, but there is usually a fusion so they have only one or two

Primary constriction Partially decondensed region that relates to the mechanisms of cell division, presenting the kinetochore, where the spindle fibers attach

Secondary constrictions Observed in at least one of the chromosomes of each species which is closely related to the organizer regions of the nucleolus presenting ribosomal RNA genes for transcription that constitute most of the nucleolus

Synapomorphies Derived character shared by more than one group

Syntenic association or synteny Presence of two or more loci genes on the same chromosome

References

Abraham A, Ninan CA (1954) The chromosomes of *Ophioglossum reticulatum* L. Curr Sci 23:213–214

Árnason U (1974) Comparative chromosome studies in *Cetacea*. Hereditas 77(1):1–36

Babcock EB (ed) (1947a) The genus *Crepis* I. In: The taxonomy, phylogeny, distribution and evolution of *Crepis*. University of California Press, Berkeley

Babcock EB (ed) (1947b) The genus I II. In: Systematic treatment. University of California Press, Berkeley

Balmus G, Trifonov VA, Biltueva LS, O'Brien PC, Alkalaeva ES, Fu B, Ferguson-Smith MA (2007) Cross-species chromosome painting among camel, cattle, pig and human: further insights into the putative Cetartiodactyla ancestral karyotype. Chromosom Res 15(4):499–514

Bantock CR, Cockayne WC (1975) Chromosomal polymorphism in *Nucella lapillus* L. Heredity 34:231–235

Bennetzen JL, Kellogg EA (1997) Do plants have a one-way ticket to genomic obesity? Plant Cell 9:1509–1514

Coelho P, Sousa P, Harris DJ, van der Meijden A (2014) Deep intraspecific divergences in the medically relevant fat-tailed scorpions (*Androctonus*, Scorpiones). Acta Trop 134:43–51

Coluzzi M, Sabatini A, Torre A, Deco MA, Petrarca V (2002) A polytene chromosome analysis of the *Anopheles gambiae* species complex. Science 298(5597):1415–1418

Correia-da-Silva M, Vasconcelos S, da Costa Soares MDL, Mayo SJ, Benko-Iseppon AM (2014) Chromosomal diversity in *Philodendron* (Araceae): taxonomic significance and a critical review. Plant Syst Evol 300(5):1111–1122

Da Cunha AB, Dobzhansky T (1954) A further study of chromosomal polymorphism in *Drosophila willistoni* in its relation to the environment. Evolution 8:119–134

Di-Nizo CB, Ventura K, Ferguson-Smith MA, O'Brien PC, Yonenaga-Yassuda Y, Silva MJ (2015) Comparative chromosome painting in six species of *Oligoryzomys* (Rodentia, Sigmodontinae) and the karyotype evolution of the genus. PLoS One 10(2):e0117579

Dutta M, Negi KS, Bandyopadhyay M (2015) Novel cytogenetic resources of wild *Allium* (Amaryllidaceae) from India. Nucleus:1–7

Eichler EE, Sankoff D (2003) Structural dynamics of eukaryotic chromosome evolution. Science 301:793–797

Enke N, Fuchs J, Gemeinholzer B (2011) Shrinking genomes? Evidence from genome size variation in Crepis (Compositae). Plant Biol 13(1):185–193. doi:10.1111/j.1438-8677.2010.00341.x

Enke N, Kunze R, Pustahija F et al (2015) Genome size shifts: karyotype evolution in Crepis section Neglectoides (Asteraceae). Plant Biol 17:775–786. doi:10.1111/plb.12318

Ferreira K, Torres GA, de Sousa SM, dos Santos AC (2010) Karyotype, meiotic behavior and pollen features of *Senna occidentalis*. Biologia 65(5):789–795

Flavell RB (1980) The molecular characterization and organization of plant chromosomal DNA sequences. Annu Rev Plant Biol 31:569–596

Fontana F, Colombo G (1974) The chromosomes of Italian sturgeons. Experientia 30(7):739–742

Futuyma DJ (ed) (2009) Evolution. Sinauer Associates, Sunderland

Goday C, Pimpinelli S (1986) Cytological analysis of chromosomes in the two species *Parascaris univalens* and *P. equorum*. Chromosoma 94(1):1–10

Gomes SSL, Saldanha CW, Neves CS, Trevizani M, Raposo NRB, Notini MM, Viccini LF (2014) Karyotype, genome size, and in vitro chromosome doubling of *Pfaffia glomerata* (Spreng.) Pedersen. Plant Cell Tissue Organ Cult (PCTOC) 118(1):45–56

Griffiths AJF, Wessler SR, Lewontin RC, Gelbart W, Suzuki DT, Miller JH (eds) (2006) Introdução à Genética. Guanabara

Guerra MS (1986) Reviewing the chromosome nomenclature of Levan et al. Revista Brasileira de Genética 9(4):741–743

Guerra M (ed) (1988) Introdução à Citogenética Geral. Guanabara Koogan, Rio de Janeiro

Guerra M (2000) Chromosome number variation and evolution in monocots. In: Monocots: systematics and evolution. CSIRO, Melbourne, pp 127–136

Guerra M (2008) Chromosome numbers in plant cytotaxonomy: concepts and implications. Cytogenet Genome Res 120:339–350

Guetat A, Rosato M, Rossello JA et al (2015) Karyotype analysis in Allium roseum L. (Alliaceae) using fluorescent in situ hybridization of rDNA sites and conventional stainings. Turk J Bot 39(5):796–807. doi:10.3906/bot-1402-48

Gurushidze M, Fuchs J, Blattner FR (2012) The evolution of genome size variation in drumstick onions (*Allium* subgenus *Melanocrommyum*). Syst Bot 37(1):96–104. doi:10.1600/036364412X616675

Harpke D, Carta A, Tomović G et al (2015) Phylogeny, karyotype evolution and taxonomy of *Crocus* series *Verni* (Iridaceae). Plant Syst Evol 301(1):309–325. doi:10.1007/s00606-014-1074-0

Hizume M, Shibata F, Matsusaki Y, Garajova Z (2002) Chromosome identification and comparative karyotypic analyses of four *Pinus* species. Theor Appl Genet 105(4):491–497

Hollingshead L (1930) Cytological investigations of hybrids and hybrid derivative *Crepis capillaris* and *Crepis tectorum*. Univ Calif Publ Agric Sci 6:55–94

Hoxmark RC (1970) The chromosome dimorphism of *Nucella lapillus* in relation to wave action. Nytt Mag Zool 18:229–238

Huang L, Wang J, Nie W, Su W, Yang F (2006) Tandem chromosome fusions in karyotypic evolution of *Muntiacus*: evidence from *M. feae* and *M. gongshanensis*. Chromosom Res 14(6):637–647

Jabbour F, Renner SS (2012) A phylogeny of Delphinieae (Ranunculaceae) shows that *Aconitum* is nested within *Delphinium* and that late miocene transitions to long life cycles in the Himalayas and Southwest China coincide with bursts in diversification. Mol Phylogenet Evol 62(3):928–942

King RC, Riley SF, Cassidy JD, White PE, Paik YK (1981) Giant polytene chromosomes from the ovaries of a *Drosophila* mutant. Science 212(4493):441–443

Klug WS, Cummings MR, Spencer CA, Palladino MA (eds) (2014) Concepts of genetics. Benjamin-Cummings Publishing Company

Kolar FR, Gosavi KVC, Chandore AN et al (2012) Comparative karyotype analysis of *Delphinium malabaricum* var. *malabaricum* (Huth) Munz. and *Delphinium malabaricum* var. *ghaticum* Billore. Cytologia 77(1):113–119

Lawlor TE (1974) Chromosomal evolution in *Peromyscus*. Evolution 28(4):689–692

Leitch IJ, Bennett MD (1997) Polyploidy in angiosperms. Trends Plant Sci 2(12):470–476

Leitch IJ, Chase MW, Bennett MD (1998) Phylogenetic analysis of DNA C-values provides evidence for a small ancestral genome size in flowering plants. Ann Bot 82:85–94

Levan A, Karl F, Avery AS (1964) Nomenclature for centromeric position on chromosomes. Hereditas 52(2):201–220

Levin DA (ed) (2002) The role of chromosomal change in plant evolution. Oxford University Press, Oxford

Lewis GP (2005) Cassieae. In: Lewis G, Schrire B, Mackinder B, Lock M (eds) Legumes of the world. Royal Botanic Gardens Kew, London, pp 111–125

Lipnerová I, Bureš P, Horová L et al (2013) Evolution of genome size in *Carex* (Cyperaceae) in relation to chromosome number and genomic base composition. Ann Bot 111(1):79–94. doi:10.1093/aob/mcs239

Makapedua DM, Barucca M, Forconi M, Antonucci N, Bizzaro D, Amici A, Canapa A (2011) Genome size, GC percentage and 5mC level in the Indonesian coelacanth *Latimeria menadoensis*. Mar Genomics 4(3):167–172

Mavrodiev EV, Chester M, Suárez-Santiago VN, Visger CJ, Rodriguez R, Susanna A, Soltis DE (2015) Multiple origins and chromosomal novelty in the allotetraploid *Tragopogon castellanus* (Asteraceae). New Phytol 206:1172–1183

Moreno NC, Amarilla LD, Las Peñas ML, Bernardello G (2015) Molecular cytogenetic insights into the evolution of the epiphytic genus *Lepismium* (Cactaceae) and related genera. Bot J Linn Soc 177(2):263–277

Pierre PM, Sousa SM, Davide LC, Machado MA, Viccini LF (2011) Karyotype analysis, DNA content and molecular screening in *Lippia alba* (Verbenaceae). An Acad Bras Cienc 83(3):993–1006

Rao SR, Pandey R, Chandel KPS (1992) Genetic stability studies in regenerated plants of *Allium tuberosum* Rottl. ex Spreng. A cytological approach. Cytologia 57(3):339–347

Resende K, Prado C, Davide L, Torres G (2014) Polyploidy and apomixis in accessions of *Senna rugosa* (G. Don) HS Irwin & Barneby. Turk J Biol 38(4):510–515

Rubes J, Musilova P, Kopecna O, Kubickova S, Cernohorska H, Kulemsina AI (2011) Comparative molecular cytogenetics in *Cetartiodactyla*. Cytogen Genome Res 137(2–4):194–207

Sacristan MD (1971) Karyotypic changes in callus cultures from haploid and diploid plants of *Crepis capillaris* (L.) Wallr. Chromosoma 33(3):273–283

Sadílek D, Nguyen P, Koç H, Kovařík F, Yağmur EA, Šťáhlavský F (2015) Molecular cytogenetics of *Androctonus scorpions*: an oasis of calm in the turbulent karyotype evolution of the diverse family Buthidae. Biol J Linn Soc 115(1):69–76

Schubert I (2007) Chromosome evolution. Curr Opin Plant Biol 10:109–115

Sharma A, Sen S (eds) (2002) Chromosome botany. CRC Press, Boca Raton

Sharma G, Sharma N (2014) Cytology as an important tool for solving evolutionary problems in angiosperms. Proc Natl Acad Sci India Sect B Biol Sci 84(1):1–7. doi:10.1007/s40011-013-0203-9

Sherman M (1946) Karyotype evolution: a cytogenetic study of seven species and six intraspecific hybrids of *Crepis*. Univ Calif Publ Bot 18:369–408

Stace CA (1980) Plant taxonomy and biosystematics. Edward Arnold, London

Stebbins GL (1958) Longevity, habitat, e release of genetic variability in the higher plants. In: Cold Spring Harbor symposia quantitative biology, vol 23, pp 365–378

Stebbins GL (1971) Chromosomal evolution in higher plants. Edward Arnold, London
Sturtevant AH (1925) The effects of unequal crossing over at the bar locus in Drosophila. Genetics 10(2):117–147
Tanomtong A, Chaveerach A, Phanjun G, Kaensa W, Khunsook AS (2005) New records of chromosomal features in Indian muntjacs (*Muntiacus muntjak*) and Fea's muntjacs (*M. feae*) of Thailand. Cytologia 70(1):71–77
Taylor KM, Hungerford DA, Snyder RL, Ulmer JFA (1968) Uniformity of karyotypes in the Camelidae. Cytogenet Genome Res 7(1):8–15
Thomas CA (1971) The genetic organization of chromosomes. Annu Rev Genet 5:237–256
Tobgy HA (1943) A cytological study of *Crepis fuliginosa*, *C. neglecta* and their F1 hybrid, and its bearing on the mechanism of phylogenetic reduction in chromosome number. J Genet 45:67–111
Trifonov VA, Stanyon R, Nesterenko AI, Fu B, Perelman PL, O'Brien PC, Yang F (2008) Multidirectional cross-species painting illuminates the history of karyotypic evolution in *Perissodactyla*. Chromosom Res 16(1):89–107
Wang AH, Sun Y, Wang FG, Schneider H, Zhai JW, Liu DM, Chen HF (2015) Identification of the relationship between Chinese *Adiantum reniforme* var. *sinense* and Canary *Adiantum reniforme*. BMC Plant Biol 15(1):36
Zarco CR (1986) A new method for estimating karyotype asymmetry. Taxon 35(3):526–530

Polyploidy: Evolution and Crop Improvement

10

Jahangir A. Dar, Zahoor A. Beigh, and Aijaz Ahmad Wani

Abstract

The condition of possessing more than two complete genomes in a cell has intrigued biologists almost a century. Many plant species including flowering plants are polyploids, and we know that it has a significant role in the evolution and crop improvement. It is well tolerated in many groups of eukaryotes. Polyploid ancestors have given rise to a number of flowering plants. Despite its widespread occurrence, the direct effect of polyploidy on evolutionary success of a species is still largely unknown. Many attractive hypotheses have been proposed in order to assign functionality to the increased content of a duplicated genome. Among these hypotheses are the proposals that genome doubling confers various advantages to polyploids which allow them to thrive well in environments that pose challenges to their diploid progenitors. Polyploidy is often accompanied with formation of improved varieties, developing sterile lines, restoring fertility in hybrids, enlargement and enhanced vigor, increasing allelic diversity and heterozygosity, etc. In genome-wide context for optimizing marker-assisted selection and crop plant improvement, all these factors need to be considered. This chapter attempts to give a brief overview of polyploidy, its origin, and role in evolution and crop improvement.

Keywords

Autopolyploidy • Allopolyploidy • Evolution • Crop improvement

J.A. Dar (✉) • Z.A. Beigh
Cytogenetics and Reproductive Biology Laboratory, Department of Botany, University of Kashmir, Srinagar 190006, J & K, India
e-mail: jahangirdar53@gmail.com

A.A. Wani
Department of Botany, University of Kashmir, Srinagar, Jammu and Kashmir, India

T.A. Bhat, A.A. Wani (eds.), *Chromosome Structure and Aberrations*,
DOI 10.1007/978-81-322-3673-3_10

10.1 Introduction

The organisms with more than two sets of chromosomes are called polyploids (Acquaah 2007; Chen 2010; Comai 2005; Ramsey and Schemske 1998). Polyploidy is widespread in nature and provides a way for adaptation and creation of new species. According to Chen et al. (2007), many crop plants have undergone polyploidy during their evolutionary process. According to Comai (2005), in every 100,000 plants, one plant is formed as polyploid by angiosperms at a significantly higher frequency. For understanding the nature of polyploidy, many studies have been carried out, and the present chapter seeks to throw light on the applications and implications of polyploidy in plant breeding and other commercial ventures. To understand polyploidy a few basic points need to be defined. The complete basic set of chromosomes is designated by "X", while the total number of chromosomes in a somatic cell is denoted by "2n". A somatic cell contains twice the number of chromosomes, while the gametes contain a haploid set only (Acquaah 2007; Otto and Whitton 2000). Three types of polyploidy, namely, autopolyploidy, allopolyploidy, and segmented allopolyploidy, have been distinguished by Stebbins (1947). In the first one, all genomes are identical and arise via genome duplication within the same species (Stebbins 1947; Lewis 1980). Allopolyploids contain two or more different genomes and can arise via hybridization of two different species linked with genome doubling (Stebbins 1947; Grant 1975). The third one, i.e., segmental allopolyploids, carries more than two incompletely distinct genomes which can lead to the formation of both bivalents and multivalents during chromosome pairing (Stebbins 1947; Levin 2002).

During the early part of the twentieth century, the phenomenon of polyploidy gained much more importance. According to Ramsey and Schemske (1998), Hugo De Vries's original mutation of *Oenothera lamarckiana* was one of the earliest examples of natural polyploidy. The occurrence of a fertile-type *Primula kewensis* from a sterile interspecific hybrid through chromosome doubling was discovered by Digby (1912), but the author failed to realize the significance of the same in the context of polyploidy (Stebbins 1971). Many crop plants like wheat, maize, sugarcane, coffee, cotton, and tobacco are polyploids either through intentional hybridization and selective breeding (e.g., some blueberry cultivars) or as a result of polyploidization event taken place in ancient times (e.g., maize) (Ramsey and Schemske 2002). In long-lived perennials that possess various vegetative means of propagation (*Fragaria*, *Rubus*, *Artemisia*, *Potamogeton*, etc.) and in those with frequent occurrences of natural interspecific hybridizations, polyploidy seems to be very favorable (Hilu 1993).

10.2 Changes in Chromosome Number

The changes either in one or a few chromosomes lead to aneuploidy. These changes in chromosome numbers are determined in relation to the somatic chromosome number (2*n*) of the species. Those aneuploid organisms which lack one

Table 10.1 Type of variations in chromosome number

Term	Type of change	Symbol
Aneuploid	One or few chromosomes extra or missing from $2n$	$2n \pm$ few
Nullisomic	One chromosome pair missing	$2n-2$
Monosomic	One chromosome missing	$2n-1$
Double monosomic	One chromosome from each of the two different chromosome pairs missing	$2n-1-1$
Trisomic	One chromosome extra	2n + 1
Double trisomic	One chromosome for each of two different chromosome pairs extra	$2n+1+1$
Tetrasomic	One chromosome pair extra	$2n+2$
Euploid	Number of genomes or copies of a single genome more or less than two	
Monoploid	One copy of a single genome	X
Haploid	Gametic chromosome complement of the species	N
Polyploid	More than two copies of one genome	
Autopolyploid	Genomes identical with each other	
Autotriploid	Three copies of one genome	$3x$
Autotetraploid	Four copies of one genome	$4x$
Autopentaploid	Five copies of one genome	$5x$
Autohexaploid	Six copies of one genome	$6x$
Allopolyploid	Two or more distinct genomes (generally each genome has two copies)	
Allotetraploid	Two distinct genomes	$2x_1 + 2x_2$
Allohexaploid	Three distinct genomes	$2x_1 + 2x_2 + 2x_3$
Allooctoploid	Four distinct genomes	$2x_1 + 2x_2 + 2x_3 + 2x_4$

Source: Singh BD (2012), Plant Breeding Principles and Methods

chromosome pair ($2n-2$) are termed as nullisomic. While those aneuploids that lack a single chromosome ($2n-1$) are known as monosomic. A double monosomic individual lacks two chromosomes which belong to two different chromosome pairs ($2n-1-1$). An aneuploid individual that contains one extra chromosome ($2n+1$) is known as trisomic and that having two extra chromosomes from two different chromosome pairs is called double trisomic ($2n+1+1$). A tetrasomic individual has one pair of extra chromosomes ($2n+2$). On the other hand, euploidy involves change in complete set of genome which is an exact multiple of the basic chromosome number of the concerned species. It is generally called polyploidy. A polyploid individual having all the genomes identical is called an autopolyploid. On the other hand, allopolyploids have two or more different genomes present. Euploids may have 3, 4, 5, 6, 7, 8, or more genomes making up their somatic chromosome number. The terminology of heteroploidy in common use is summarized in Table 10.1.

10.3 Origin of Polyploidy

Polyploidy is originated by different means. Somatic doubling during mitosis, nonreduction in meiosis leading to the formation of unreduced gametes, polyspermy (fertilization of the egg by two male nuclei), and endoreduplication (replication of the DNA but no cytokinesis) are some of the basic processes which give rise to polyploidy. According to Grant (1981), some authors have reported endoreduplication and somatic doubling more similar and have not viewed these as separate mechanisms. Chromosome doubling can occur either in the zygote or in some apical meristems to produce complete polyploids and polyploidy chimeras, respectively. Ramsey and Schemske (1998) have reported somatic polyploidy in some nonmeristematic plant tissues (e.g., tetraploid and octoploid cells in the cortex and pith of *Vicia faba*). According to Grant (1981), mitotic nondisjunction is the main cause of somatic doubling. Somatic doubling can occur in purely vegetative tissues, in branches that may produce flowers, or in early embryonic divisions (Grant 1981). Chromosome doubling in the zygotes was best described from heat shock experiments in which young embryos were briefly exposed to high temperatures (Lewis 1980).

10.3.1 Origin of Autopolyploidy

Autoploypoidy can be defined as the individuals with multiple sets of chromosomes derived from a single species. Autopolyploids can occur spontaneously in nature in low frequency and can be artificially induced by various means such as heat and chemical treatments, decapitation, and selection from twin seedlings. In autopolyploidy, the chromosomes fail to separate due to meiotic failure resulting in gametes with twice as many chromosomes as normal ($2n$). Autopolyploids can be formed by unreduced $2n$ gametes which are produced due to gametic nonreduction or meiotic nuclear restitution during micro- and megasporogenesis. Figure 10.1 shows the origin of autopolyploidy from two nonreduced gametes.

10.3.2 Origin of Allopolyploidy

The polyploids with chromosomes derived from different species are called allopolyploids. The fusion of reduced $1n$ gamete with unreduced $2n$ gamete gives rise to $3n$ zygote followed by the subsequent fusion of $1n$ reduced gamete with $3n$ gamete in the next generation giving rise to a tetraploid individual. This two-step process of allopolyploid production is sometimes referred to as a triploid bridge. The diagrammatic representation of allopolyploid formation is given in Fig. 10.2.

Environment and genotype have the influence on the formation of nonreduced gametes. For example, an increase in the number of nonreduced gametes in *Gilia*

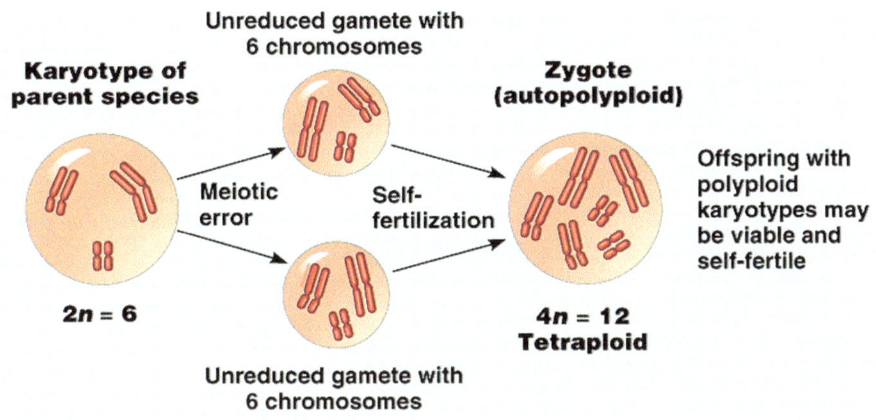

Fig. 10.1 Showing the origin of autopolyploid from unreduced gametes (*Source:* Campbell's Biology, page 454, 5th Edition)

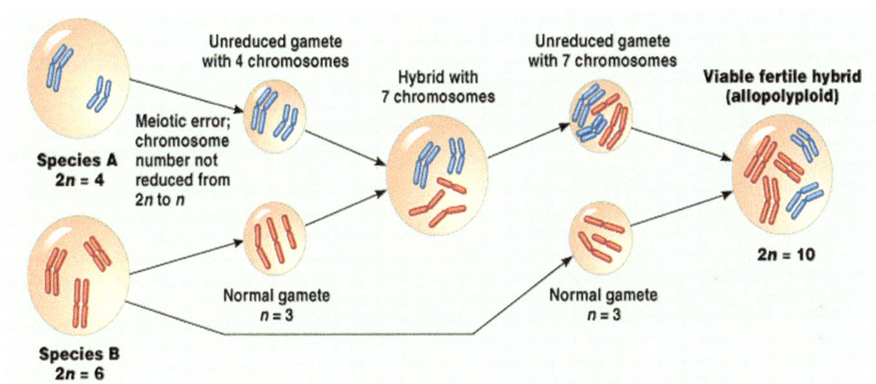

Fig. 10.2 Showing the origin of allopolyploid from unreduced and reduced gametes (*Source*: Campbell's Biology, page 454, 5th Edition)

has been shown to be favored by adverse growing conditions. In case of maize, the gene *elongate* on chromosome 3 was found to increase the proportion of diploid eggs which is an example of genotype in modulating the nonreduced gamete production (Grant 1981; Lewis 1980). The rapid screening techniques like flow cytometry, chromosome pairing, and other genomic techniques are helpful in studies on unreduced gametes in both plants and animals (Mable 2003). According to Ramsey and Schemske (1998), the contribution of polyspermy as a mechanism of polyploidy formation is rare except in some orchids. Endoreduplication has been known to occur in endosperm and the cotyledons of developing seeds, leaves, and stems of bolting plants (Larkins et al. 2001).

10.4 Methods for Inducing Polyploidy

In the1930s it was discovered that colchicine inhibits the formation of spindle fibers and temporarily arrests chromosomes at the anaphase stage (Blakeslee and Avery 1937). At this point, the chromosomes have replicated, but cell division has not yet taken place resulting in the formation of polyploidy cells. Other mitotic inhibitors, namely, oryzalin, trifluralin, amiprophos-methyl, and N_2O gas, have also been identified and used as doubling agents (Bouvier et al. 1994; Van Tuyl et al. 1992; Taylor et al. 1976). There are various methods for applying these doubling agents. One of the easiest and most effective methods is to work with a large number of seedlings having small, actively growing meristems. Seedlings can be soaked or the apical meristems can be submerged with different concentrations, durations, or frequencies of a given doubling agent. Shoots of older plants can be treated, but it is often less successful and results in a greater percentage of cytochimeras. Treatment of smaller axillary or subaxillary meristems is sometimes more effective. Chemical solutions can be applied to buds using cotton, agar, or lanolin or by dipping branch tips into a solution for a few hours or days. Surfactants, wetting agents, and other carriers (dimethyl sulfoxide) are sometimes used to enhance efficacy. Heat or cold treatment, X-ray, or gamma ray irradiation may also induce polyploidy in low frequencies. Triploid branches have been produced in *Datura* by cold treatment. Exposure of maize plants or ears to high temperature (38–45 °C) at the time of first zygotic division produces 2–5 % tetraploid progeny (Randolph 1941). Similarly heat treatments in barley, wheat, rye, and some other crop species have been successfully used for inducing polyploidy.

10.5 Role of Polyploidy in Plant Evolution

In comparison to the gradual evolutionary process whereby species evolve by small spontaneous mutations accumulated over time in the population, new species of plants can also arise rapidly. The most common mechanism for abrupt speciation is through the formation of natural polyploids. Once a tetraploid arises in a population, it can generally hybridize with other tetraploids. However, these tetraploids are reproductively isolated from their parental species. Tetraploids that cross with diploids of the parental species will result in triploids that are typically sterile. This phenomenon provides reproductive barrier between the polyploids and the parental species which is a driving force for speciation. Various estimates suggest that about 47–70 % flowering plants are of polyploidy origin (Grant 1971; Goldblatt 1980; Ramsey and Schemske 1998). For example, the plants in the rosaceous subfamily Maloideae (*Malus*, *Pyrus*, *Photinia*, etc.) are believed to have originated from ancient allopolyploids since they have $n = 17$ basic chromosome number, whereas plants in other subfamilies of Rosaceae have $n = 8$ or 9 (Rowley 1993). In many genera, different species will have different ploidy levels representing a series of polyploids. In *Chrysanthemum* different species have chromosome numbers of $2n =$ 18, 36, 54, 72, 90, and 198 – all multiples of a basic chromosome number of 9.

Polyploids have adaptive and evolutionary advantages due to certain factors. They are significantly more heterozygous than their diploid counterparts. The heterozygosity can be a key factor in growth, performance, and adaptability of a polyploidy plant. Allopolyploids can contribute to heterosis or hybrid vigor due to dissimilarity in genes. All polyploid individuals have a certain amount of genetic redundancy; extra copies of genes can mutate and diverge resulting in new traits without compromising essential functions. Polyploid populations often reveal extensive genomic rearrangement including the origin of novel DNA regions (Arnold 1997; Song et al. 1995; Wendel 2000). Ancient polyploids can eventually undergo such changes to the extent that they effectively become diploidized where diploid gene ratios are restored.

Polyploid plants also tend to be more self fertile and apomictic. Since polyploids usually arise at a low frequency, greater self fertility and apomixes would help to compensate for their minority disadvantage (Briggs and Walters 1977) and would provide further benefits in areas where breeding systems are compromised in stressful environments. Furthermore, inbreeding is less harmful for allopolyploid plants due to their greater heterozygous nature. One question that frequently arises is whether polyploids inherently have greater stress tolerance or not. For example, it has often been observed that unequal number of polyploids is found in stressful conditions like cold and dry regions. Some argue that this is a spurious correlation (Sanford 1983) or possibly the result of intermixing of species and formation of allopolyploids during glacial periods (Stebbins 1984). However, polyploids may also have positive characteristics that provide some benefits helpful in adaptation. Molecular studies have confirmed that allopolyploids exhibit enzyme multiplicity (Soltis and Soltis 1993). Since allopolyploids represent a fusion of two different genomes, these polyploids can potentially produce all of the enzymes produced by each parent as well as new hybrid enzymes. This enzyme multiplicity may give polyploids with greater biochemical flexibility, possibly extending the range of environments in which the plant can grow (Roose and Gottlieb 1976). Other changes in expression of genes, altered regulatory interactions, and rapid genetic and epigenetic changes could further contribute to increased variation and new phenotypes (Osborn et al. 2003).

10.5.1 Autopolyploidy and Evolution

Autopolyploidy has played a limited role in evolution of plant species. Some of our present-day crop species are considered to be autopolyploids. Autotetraploids appear to have been more successful as crops than other forms of autopolyploidy. In addition many forage grasses and several ornamentals are most likely autopolyploids. Recent studies using genomic in situ hybridization (GISH), however, have revealed peanut and coffee to be allopolyploids. GISH is a powerful tool for investigation of genome organization and evolutionary relationships. The diploid progenitors (parental species) of *Arachis hypogaea* are *A. villosa* and *A. ipaensis* (Raina and Mukai 1999). Similarly the most likely diploid progenitors of *Coffea arabica* are the wild species *C. congensis* and *C. eugenioides* (Raina et al. 1998). A similar analysis of other putative autotetraploids may reveal them to be allopolyploid in nature. Molecular analysis like genome sequencing and comparative genomic studies reveal that most species of angiosperms and vertebrates have experienced whole genome duplications, followed by loss of most of the duplicated regions of the genome. In this process, their genomes have retained considerable amounts of duplications, which have expanded the range of genetic diversity in these species. For example, *Oryza sativa* and *Arabidopsis thaliana* have experienced whole genome duplications three times at approximately 70 million years ago, 65 million years ago, and 40 million years ago. A large part of the duplicated genome has been lost since then, but a large number of duplicated genes exist even today. These duplicated genes have diverged to various degrees and become subfunctional, i.e., they show reduced levels of expression as confirmed by transcriptome analyses.

10.5.2 Allopolyploidy and Evolution

Allopolyploidy has played an important role in evolution. Allopolyploidy occurs in various genera of plants and has enjoyed considerable success in natural populations. It is expected that one third of the angiosperms are polyploids, and a huge number of them are allopolyploids. Allopolyploids have been more successful as crop species than autopolyploids. Some of the present-day allopolyploid crop species are given in Table 10.2.

Some naturally occurring allopolyploid crops include wheat, cotton, tobacco, mustard, oat, etc. Interspecific crossing followed by chromosome doubling in nature has resulted in the origin of allopolyploid crop species. The evolutionary origins of some natural allopolyploid crops are described below:

(i) Bread wheat (*Triticum aestivum*)

Evolutionary origin of bread wheat has been the most extensively investigated. Identity of the diploid species contributing the three different genomes (A, B, and D) of *Triticum aestivum* has been investigated by many workers more notably by Sears, Kihara, and others. It is generally accepted that the genome A present in diploid wheat is similar to those present in tetraploid and hexaploid wheat. Further, the genome B of tetraploid emmer wheat is found similar to that in hexaploid wheat. This is evident from chromosome pairing in crosses among diploid, tetraploid, and hexaploid wheat. Hybrid between diploid and tetraploid wheat shows 7^{II} and 7^{I}, while those between tetraploid and hexaploid wheat shows about 14^{II} and 7^{I}. It is believed that A genome of wheat has come from *Triticum monococcum* ($2n = 14$), D genome from *Triticum tauschii* (2n = 14), and B genome from unknown source probably from an extinct species ($2n = 14$) (Fig. 10.3).

Table 10.2 Some allopolyploid crop species and their gametic chromosome numbers

Scientific name	Common name	Gametic ch. no.	Cultivated/wild
Avena strigosa	Sand oats	7	Wild
A. barbata	Slender wild oats	14	Wild
A. sativa	Cultivated oats	21	Cultivated
A. byzantine	Cultivated red oats	21	Cultivated
Brassica nigra	Black sarson	8 (B)[a]	Cultivated
B. oleracea	Cabbage, cauliflower, etc.	9 (C)	Cultivated
B. campestris	Turnip rape	10 (A)	Cultivated
B. carinata	Abyssinian cabbage	17	Wild
B. juncea	Rai, Indian mustard	18	Cultivated
B. napus	Rape	19	Cultivated
Gossypium arboreum	Asiatic (desi) cotton	13 (A_2)	Cultivated
G. herbaceum	Asiatic cotton	13 (A_1)	Cultivated
G. thurberi	Wild American cotton	13 (D_1)	Wild
G. barbadense	Sea island (Egyptian) cotton	26 (A_2D_2)	Cultivated
G. hirsutum	American upland cotton	26 (A_1D_1)	Cultivated
Hordeum vulgare	Cultivated barley	7	Cultivated
H. jubatum	Squirrel tail barley	14	Wild
H. nodosum	Foxtail barley	21	Wild
Medicago hispida	California burclover	7	Cultivated
M. lupulina	Black medic	8, 16	Cultivated
M. falcate	Yellow alfalfa	8, 16	Cultivated
Nicotiana sylvestris	Wild tobacco	12	Wild
N. tomentosa	Wild tobacco	12	Wild
N. tabacum	Cultivated tobacco	24	Cultivated
N. rustica	Cultivated tobacco	24	Cultivated
N. bigelovii	Wild tobacco	24	Wild
N. debneyi	Wild tobacco	24	Wild
Prunus americana	American plum	8	Cultivated
P. avium	Sweet cherry	8	Cultivated
P. persica	Peach	8	Cultivated
P. cerasus	Sour cherry	16	Cultivated
P. domestica	European plum	16	Cultivated
Saccharum officinarum	Noble canes	40	Cultivated
S. barberi	Indian canes	41, 45, 46, 58, 62	Cultivated
S. sinensis	Indian canes	58, 59	Cultivated
S. spontaneum	Wild canes	20–64	Wild

(continued)

Table 10.2 (continued)

Scientific name	Common name	Gametic ch. no.	Cultivated/wild
S. robustum	Wild canes	30–74	Wild
Sorghum versicolor	Wild sorghum	5	Wild
S. bicolor	Jowar	10	Cultivated
S. halepense	Johnson grass	20	Cultivated
Trifolium pratense	Red clover	7	Cultivated
T. alexandrinum	Berseem clover	8	Cultivated
T. repens	White clover	16	Cultivated
T. medium	Zigzag clover	40, 48, 42, 49	Cultivated
Triticcum monococcum	Wild einkorn	7 (A)	Wild
T. turgidum var. dicoccoides	Wild emmer	14 (AB)	Wild
T. turgidum var. dicoccum	Emmer wheat	14 (AB)	Cultivated
T. turgidum var. turgidum	Solid stem wheat	14 (AB)	Cultivated
T. turgidum var. carthlicum	Persian wheat	14 (AB)	Cultivated
T. turgidum var. polonicum	Polish wheat	14 (AB)	Cultivated
T. turgidum var. durum	Durum wheat	14 (AB)	Cultivated
T. timopheevii	–	14 (AG)	Cultivated
T. timopheevii var. americanum	–	14 (AG)	Wild
T. aestivum var. spelta	Spelt wheat	21 (ABD)	Cultivated
T. aestivum var. aestivum	Common bread wheat	21 (ABD)	Cultivated
T. aestivum var. macha	–	21 (ABD)	Wild
T. aestivum var. compactum	Club wheat	21 (ABD)	Cultivated
T. aestivum var. sphaerococcum	Indian dwarf wheat	21 (ABD)	Wild
T. zhukovskyi	–	21 (AAG)	Cultivated

[a]Letters within parentheses denote the genomes present in the species
Source: Singh BD (2012), Plant Breeding, Principles and Methods

(ii) Tobacco (*Nicotiana tabacum*)

The genus *Nicotiana* comprises about 76 currently recognized naturally occurring species that are subdivided into 13 sections (Knapp et al. 2004). *Nicotiana tabacum* ($n = 14$) is a classic amphidiploid species originated from a hybridization event between *Nicotiana sylvestris* and *Nicotiana tomentosa*, both the species are diploid with $n = 12$ (Fig. 10.4). It has been reported that the maternal parent and the other donor of the S genome is *Nicotiana sylvestris* (Bland et al. 1985; Olmstead and Palmer 1991; Aoki and Ito 2000; Yukawa et al. 2006), whereas the section Tomentosae (*Nicotiana tomentosiformis*, *Nicotiana otophora*, or an introgressive hybrid between the two) has contributed the T genome (Kenton et al. 1993; Riechers and Timko 1999; Lim et al. 2000; Kitamura et al. 2001; Ren and Timko 2001).

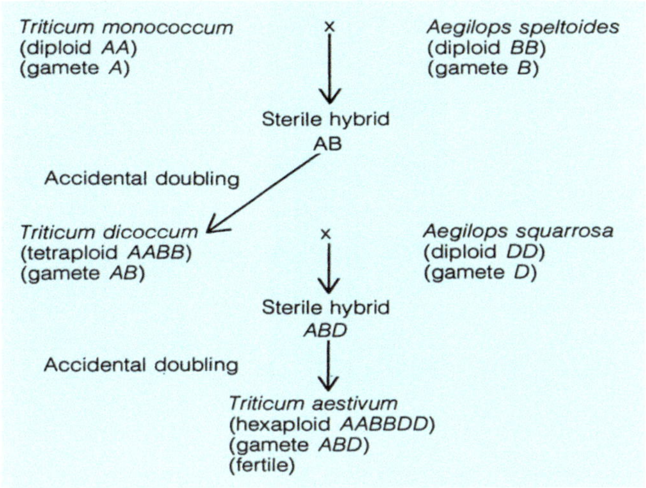

Fig. 10.3 Evolution of bread wheat (*Triticum aestivum*) (Source: http://www.ibri.org/Books/Pun_Evolution/Figures/Fig03-12.gif)

(iii) Cotton (*Gossypium hirsutum*)

All the diploid species of genus *Gossypium* have haploid chromosome number of 13 and fall into seven different genome types, designed A–G based on chromosome pairing relationships (Beasley 1942; Endrizzi et al. 1984). A total of five tetraploid ($n = 2x = 26$) species are recognized in *Gossypium*. According to Kimber (1961), all tetraploid species show disomic chromosome pairing. Chromosome pairing in interspecific crosses between diploid and tetraploid cotton suggests that tetraploids contain two different genomes, which resemble the A genome of *G. hirsutum* ($n = 13$) and D genome of *G. raimondii* ($n = 13$), respectively. Both the A and D genome species diverged from a common ancestor about 6–11 million years ago (Wendeil 1989). The putative A x D polyploidization event occurred in the New World, about 1.1–1.9 million years ago, in which A genome donor which is native to the old world served as the female parent (Wendeil 1989; Wendeil and Albert 1992). The five allotetraploid species (*G. hirsutum*, *G. barbadense*, *G. darwini*, *G. mustelinum*, and *G. tomentosum*) are thought to have originated by diversification at the polyploidy level (Fig. 10.5).

(iv) Amphidiploid *Brassica* species

An interesting example of the role of allopolyploidy in the evolution of different *Brassica* species is presented in the *Brassica* triangle (Morinaga 1934) (Fig. 10.6). As per the scheme, *Brassica juncea* ($n = 18$) is an amphidiploid from an interspecific cross between *Brassica nigra* ($n = 8$) and *Brassica campestris* ($n = 10$), whereas an interspecific cross between *Brassica oleracea* ($n = 9$) and *Brassica campestris* ($n = 10$) has given rise to amphidiploid *Brassica napus* ($n = 19$). On the other hand *Brassica carinata* ($n = 17$) is a result of an interspecific cross between *Brassica nigra* ($n = 8$) and *Brassica oleracea* ($n = 9$).

Fig. 10.4 Evolution of *Nicotiana tabacum*

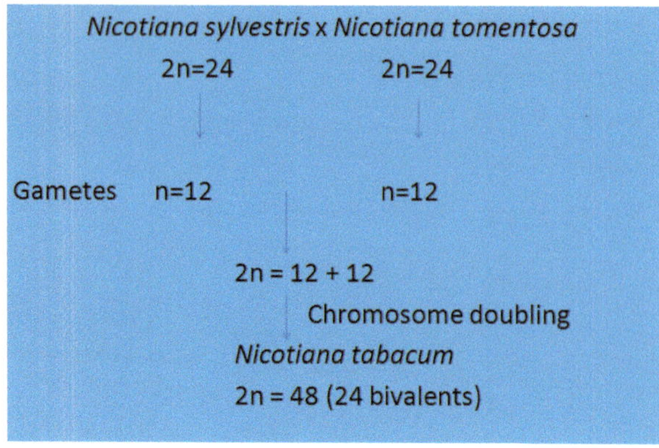

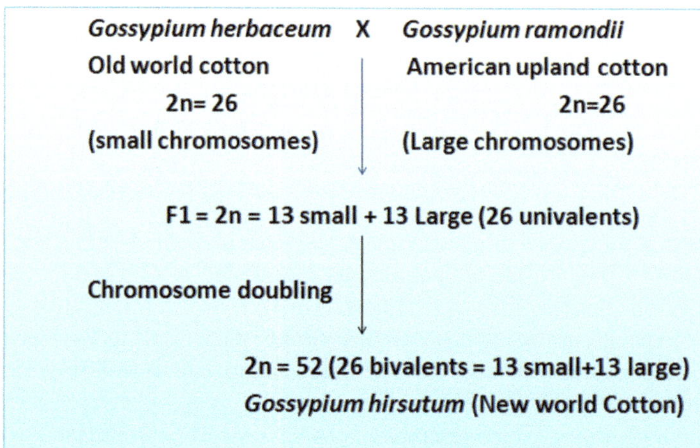

Fig. 10.5 Evolution of cotton (*Gossypium hirsutum*)

10.6 Polyploidy and Crop Improvement

Polyploidy has found some practical applications in improvement of crops, including the development of commercial varieties in India as well as in some other countries, utilization as a bridging species, creation of new crop species, widening the genetic base of existing polyploids, etc. Some important examples are discussed below.

(a) *Autotetraploids (triploid sugar beet and watermelon):* Triploid sugar beets have larger roots as compared to diploids, but they maintain the sugar content of diploids and thus yield more sugar per unit area. Such triploids are produced by inter planting diploid and tetraploid plants. Autotetraploid sugar beets have

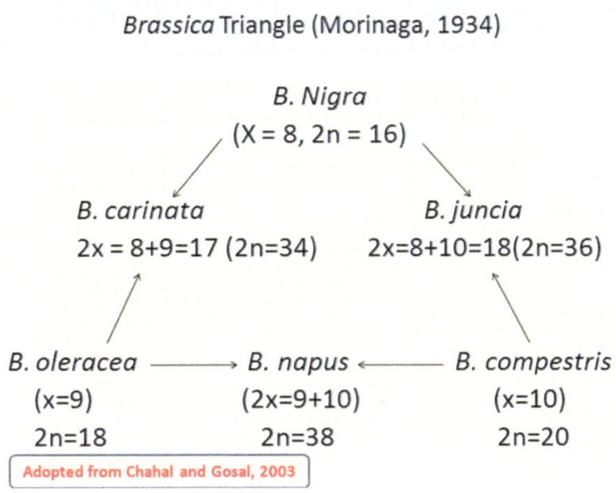

Fig. 10.6 Evolution of *Brassica* species and the relationships between diploid and naturally occurring amphidiploid species of *Brassica*

smaller roots as compared to diploid. Triploids can be produced when tetraploid and diploid are crossed or when something goes erroneous in meiosis and unreduced gametes are produced (with 2*n* chromosomes) which unite with gametes carrying haploid (*n*) chromosomes. Triploidy can also be beneficial in watermelons. Diploid watermelons (*Citrullus lanatus*) have 22 chromosomes per somatic cell and are fully fertile and produce a huge number of seeds per fruit. However, natural parthenogenesis is not known for watermelons. Hence Kihara's (1951) technique of producing seedless triploid watermelons can be utilized. The autotetraploid lines are planted alternately with diploids in isolation. Tetraploids are used as seed parent. The seeds produced in triploids from tetraploids are viable. However, when diploid is used as female or seed parent, then the program of triploid seed production is unproductive. Because of meiotic abnormalities, triploids cannot produce true seeds, but rudimentary structures similar to seeds of cucumber (small and white). For raising a commercial successful crop of triploid seedless watermelon, it is necessary to interplant diploid variety because fruit setting on triploids depends upon stimulus provided by the pollen.

(b) *Autotetraploids:* The only successfully developed autotetraploid among the grain crops is the rye (*Secale cereale*). Tetraploid rye has better qualities than diploids like larger kernels, superior ability to emerge under adverse conditions, and higher protein content. It is grown in Sweden and Germany. Similarly tetraploid grapes with larger fruits and fewer seeds per berry than diploids have been developed in California, USA. Tetraploid strains of red clover grown in Sweden have given higher hay yields than corresponding diploids among the forage crops. Pusa giant berseem (a variety of Egyptian clover) has been released in India for higher fodder yield. In cases of ornamental plants like phlox, dahlia, snapdragon, etc., induced autopolyploidy has been most successful as they have bigger flower size, longer blooming period, and relatively longer lasting flowers (Kehr 1996). Autotetraploidy has also been successful in crops like banana, maize, potato, and turnip.

(c) *Overcoming barriers to hybridization*: Due to differences in ploidy levels between prospective parents, the desirable crosses are difficult to obtain. Such interploid barriers appear to arise from abnormal endosperm formation. In species where there is a block at interploid level, seeds will often only develop normally if there is a 1 paternal: 2 maternal ratio in the genomic makeup of the endosperm, which would be the normal case for two diploid parents (Ramsey and Schemske 1998). Seeds that do not meet this criterion are often immature or aborted. In some cases this ratio is not accurate, but the larger the variation, the lower the viability of the seeds (Sanford 1983). In cases where the blocks due to the difference in ploidy level exist, barriers to hybridization may be overcome by manipulating the ploidy levels to match prior to hybridization.

(d) *Developing sterile cultivars:* A significant threat to certain ecosystems is the introduction of invasive species. An ideal approach for addressing this problem is the development of sterile forms of important nursery crops. There are various methods for developing sterile plants. Among them one of the rapid and efficient approaches is to create polyploidy. In most cases, these sterile plants function normally except reproduction particularly meiosis. In spite of these complexities, autotetraploids of some species can produce seeds that are fertile. In such a situation, triploids can be created by hybridizing tetraploids with diploids. In some species triploid development can be complicated due to the interploid block that prevents the normal development of triploid embryo. However, embryo culture is one of the techniques that can be used to produce sterile triploid plants. Another approach for triploid development is the regeneration of plants from endosperm. In most angiospermic seeds, the embryo is diploid, and the adjoining endosperm originates from the fusion of three haploid nuclei (one from male gametophyte and two from female) resulting in triploid tissue. This tissue can be excised from developing seeds and cultured in vitro to eventually give rise to regenerated embryos and plantlets. This method has been successful in various plants like citrus, kiwifruit, loquat, etc.

(e) *Restoring fertility in wide hybrids:* It is not necessary for hybrids between different taxa to be sterile. This often occurs due to failure of the chromosomes to pair correctly during meiosis – referred to as chromosomal sterility. The fertility can be restored by doubling the chromosomes of a wide hybrid. This approach has been used successfully in *Rhododendron* and *Chitalpa tashkentensis* (Contreras 2006; Olsen 2006). However, in some cases this technique has been successful in restoring fertility, as was the case with tetraploid hybrids of *Alstroemeria aurea* × *A. caryophyllaceae* (Lu and Bridgen 1997).

(f) *Enhancing pest resistance and stress tolerance:* Polyploids have played an important role in adaptability and resistance to biotic and abiotic stresses (Levin 1983). In some cases polyploids have demonstrated greater resistance to pests and pathogens, greater nutrient uptake efficiency, better drought resistance, and superior cold tolerance. There are a number of strategies for inducing polyploids as a means of enhancing adaptability. The expression and concentration of certain secondary metabolites and defense chemicals can be enhanced by increasing the chromosome number and related gene dose. However, this is not always true, and little is generally known about the relationship between gene dose, gene silencing, and expression secondary metabolites. An important method would be to develop allopolyploids between plants with diverse endogenous secondary metabolites. A unique and valuable characteristic of allopolyploids is that they often produce all the enzymes and metabolites (including defense chemicals) of both parents. This could be particularly effective for combining the characteristics of pest resistance of two species and contributing to a much broader and more horizontal form of pest resistance. The same approach may be useful for enhancing tolerance to certain environmental stresses.

(g) *Increased allelic diversity and heterozygosity:* Polyploidy has played an important role in increasing the allelic copy number and heterozygosity leading to novel phenotypes. Allelic diversity also increases during allopolyploidy, when two (or more) different genomes are present in a common nucleus. According to Osborn et al. (2003), the oil seed production in *B. napus* is positively affected by intergenomic heterozygosity. The QTL for seed yield and other traits in other populations of *B. napus* is also affected by intergenomic heterozygosity (Udall et al. 2006; Quijada et al. 2006). The tetraploid cotton also dominates the global market in terms of fiber production because they produce longer, finer, and stronger fiber than do their diploid relatives. According to Jiang et al. (1998), several QTL located on the D genome suggested that D genome loci had been used for the synthesis of fiber subsequent to polyploidy formation.

(h) *Creation of new crop species:* New crop species can be developed by polyploidy as triticale as the best example which is an allopolyploid between *Triticum aestivum* and *Secale cereal*. Poland, Germany, and France mainly cultivate the triticale varieties. An induced polyploid variety *Raphanobrassica* was of no use as the desired traits were not obtained from the cross. Another new autotetraploid variety was produced in kiwi (*Actinidia chinensis*) with the help of colchicine treatment, highlighting the considerable potential of this method to produce new cultivars with satisfactory fruit size (Wu et al. 2012).

10.7 Conclusions

Despite the prevalent occurrence of polyploidy in nature and the occurrence of its footprints in all angiospermic genomes, the question of effects of polyploidy on the evolutionary route of a species is still unclear. Earlier questions about the role of polyploidy in response to environmental stress or whether genome doubling is advantageous or disadvantageous to evolutionary success are being revisited using current genomic tools. Studies based on molecular levels are evident for genomic change on numerous levels of regulation related to polyploidization. However, the effects of polyploidy on fitness under different environmental conditions are not known still in many cases, and there is little evidence that observed transcriptional and genomic changes actually lead to faster evolution or greater adaptation in natural populations. Polyploids are looking generally different from their progenitors in morphological, ecological, physiological, and cytological characteristics that can contribute both to exploitation of a new niche and to reproductive isolation. Therefore, polyploidy is a major mechanism for adaptation and speciation in plants. The development of new crops and the interspecific gene transfer and also the origin of new crops can be traced with the help of polyploidy breeding. Thus, polyploidy is an interesting field of study to demonstrate the evolution of crop plants and utilize their variability in the field of crop breeding.

References

Acquaah G (2007) Principles of plant genetics and breeding. Wiley-Blackwell, Malden

Aoki S, Ito M (2000) Molecular phylogeny of *Nicotiana* (Solanaceae) based on the nucleotide sequence of the matK gene. Plant Biol 2(3):316–324

Arnold ML (1997) Natural hybridization and evolution. Oxford University Press, New York

Beasley YO (1942) Meiotic chromosome behavior in species hybrids, haploids, and induced polyploids of *Gossypium*. Genetics 27:25–54

Blakeslee AF, Avery AG (1937) Method of inducing doubling of chromosomes in plants by treatment with colchicine. J Hered 28:393–411

Bland MM, Matzinger DF, Levings CS (1985) Comparison of the mitochondrial genome of *Nicotiana tabacum* with its progenitor species. Theor Appl Genet 69:535–541

Bouvier L, Fillon FR, Lespinasse Y (1994) Oryzalin as an efficient agent for chromosome doubling of haploid apple shoots in vitro. Plant Breed 113:343–346

Briggs D, Walters SM (1977) Plant variation and evolution, 3rd edn. Cambridge University Press, Cambridge

Campbell NA, Jane BR, Mitchell LG (1999) Biology, 5th edn. Benjamin-Cummings Pub. Co, San Francisco. pp 454

Chahal GS, Gosal SS (2003) Principles and procedures of plant breeding – biotechnological and conventional approaches. Narosa Publishing House, New Delhi

Chen Z (2010) Molecular mechanisms of polyploidy and hybrid vigor. Trends Plant Sci 15:57–71

Chen L, Lou Q, Zhuang Y, Chen J, Zhang X, Wolukau JN (2007) Cytological diploidization and rapid genome changes of the newly synthesized allotetraploids *Cucumis* × *hytivus*. Planta 225:603–614

Comai L (2005) The advantages and disadvantages of being polyploid. Nat Rev Genet 6:836–846

Contreras RN (2006) Using polyploidy to restore fertility in azaleodendrons and investigating parentage of these wide hybrids. MS thesis, North Carolina State University, Raleigh

Digby L (1912) The cytology of *Primula kewensis* and of other related *Primula* hybrids. Ann Bot 26:357–388

Endrizzi JE, Turcotte EL, Kohel RJ (1984) Qualitative genetics, cytology, and cytogenetics. In: Kohel RJ, Lewis CF (eds) Cotton. American Society of Agronomy, Madison, pp 81–129

Goldblatt P (1980) Polyploidy in angiosperms: monocotyledons. In: Lewis WH (ed) Polyploidy. Plenum Press, New York, pp 219–239

Grant B (1971) Plant speciation. Columbia University Press, New York

Grant V (1975) Genetics of flowering plants. Columbia University Press, New York

Grant V (1981) Plant speciation, 2nd edn. Columbia University Press, New York

Hilu KW (1993) Polyploidy and the evolution of domesticated plants. Am J Bot 80(12):1494–1499

Jiang Y, Scarpa A, Zhang L, Stone S, Feliciano E, Ferro-Novick S (1998) A high copy suppressor screen reveals genetic interactions between BET3 and a new gene. Evidence for a novel complex in ER-to-Golgi transport. Genetics 149(2):833–841

Kehr AE (1996) Woody plant polyploidy. Am Nurserym 183(3):38–47

Kenton A, Parokonny AS, Gleba YY, Bennett MD (1993) Characterization of the *Nicotiana tabacum* L. genome by nuclear cytogenetics. Mol Gen Genet 240:159–169

Kihara H (1951) Triploid watermelons. Proc Am Soc Hortic Sci 58:217–230

Kimber G (1961) Basis of the diploid-like meiotic behavior of polyploid cotton. Nature 191:98–99

Kitamura S, Inoue M, Shikazono N, Tanaka A (2001) Relationships among *Nicotiana* species revealed by the 5S rDNA spacer sequence and fluorescence in situ hybridization. Theor Appl Genet 103:678–686

Knapp S, Chase MW, Clarkson JJ (2004) Nomenclatural changes and a new section classification in *Nicotiana* (Solanaceae). Taxon 53:73–82

Larkins BA, Dilkes BP, Dante RA, Coelho CM, Woo YM, Liu Y (2001) Investigating the hows and whys of DNA endoreduplication. J Exp Bot 52:183–192

Levin DA (1983) Polyploidy and novelty in flowering plants, Am Nat 122(1):1–24. In Levin DA (2002) The role of chromosomal change in plant evolution. Oxford University Press, Oxford

Levin D (2002) The role of chromosomal change in plant evolution. Oxford University, Oxford

Lewis W (1980) Polyploidy in species populations. In: Lewis W (ed) Polyploidy: biological relevance. Plenum, New York, pp 103–144

Lim CR, Kimata Y, Ohdate H, Kokubo T, Kikuchi N, Horigome T, Kohno K (2000) The *Saccharomyces cerevisiae* RuvB-like protein, Tih2p, is required for cell cycle progression and RNA polymerase II-directed transcription. J Biol Chem 275(29):22409–22417

Lu C, Bridgen MP (1997) Chromosome doubling and fertility study of Alstroemeria aurea A. caryophyllaceae. Euphytica 94:75–81

Mable BK (2003) Breaking down taxonomic barriers in polyploidy research. Trends Plant Sci 8(12):582–590

Morinaga T (1934) Interspecific hybridization in Brassica. VI. The cytology of F1 hybrids of *B. Juncea* and *B. nigra*. Cytologia 6:62–67

Olmstead RG, Palmer JD (1991) Chloroplast DNA and systematics of the Solanaceae. In: Hawkes JG, Lester RN, Nee M, Estrada N (eds) Solanaceae III: taxonomy, chemistry, evolution. Royal Botanic Gardens, Kew, London, pp 161–168

Olsen KB (2006) Productivity impacts of offshoring and outsourcing: a review. OECD Directorate for Science, Technology and Industry (STI), Paris, pp 1–35

Osborn TC, Pires JC, Birchler JA, Auger DL, Chen ZJ, Lee HS, Comai L, Madlung A, Doerge RW, Colot V, Martienssen RA (2003) Understanding mechanisms of novel gene expression in polyploids. Trends Genet 19:141–147

Otto SP, Whitton J (2000) Polyploid incidence and evolution. Annu Rev Genet 34:401–437

Quijada PA, Joshua A, Udall JA, Lambert B, Osborn TC (2006) Quantitative trait analysis of seed yield and other complex traits in hybrid spring rapeseed (*Brassica napus* L.): 1. Identification of genomic regions from winter germplasm. Theor Appl Genet 113:549–561

Raina SN, Mukai (1999) Genomic in situ hybridization in *Arachis* (Fabaceae) identifies the diploid progenitors of (*A. hypogaea* and wild *A. monticola*) peanut species. Plant Syst Evol 214:251–262

Raina SN, Mukai Y, Yamamoto M (1998) Insitu hybridization identifies the diploid progenitor species of *Coffea arabica* (Rubiaceae). Theor Appl Genet 97:1204–1209

Ramsey J, Schemske DW (1998) Pathways, mechanisms, and rates of polyploid formation in flowering plants. Annu Rev Ecol Syst 29:467–501

Ramsey J, Schemske DW (2002) Neopolyploidy in flowering plants. Annu Rev Ecol Syst 33:589–639

Randolph LF (1941) An evaluation of induced polyploidy as a method of breeding crop plants. Am Soc Nat 75(759):347–363

Ren N, Timko MP (2001) AFLP analysis of genetic polymorphism and evolutionary relationships among cultivated and wild *Nicotiana* species. Genome 44:559–571

Riechers DE, Timko MP (1999) Structure and expression of the gene family encoding putrescine N-methyltransferase in *Nicotiana tabacum*: new clues to the evolutionary origin of cultivated tobacco. Plant Mol Biol 41:387–401

Roose ML, Gottlieb LD (1976) Genetic and biochemical consequences of polyploidy in *Tragopogon*. Evolution 30:818–830

Rowley GD (1993) Rosaceae: the rose family. In: Heywood VH (ed) Flowering plants of the world. Batsford Pub, London, pp 141–144

Sanford JC (1983) Ploidy manipulations. In: Moore JN, Janick J (eds) Methods in fruit breeding. Purdue University Press, West Lafayette, pp 100–123

Singh BD (2012) Plant breeding principles and methods, 9th edn. Kalyani Publishers, New Delhi, pp 641–667

Soltis DE, Soltis PS (1993) Molecular data and the dynamic nature of polyploidy. Crit Rev Plant Sci 12:243–273

Song K, Lu P, Tank K, Osborn TC (1995) Rapid genome change in synthetic polyploids of *Brassica* and its implications for polyploid evolution. Proc Natl Acad Sci U S A 92:7719–7723

Stebbins GL (1947) Types of polyploids: their classification and significance. Adv Genet 1:403–429

Stebbins GL (1971) Processes of organic evolution. Prentice- Hall, Englewood Cliffs

Stebbins GL (1984) Polyploidy and the distribution of the arctic-alpine flora: new evidence and a new approach. Bot Helv 94:1–13

Taylor NL, Anderson MK, Wuesenberry KH, Watson L (1976) Doubling the chromosome number of *Trifolium* species using nitrous oxide. Crop Sci 16:516–518

Udall JA, Quijada PA, Lambert BA, Osborn TC (2006) Quantitative trait analysis of seed yield and other complex traits in hybrid spring rapeseed (*Brassica napus* L.): 2. Identification of alleles from unadapted germplasm. Theor Appl Genet 113:597–609

Van Tuyl JM, Meijer B, Van Dien MP (1992) The use of oryzalin as an alternative for colchicine in in-vitro chromosome doubling of *Lilium* and *Nerine*. Acta Hortic 325:625–629

Wendeil F (1989) New world cottons contain old world cytoplasm. Proc Natl Acad Sci USA 86:4132–4136

Wendeil F, Albert VA (1992) Phylogenetics of the cotton genus (*Gossypium*): character-state weighted parsimony analysis of chloroplast-DNA restriction site data and its systematic and biogeographic implications. Syst Bot 17:115–143

Wendel JF (2000) Genome evolution in polyploids. Plant Mol Biol 42:225–249

Wu J, Ferguson AR, Murray BG, Yilin J, Paul MD, Jingli Z (2012) Induced polyploidy dramatically increases the size and alters the shape of fruit in *Actinidia chinensis*. Ann Bot 109:169–179

Yukawa M, Tsudzuki T, Sugiura M (2006) The chloroplast genome of *Nicotiana sylvestris* and *Nicotiana tomentosiformis*: complete sequencing confirms that the *Nicotiana sylvestris* progenitor is the maternal genome donor *of Nicotiana tabacum*. Mol Genet Genomics 275:367–373

Links

http://www.ibri.org/Books/Pun_Evolution/Figures/Fig03-12.gif

Mutagenic Effects on Meiosis in Legumes and a Practical Case Study of *Vicia faba* L.

11

Tariq Ahmad Bhat and Aijaz Ahmad Wani

Abstract

Mutagens typically physical (ionising radiations, e.g. particulate (α-ray, β-ray and thermal neutrons) and non-particulate (X-ray and ɣ ray); non-ionising radiation, e.g. UV ray) and chemical (EMS, dES, NMG, MMS, EO, hydroxyl amine, nitrous acid, 5-bromouracil, 2-aminopurine and ethidium bromide, among others) are widely used in plant species of interest with an objective to create genetic variations by widening the gene pool and to induce gene mutation of commercial importance (superior qualitative trait(s) and enhancement in raw and value-added product(s)). Gene mutation is of global significance, and successful mutagenesis experiment depends on the sensitivity of the genotype(s) to the administered doses of the mutagen(s) employed. Assessment of LD_{50}, lethality, injury, mitotic and meiotic aberration frequency (key components to determine sensitivity of a species) is prerequisite for determining sublethal doses for monitoring successful mutation breeding experiments. The chapter gives a comparative observation of ethyl methane sulphonate (EMS), methyl methane sulphonate (MMS) and gamma irradiation on cytological and developmental parameters, i.e. meiotic features, pollen sterility, seed germination and seedling survival in *Vicia faba* L. The present study on the *Vicia faba* L. var. minor and major in M_1 generation showed that all the mutagens used elicit numerous chromosomal aberrations in meiosis and decrease in seed germination, pollen fertility and seedling survival. The combined treatments induced more chromosomal aberrations than the individual doses of mutagens which represents that combined treatments could

T.A. Bhat (✉)
Department of Education, Government of Jammu and Kashmir, Srinagar, India
e-mail: bhattariq110@gmail.com

A.A. Wani
Department of Botany, University of Kashmir, Srinagar, Jammu and Kashmir, India

T.A. Bhat, A.A. Wani (eds.), *Chromosome Structure and Aberrations*,
DOI 10.1007/978-81-322-3673-3_11

be more effective in creating more favourable variability than individual doses in both the varieties, i.e. minor and major of *Vicia faba* L.

Keywords
Mutagens • Chromosomal abnormalities • Chromosome analysis • Seed germination • Pollen fertility • Seedling survival • Phenotypic mutants • Breeding experiments

11.1 Description of *Vicia faba* L.

11.1.1 Origin

Faba bean is believed to be originated from Central Asian, Mediterranean and South American centres. According to Cubero (1973), the faba bean originated from four centres of eastern origin: (1) Europe (2) North African coast to Spain, (3) Nile to Ethiopia and (4) Mesopotamia to India. Secondary centres of diversity are postulated in Afghanistan and Ethiopia. The *Vicia narbonensis* L. and *Vicia galilaea* Plitmann and Zohary are taxonomically closely related to the cultivated crop having 2n = 12 chromosomes. Ladizinsky (1975) believed its origin to be Central Asia. China is the biggest centre of diversity of *Vicia faba* L. which produces 60 % of total world production (FAO 1994).

11.1.2 Taxonomy, Morphology and Floral Biology

Vicia faba L. is an unbranched annual herb with upright shoot system 0.3–2 m tall with 1 or more hollow stems from the base (Bond et al. 1985; Duke 1981; Health et al. 1994). The leaves are compound having 2–6 leaflets without tendrils or with rudimentary tendrils (Kay 1979; Bond et al. 1985). Flowers are large with white and dark purple markings, produced from short pedicels with 1–5 in number per cluster on each axillary raceme between the fifth and tenth node; the cross pollination is predominant and bumble bees act as main pollinator. The plant shows tap root system with profusely branched secondary roots (Bond et al. 1985).

The systematic position of *Vicia faba* L. under Bentham and Hooker's system of classification is as follows:

Division – Phanerogams
  Class – Dicotyledons
    Subclass – Polypetalae
      Series – Calyciflorae
        Order – Rosales
          Family – Leguminosae
            Subfamily – Papilionaceae (Fabaceae)
              Genus – Vicia
                Species – Faba

11.1.3 Ecology

Faba bean develops well in cool season. It grows in warm temperate and subtropical areas as a winter annual herb. It tolerates the temperature up to −15 °C (Robertson et al. 1996). It can be grown anywhere particularly in wetter portions of cereal-growing areas. It grows in a range of soil but grows best on rich loams with moderate moisture (Duke 1981). They are least drought resistant among the legumes, but cultivars with high water use efficiency have been developed at ICARDA (Robertson et al. 1996). Faba bean is more tolerant to acidic soils than most of the legumes. Its production is best in optimum temperatures which range from 18 to 27 °C (Duke 1981). The ideal rainfall which favours its best production is 650–1000 mm per annum (Kay 1979). The life cycle ranges from 90 to 220 days depending upon the cultivars and climatic conditions (Bond et al. 1985).

11.1.4 Yield and Economics

The attack of parasites and pathogens such as *Botrytis fabae*, *Ascochyta fabae*, *Uromyces fabae*, *Orobanche crenata* and *Aphis fabae*, flower drop, seed abortion, lack of adequate pollination and reduced seed setting can be major constraints to yield. The yield is also influenced by drought, high temperature, deficiency of nutrients, salinity and water logging. China was the largest producer with estimated annual production of 2.4–2.6 million MT (1161–1447 kg/ha) from 1979 to 1994 (FAO 1994). Argentina reported the production of more than 9000 kg/ha, followed by Switzerland (3350–4375 kg/ha), France (3000–3900 kg/ha) and Belgium (3350–3750 kg/ha) from 1992 to 1994 (FAO 1994). In Southern United States and US Pacific Coast, the faba bean has been in use for human consumption, feed for cattle particularly to horse (popularly called as horse beans) and green manuring. The immature seeds of faba bean are used as green vegetable.

The major agro-ecological regions of faba bean are Northern Europe, Mediterranean, the Nile valley, Ethiopia, Central Asia, East Asia, Oceana, Latin America and North America (Bond et al. 1985). The production of faba bean increased manyfold in Australia, EEC, West Asia and North Africa, but there is 25 % decrease in area meant for sowing of faba bean in China and decrease in production in Africa from 1982 to 1992 (Oram and Agacoili 1994). Faba bean is the second abundant food legume in Europe (Picard et al. 1988).

11.1.5 Uses

The faba bean is a source of protein and carbohydrate from the very ancient times. It plays a vital role in crop rotation, improving physical conditions of soil, disease control and weed population. It is cultivated as a vegetable and used green or dried and fresh or canned and for stock feed and is considered superior to other legumes in food value. Faba bean is used as human food as well as animal feed both in developing countries and in industrialised countries. It is used as breakfast in the Middle

East, Mediterranean region, China and Ethiopia (Bond et al. 1985). The most popular dishes of faba bean are medamis (stewed beans), falafel (deep-fried cotyledon paste with some vegetables and spices), bissara (cotyledon paste poured onto plates) and nabet soup (boiled germinated beans) (Jambunathan et al. 1994). Roasted seeds are eaten in India (Duke 1981). Straw from faba bean harvest fetches a premium in Egypt and Sudan and is considered as a cash crop (Bond et al. 1985). The straw can also be used for brick making and as a fuel in parts of Sudan and Ethiopia.

Broad bean has been considered as meat and milk substitute. Large seeded cultivars are grown in gardens and are used as vegetable and one of the most important winter crops for human consumption in the Middle East. The green seeds are consumed as a cooked vegetable. The dry beans are used as food and feed to livestock. Broad beans are very nutritious, containing 23 % proteins. The whole dry seeds contain (per 100 g) 344 cal, 10.1 % moisture, 26.2 g protein, 1.3 g fat, 59.4 g total carbohydrate, 6.8 g fibre, 3.0 g ash, 104 mg ca, 301 mg p, 6.7 mg Fe, 8 mg riboflavin, 2.1 mg niacin, 162 tryptophan and 16 mg ascorbic acid. The faba bean is said to be used for diuretic, expectorant and tonic.

Broad bean proteins consist of globulins, albumins and glutelins. They also contain two minor proteoses and two globulins, i.e. legumin and vicilim. The favaism which is a paralytic condition has been reported in certain tropical countries so the faba bean should be used with care and cautious manner particularly when it is part of regular diet.

11.2 Cytology

The *Vicia faba* L. has six pairs of comparatively bigger chromosomes which are treated as excellent cytogenetic material for studying chromosomal aberrations. Root tips and pollen mother cells of *Vicia faba* is the best experimental material for assessing the chromosomal damage.

11.2.1 Cytological Analysis

Estimation of cytological abnormalities and their magnitude during mitosis or meiosis is the most important index for evaluating the effect of a mutagen. It is also a dependable parameter to assess radio sensitivity of plants to both physical and chemical mutagens. The chromosomal aberrations induced by mutagens have been reported by many researchers in several plants such as in pea (Kallo,1972), triticale (Pushpalatha et al. 1992), lentil (Reddy and Annadurai 1992), fenugreek (Anis and Wani 1997; Parveen et al. 2006), *Cicer* (Ganai et al. 2005), *Capsicum annuum* (Anis et al. 2000; Gulfishan et al. 2010, 2011, 2012) and broad bean (Bhat et al. 2005a, b, c, 2006a, b, c, d, 2007a, b, c, d).

11.2.2 Cytological Observations in M_1 Generation

The control plants of var. minor and var. major revealed six perfect bivalents (2n = 12) at diakinesis and metaphase which showed equal separation (6:6) at anaphase I1. Telophase II, anaphase II and telophase II were normal giving rise to normal tetrad.

The microsporogenesis of plants raised from tetrad seeds was highly disturbed. The meiotic studies showed that although the types of chromosomal aberrations were more or less similar in both the varieties, the frequencies of these chromosomal aberrations were different.

The mutagenic treatments induced various chromosomal aberrations during microsporogenesis in M_1 generation. The most frequent aberrations were stickiness, univalents, multivalents, precocious separation, non-orientation of bivalents and fragments at metaphase I/II and laggards, bridges, unequal separation, non-disjunction and cytomixis at anaphase I/II and laggards I/II. The dominant meiotic aberrations at telophase I/II were disturbed polarity, micronuclei, multinucleate condition and cytomixis. Representative cytological features are shown in the plates. The data recorded in the tables showed that almost all types of chromosomal aberrations were dose dependent in both the varieties.

The analysis of cytological parameters of mitosis and meiosis is considered as one of the dependable indices to measure the potency of a mutagen. Therefore, investigations on disturbances in meiotic behaviour indicating mutational genetic load form an integral part of most of mutation studies and ascertain the most effective mutagen for a given crop to realise maximum results.

In the present investigation, a vast array of meiotic aberrations were observed in the pollen mother cells obtained from plants raised from the seeds treated with different concentrations of EMS, MMS, DES, SA and gamma rays. The different types of chromosomal aberrations, viz. univalents, multivalents, stickiness, precocious separation, chromatin bridges, laggards, unequal separation, disturbed polarity, micronuclei and cytomixis, were observed in the present investigation. Similar results were also reported by many workers in different plants after treatments with physical and chemical mutagens, viz. Anis and Wani (1997) in *Trigonella foenum-graecum*; Dhamyanthi and Reddi (2000) and Gulfishan et al. (2013) in *Capsicum annuum*; Singh (2003) in *Vigna radiata*; Rao and Laxmi (1980) and Katiyar (1978) in *Capsicum annuum* L.; Singh and Chaudhary (2005) in chilli; Bhat et al. (2005a, 2007a, b, c, d),Sharma et al. (2009) and Bhat and Wani (2015) in *Vicia faba*; Kumar and Singh (2003) in *Hordium* vulgare; Kumar and Dubey (1998a) in *Lathyrus sativus*; Alka *et al.* (2012) *in Linum usitatissimum L.*; and Aslam *et al.* (2012) in *Cichorium intybus* L.

11.2.2.1 Univalents

The frequency of univalents ranged from 1 to 12 per PMC at metaphase I. The univalents ranged from 0.74 to 1.70 % (EMS), 0.91 to 1.85 % (ϒ-rays), 0.92 to 1.90 % (MMS), 0.91 to 2.01 % (ϒ-rays + EMS) and 1.09 to 2.23 % (ϒ-rays + MMS) in var. minor, while the frequency of univalents was from 0.54 to 1.45 % (EMS), 0.74 to 1.66 % (ϒ-rays), 0.73 to 1.65 % (MMS), 0.90 to 1.82 % (ϒ-rays + EMS) and

92 to 2.01 % (ϒ-rays + MMS) in var. major. The combined mutagenic treatments induced more univalents than the individual mutagenic treatments. Moreover, the maximum univalents were recorded at the highest dose/concentration of each individual as well as in combined mutagenic treatments in both the varieties. However, var. minor recorded more frequency of univalents than var. major.

The frequency of univalents ranged from 1 to 12 per PMC, and these were later found as laggards at anaphase and telophase stages. The univalents were also reported by Saha and Datta (2000) in *Nigella sativa*, Kumar et al. (2003) in *Lens culinaris*, Sengupta and Datta (2004) in *Sesamum indicum* L., Bhat et al. (2006a, b, c, d) in *Vicia faba* L. and Aslam et al. (2012) in *Cichorium intybus* L. It seems more likely that mutagens induced univalent formation through cryptic structural changes in some of the chromosomes which restricted pairing between homologous chromosomes and in this way reduced chiasma frequency.

According to Zeerak (1992), the failure of pairing among homologous chromosomes might be due to structural changes in chromosomes and gene mutations induced by mutagens resulting in the occurrence of univalents. Previously it has been reported that the presence of univalents at metaphase I might be due to asynapsis (lack of chromosome pairing during the late prophase I) so that the homologous chromosomes failed to pair (Kaltsikes 1973; Gustafsson 1983) or desynapsis (inability to retain chiasmata in synapsed homologous chromosomes) resulting in premature separation of bivalents so that the separated chromosomes will not be able to orient themselves at equatorial plate. Koduru and Rao (1981) are also of the opinion that the univalents occur due to asynaptic or desynaptic genes in prophase I. Gottschalk and Kleine (1976) explained that the chromosome pairing is under the control of two groups of genes, viz. *As* and *Ds* which when present in recessive state, may cause chromosome pairing to fail. The emergence of univalents could also be due to precocious chiasma terminalisation (Gottschalk and Kaul 1980b; Sidhu 2008). Thus, reduced chiasma frequency as a result of increased heterology may be one of the reasons of increased number of univalents with increasing concentration of mutagens (Gottschalk and Kaul 1980a; Jabee and Ansari 2005). The absence or highly reduced number of univalents in M_3 generation was due to the fact that desynapsis or asynapsis did not occur due to the ceasing effect of mutagens and the normal pairing of bivalents. It may also be due to repair mechanism in the case of damaged DNA. Some of these univalents were later on found to be laggards at anaphase and telophase stages. Thus, gamma rays, MMS and DES might have induced genic disturbances due to mutagenic activity and hence the disturbances in homology and pairing of homologous chromosomes.

11.2.2.2 Multivalents

Multivalents such as trivalents, tetravalents, hexavalents, heptavalents and octavalents were observed at the metaphase I in the treated populations. The multivalents showed dose-dependent increase in all the individuals and combined mutagenic treatments in both the varieties. The frequency of PMCs with multivalents ranged from 1.11 to 2.05 % (EMS), 1.30 to 2.03 % (ϒ-rays), 1.28 to 2.05 % (MMS), 1.30 to 2.40 % (ϒ-rays + EMS) and 1.30 to 2.60 % (ϒ-rays + MMS) in var. minor, while the range was from 0.90 to 1.81 % (EMS), 0.92 to 1.84 % (ϒ-rays) 0.91 to 2.02 %(MMS), 1.08 to 2.18 % (ϒ-rays + EMS) and 1.10 to 2.19 % (ϒ-rays + MMS) in var. major. The combined mutagenic treatments showed higher frequency of chromosomal aberrations than the individual mutagenic treatments. Moreover, the highest dose of each individual and combined mutagenic treatment showed the higher frequency of multivalents in both the varieties. The maximum multivalent formation was 2.60 % and 2.19 % at 20 Kr + 0.04 %MMS in var. minor and var. major, respectively.

Multivalent formation as observed in the present investigation has also been reported by many workers in several crops such as Zeerak (1992) in *Lycopersicon esculentum*, Siddique and Ansari (2005) in *Solanum melongena* L., Kumar and Rai (2007a) in *Zea mays* L., Kumar and Tripathi (2004) and Kumar and Gupta (2009) in *Capsicum annuum*, etc.

The multivalent formation was due to the breakage in chromosomes caused by these mutagens and their reunion through reciprocal translocations. Chughtai and Hasan (1979) recorded the multivalents with increasing dosage of EMS, MES and MMS in *Lens esculenta* and suggested that translocation might have been produced due to terminal affinities of broken chromosomes. Zeerak (1992), Vandana and Kumar (1996), Kumar and Sinha (1991), Anis and Wani (1997) and Kumar and Gupta (2009) attributed the multivalent formation to irregular pairing and breakage followed by translocation and inversion. According to Lea (1955) and Srivastava (1979), the broken ends of the chromosomes when fused at random may bring about unequal changes making up the multivalents.

11.2.2.3 Stickiness

Stickiness or clumping of chromosomes at metaphase I/II was the most common meiotic aberration. Chromosomes were clumped either in one or different groups. The frequency of PMCs showing stickiness ranged from 0.92 to 1.86 %, 0.91 to 2.03 % (ϒ + rays), 0.92 to 2.05 % (MMS), 0.91 to 2.40 % (ϒ-rays + EMS) and 0.91 to 2.60 % (ϒ-rays + MMS) in var. minor, while the frequency of PMCs showing stickiness ranged from 0.72 to 1.63 % (EMS), 0.74 to 1.84 % (ϒ-rays), 0.73 to 2.02 % (MMS), 0.72 to 2.18 % (ϒ-rays + EMS) and 0.73 to 2.38 % (ϒ-rays + MMS) in var. major. The combined mutagenic treatments showed the more frequency of PMCs with stickiness than the individual mutagenic treatments that showed the higher frequency of PMCs with stickiness in both the varieties. The maximum PMCs with stickiness were 2.60 % and 2.38 % at 20 Kr + 0.04 %MMS in var. minor and var. major, respectively.

Stickiness could be due to depolymerisation of nucleic acid caused by mutagenic treatment or due to partial dissociation of the nucleoproteins and alteration in their pattern of organisation (Evans 1962). Jabee and Ansari (2005) suggested that chromosomal breakage may cause stickiness among the chromosomes. It may also be due to genetic and environmental factors (Rao et al. 1990; Nirmala and Rao 1996; Baptista-Giacomelli et al. 2000a). Stickiness could arise due to depolymerisation of nucleic acid caused by mutagenic treatment (Tarar and Dnyansagar 1980; Kumar et al. 2003; Kumar and Tripathi 2003; Jabee et al. 2008) or due to dissociation of nucleoproteins and alteration in their pattern of organisation (Katiyar 1978; Myers et al. 1992; Kumar et al. 2003; Kumar and Rai 2007c). In the present case, the mutagens seem to be responsible for induced stickiness, which is caused by the chromosomal condensation during active cell divisional stages by target proteins, gene mutation or direct action of the mutagens on the proteins or by abnormal functioning of target proteins caused by disturbance in the chromosomes during the course of their condensation from prophase I to metaphase II.

11.2.3 Precocious Movement

Precocious separation at metaphase I/II was also one of the most common cytological aberrations. The bivalents which showed precocious separation ranged from 1 to 3. The frequency of PMCs with precocious separation ranged from 0.55 to 1.50 % (EMS), 0.73 to 1.70 % (ϒ-rays), 0.73 to 1.90 % (MMS), 0.73 to 2.01 % (ϒ-rays + EMS) and 0.91 to 2.23 % (ϒ-rays + MMS) in var. minor, while the frequency of PMCs with precocious separation ranged from 0.40 to 1.30 % (EMS), 0.40 to 1.47 % (ϒ-rays), 0.36 to 1.84 % (MMS), 0.36 to 2.00 % (ϒ-rays + EMS) and 0.73 to 2.19 % (ϒ-rays + MMS) in var. major. The maximum frequency of PMCs with precocious separation was 2.23 % and 2.19 % at 20Kr + 0.04 %MMS in var. minor and var. major, respectively. The combined mutagenic treatments induced more frequency of PMCs with precocious separation than the individual treatments. Moreover, the highest dose/concentration of both individual and combined treatments showed the highest frequency of PMCs with precocious separation.

Precocious movement of chromosomes was frequently found in present investigation at metaphase I and II stages. Similar was reported by the work of Das and Roy (1989) in *Solanum sisymbriifolium*, Pagliarini (1990) in *Aptenia cordifolia*, Anis and Wani (1997) and Siddiqui et al. (2007) in *Trigonella foenum-graecum*, Kumar and Tripathi (2004) in *Capsicum annuum*, Kumar and Rai (2007c) in *Glycine max*, Malik and Shrivastava (2007) in *Carthamus*, Bhat et al. (2007a, b, c, d) and Khan et al. (1998, 2007a, b) in *Vicia faba*, Defani-Scoarize et al. (1995) and Kumar and Rai (2007a) in *Zea mays*, etc.

The precocious movement of chromosomes might have been caused by the early terminalisation or stickiness of chromosomes and/or movement of chromosomes ahead of the rest during anaphase (Premjit and Grover 1985). It may be due to the abnormal homology for chromosome pairing and spindle mechanism (Agarwal and Ansari 2001), either because of the abnormal spindle activity (Kumar and Gupta 2009) or due to the reunion of chromatids during meiotic prophase (Rees and Thompson 1955; Lewis and John 1966; Newmann 1966).

The desynapsis is caused by precocious separation of univalents (Bose and Saha 1970; Kaul and Nirmala 1993; Kumar and Rai 2007a) or asynapsis (Roy et al. 1971). Moreover, structural differences of homologous pairs followed with disturbed spindle mechanism might have resulted in haphazard movement of chromosomes, some of them being precocious.

11.2.4 Stray Chromosomes

Stray chromosomes were observed at metaphase I which may be caused by abnormal spindle function and clumping of chromosomes. Similar results were also reported in *Vicia faba* L. (Bhat et al. 2007a, b, c, d; Gulfishan et al. 2010) and in *Hordeum vulgare* L. (Jafri et al. 2012).

11.2.5 Non-orientation

Non-orientation of bivalents was observed in all the mutagenic treatments in both the varieties. The frequency of PMCs with non-orientation of bivalents ranged from 0.92 to 1.30 %(EMS),1.10 to 1.85 %(ϒ-rays), 1.10 to 2.05 % (MMS), 1.10 to 2.20 % (ϒ-rays + EMS) and 1.09 to 2.41 % (ϒ-rays + MMS) in var. minor, while the frequency of PMCs showing non-orientation of bivalents ranged from 0.72 to 1.45 %(EMS), 0.74 to 1.84 %(ϒ-rays), 0.73 to 2.20 % (MMS), 0.72 to 2.37 % (ϒ-rays + EMS) and 1.10 to 2.19 % (ϒ-rays + MMS) in var. major. The maximum frequency of PMCs with non-disjunction of bivalents was 2.41 % at 20 Kr + 0.04 % MMS and 2.37 % at 20 Kr + 0.4 % EMS in var. minor and var. major, respectively. The combined mutagenic treatments induced more frequency of PMCs with non-orientation of bivalents than the individual mutagenic treatments. Moreover, the highest dose/concentration of each individual and combined mutagenic treatment induced the highest frequency of PMCs with non-orientation of bivalents in both the varieties.

11.2.6 Fragments

The fragments were observed in almost all the treatments with a few exceptions (0.1 %EMS and 0.2 %EMS in var. minor and 0.1 % EMS, 0.2 %EMS, 10 Kr rays, 0.01 % MMS, 10 Kr + 0.1 % EMS and 10 Kr + 0.01 % MMS in var. major). The frequency of PMCs with fragments ranged from 0.54 to 0.74 % (EMS), 0.40 to 0.92 % (ϒ-rays), 0.20 to 1.12 % (MMS), 0.20 to 1.9 % (ϒ-rays + EMS) and 0.20 to 1.30 % (ϒ-rays + MMS) in var. minor, while the frequency of PMCs with fragments ranged from 0.40 to 0.54 % (EMS), 0.36 to 0.55 % (ϒ-rays), 0.55 to 0.92 % (MMS), 0.54 to 0.91 % (ϒ-rays + EMS) and 0.55 to 1.09 % (ϒ-rays + MMS) in var. major. The highest frequency of pollen mother cells with fragments was 1.30 % at 20 Kr ϒ-rays + 0.04 % MMS and 1.10 % at 20Kr ϒ-rays + 0.03 %MMS in var. minor and var. major, respectively.

11.3 Chromosomal Aberrations at Anaphase I and II

11.3.1 Laggards

The PMCs with laggards were observed in almost all the treatments with a few exceptions (0.1 % EMS in var. minor and 0.1 % EMS and 10 Kr ϒ-rays in var. major). The frequency of lagging chromosomes ranged from 1 to 4 per PMC. These laggards were present either as univalents or as whole bivalents at anaphase I/II. The frequency of PMCs with laggards at anaphase I/II ranged from 0.73 to 1.30 % (EMS), 0.20 to 1.50 % (ϒ-rays), 0.40 to 1.70 % (MMS), 0.40 to 1.83 % (ϒ-rays + EMS) and 0.54 to 2.04 % (ϒ-rays + MMS) in var. minor, while the frequency of PMCs with laggards ranged from 0.54 to 1.10 % (EMS), 0.72 to 1.29 % (ϒ-rays), 0.18 to 1.47 % (MMS), 0.18 to 1.64 % (ϒ-rays + EMS) and 0.36 to 1.83 % (ϒ-rays + MMS) in var. major. The highest frequency of PMCs with laggards was 2.04 % and 1.83 % at20 Kr + 0.04 % MMS in var. minor and var. major, respectively.

The occurrence of laggards may be explained on the basis of abnormal spindle formation and chromosomal breakage. The laggards observed in the present study might be due to delayed terminalisation or stickiness of chromosome ends or because of failure of chromosomal movement (Permjit and Grover 1985; Jayabalan and Rao 1987b; Soheir et al. 1989).

Tarar and Dnyansagar (1980) and Das and Roy (1989) are of the opinion that due to the effects of mutagens, the spindle fibres failed to carry the respective chromosome to the polar regions, and resultantly the lagging chromosomes appeared at anaphase I. According to Pagliarini (1990), laggards may be the result of late chiasma terminalisation. Kumar and Rai (2007a, 2009) also support the opinion that laggards might have appeared due to improper spindle functioning. Kumar and Gupta (2009) reported that fragments which appeared on the breakage of bridges, as a result of spindle fibre functioning to pull the chromosomes towards the poles, formed laggards.

11.3.2 Bridges

Bridges at anaphase stages were frequently observed in all the treatments in both the varieties but were more frequent in combination treatments than the individual mutagenic treatments. The frequency of PMCs with bridges ranged from 1.11 to 1.70 % (EMS), 1.30 to 1.85 % (ϒ-rays) and 1.30 to 2.05 % (MMS) in var. minor, while the frequency of PMCs with bridges ranged from 0.90 to 1.45 % (EMS), 1.11 to 1.66 %(ϒ-rays), 1.09 to 1.84 % (MMS), 1.08 to 2.00 % (ϒ-rays + EMS) and 1.10 to 2.19 % (ϒ-rays + MMS) in var. major. The maximum frequency of PMCs with bridges was 2.14 % and 2.19 % at 20 Kr + 0.04 % MMS in var. minor and var. major, respectively.

Bridges with or without fragments at anaphase stages were observed in the present investigation. The bridges were also observed by Reddy and Annadurai (1992) in *Lens culinaris* by different mutagens; Khan et al. (1998a, 2007b) in *Vicia faba* by caffeine, DES and 8-HQ; Kumar et al. (2003) in *Lens culinaris* by combined treatment of gamma rays and EMS; Jabee and Ansari (2005) in *Cicer arietinum* by hydrazine sulphate (HS); Kumar and Rai (2007a) in *Zea mays* by EMS; Siddiqui et al. (2007) in *Trigonella foenum-graecum* by sodium azide (NaN3); and Singh and Chaudhary (2005) and Kumar and Gupta (2009) in *Capsicum annuum* L.

In the present study, occurrence of bridge with fragment may be due to paracentric inversion. The present findings are in agreement with the earlier results of Jayabalan and Rao (1987a, b) in tomato and Mitra and Bhowmik (1996) in *Nigella sativa.* Saylor and Smith (1966) suggested that the formation of chromatin bridges might be due to the failure of chiasmata in a bivalent to terminalise and the chromosomes get stretched between the poles. Bhattacherjee (1953) attributed bridge formation to interlocking of bivalent chromosomes and Sinha and Godward (1972) to paracentric inversions. The occurrence of breaks at the same locus and their lateral fusion leads to the formation of dicentric chromosome which is pulled equally to both the poles forming a bridge (Anis et al. 1998). Bridges can also be attributed to the general stickiness of chromosomes at metaphase which further led to their inability to separate or to the breakage and reunion of chromosomes (Ahmad 1993; Anis and Wani 1997; Kumar and Gupta 2009).

11.3.3 Unequal Separation

Unequal separation of chromosomes at anaphase I was noticed in all the treatments in both the varieties without any exception. The chromosomes segregated mostly in the ratio of 7:5 and 8:4. The combined treatments showed more frequency of PMCs with unequal separation than the individual mutagenic treatments. However, dose-dependent increase was observed in all the mutagens in both the varieties. The frequency of PMCs with unequal separation ranged from 0.40 to 0.93 % (EMS), 0.40 to 1.11 % (ϒ-rays), 0.40 to 1.30 % (MMS), 0.73 to 1.46 % (ϒ-rays + EMS) and 0.91 to 1.85 % (ϒ-rays + MMS) in var. minor, while the frequency of PMCs with unequal separation ranged from 0.20 to 0.72 % (EMS), 0.20 to 0.92 % (ϒ-rays), 0.36 to 1.10 % (MMS), 0.54 to 1.27 % (ϒ-rays + EMS) and 0.73 to 1.64 % (ϒ-rays + MMS) in var. major. The maximum frequency of PMCs with unequal separation was 1.85 % and 1.64 % observed at 20 Kr + 0.04 % MMS in var. minor and var. major, respectively.

Unequal separation of chromosomes towards poles at anaphase due to non-disjunction of homologous chromosomes at metaphase as observed during the present study was due to the stickiness of chromosomes and could result in unequal distribution of chromosomes in the daughter nuclei (Anis and Wani 1997). Mitra and Bhowmik (1996) reported that unequal separation of chromosomes was caused by spindle irregularities. Random movement of univalents to any one of the poles leads to unequal separation of chromosomes (Kumar and Singh 2003). Kumar and Rai (2007c) reported that unequal separation of chromosomes in meiosis I and II might be the result of the non-oriented bivalent formation due to spindle dysfunction or due to the formation of univalents at diakinesis or metaphase. It might have also occurred due to early or delayed separation of bivalents. Micronuclei generally arise from fragments and lagging chromosomes which fail to reach the poles and get included in the daughter nuclei (Kumar and Duby 1998b). Laxmi et al. (1975) suggested that irregular distribution of acentric fragments or laggards results in the formation of micronuclei at telophase resulting in variation in number and size of pollen grains obtained from the pollen mother cell. Micronuclei at dyad or tetrad stage of PMCs in mutagen-treated population might have resulted due to non-orientation of chromosomes and laggards since they were of frequent occurrence. Micronuclei lead to the loss of genetic material. Their presence, therefore, suggests that the resultant product of meiotic division is deficient in one or the other chromosome. This usually leads to the formation of sterile pollen grains.

11.3.4 Disturbed Polarity

The disturbed polarity was observed at telophase II in all treatments in both the varieties. However, the frequency was observed more in combined mutagenic treatments than the individual mutagenic treatments. The frequency of PMCs which showed disturbed polarity ranged from 1.39 to 2.24 % (EMS), 1.30 to 2.40 % (ϒ-rays), 1.50 to 2.42 % (MMS), 1.47 to 2.60 % (ϒ-rays + EMS) and 1.45 to 2.80 % (ϒ-rays + MMS) in var. minor, whereas the frequency of PMCs with disturbed polarity ranged from 0.90 to 1.81 % (EMS), 0.92 to 2.02 % (ϒ-rays), 1.09 to 2.02 % (MMS), 1.08 to 2.18 % (ϒ-rays + EMS) and 1.10 to 2.38 % (ϒ-rays + MMS) in var. major. The maximum frequency of PMCs with disturbed polarity ranged from 2.80 % and 2.38 % at 20 Kr + 0.04 % MMs in var. minor and var. major, respectively.

Disturbed polarity at telophase stages was recorded in certain percentage of the PMCs which could be due to spindle disturbance. Disturbed polarity was also reported by Sharma et al. (2004) in chickpea and Bhat et al. (2006a, b, c, d) in *Vicia faba* L.

11.3.5 Cytomixis

The cytomixis was observed in almost all the treatments in both varieties with some exceptions, viz. 0.1 % EMS in var. minor and 0.1 %, 0.2 % EMS and 10 Kr rays in var. major. The frequency of PMCs with cytomixis ranged from 0.36 to 0.60 % (EMS), 0.20 to 0.74 % (ϒ-rays), 0.36 to 0.93 % (MMS), 0.40 to 1.09 % (ϒ-rays + EMS) and 0.54 to 1.50 % (ϒ-rays + MMS) in var. minor, while in var. major 0.20 to 0.40 % (EMS), 0.36 to 0.55 % (MMS), 0.18 to 0.73 % (ϒ-rays), 0.18 to 0.91 % (ϒ-rays + EMS) and 0.36 to 1.28 % (ϒ-rays + MMS). The maximum frequency of PMCs with cytomixis was 1.50 % and 1.28 % at 20Kr + 0.04 % MMS in var. minor and var. major, respectively.

In the present investigation, very low frequency of pollen mother cells showed cytomixis at various stages of meiosis. Similar observation was reported by several workers during microsporogenesis (Sudan and Wafai 1987; Bahl and Tyagi 1988; Kaul 1990; Kumar and Sharma 2002; Bhat et al. 2006a, b, c, d; Kumar and Rai 2007a, b, c).

Among the factors proposed to cause the cytomixis are (i) influence of genes (Kaul and Nirmala 1993), (ii) abnormal formation of the cell wall during premeiotic divisions (Kamra 1960), (iii) action of agents such as colchicines (Sinha 1988), (iv) action of chemical mutagens (Bhat et al. 2006a, b, c, d), (v) rotenone (Amer and Mikhael 1986), (vi) changes in the biochemical process that involve microsporogenesis modifying the micro-environment of affected anthers (Kaul 1990), (vii) effect of gamma radiation resulting in an unbalanced and sterile genetic system (Amma et al. 1990) and (viii) due to the presence of a male-sterile mutant gene and its frequency altered by environmental factors (Nirmala and Kaul 1994) and environmental stress and pollution (Haroun et al. 2004). In the present investigation, the occurrence of cytomixis may be due to action of mutagens used (Sinha 1988).

Cytomixis may lead to the production of aneuploid plants with certain morphological peculiarities (Sheidai et al. 1993) or produce gametes with no reduction in chromosomal number as reported in *Aegilops* (Sheidai et al. 2002) having evolutionary importance which leads to the plants with polyploidy chromosomal level. The change in chromosomal level is also responsible for reduction in pollen fertility.

Cytomixis was first observed by Korniche (1901) in pollen mother cells (PMCs) of *Crocus sativus* and then by Gates (1911), who studied the PMCs of *Oenothera gigas* and defined it as process of transmigration of chromatin from one cell to the adjoining cell and coined the term cytomixis. The cytomixis is a common phenomenon reported more commonly during gametogenesis in several plants (Mc Clintock 1929; Bahl and Tyagi 1988; Koul 1990; Yen et al. 1993; Kumar and Sharma 2002; Haroun et al. 2004; Bhat et al. 2006b). In addition, the cytomixis has also been observed in somatic cells (Bowes 1973; George and Geethamma 1985) and in the interphase between mitotic and meiotic cells (Cooper 1952). In legumes, intermigration of chromatin has been reported in many genera, e.g. *Vigna* (Sen and Bhattacharya 1988), *Lathyrus* (Kumar and Sinha 1991; Seijo 1996), *Pisum* (Gottschalk 1970; Nirmala and Kaul 1994) and *Vicia faba* L (Bhat et al. 2006b).

The connection between the cells adjacent to each other is determined by the frequency and intensity of cytomictic flow. The pollen mother cells showed two types of connections, i.e. direct fusion and cytoplasmic channels. The PMCs showed cytomixis at various stages of cell division, and mostly the cells showed single cytoplasmic channel; the multiple cytoplasmic channels were also observed though rarely. The first phase of meiosis showed less abundant cytoplasmic channels than the second phase, but the cells at different phases of cell division were also evident. Some PMCs showed cytomixis to other PMCs both by cytoplasmic channel and direct fusion simultaneously. Similar observations have been reported by many workers (Bahl and Tyagi 1988; Seijo 1996 and Haroun et al. 2004).

The breadth of cytomictic channels were of varying dimensions showing transfer of chromosomes from one cell to other cell/cells at different stages of meiosis. The multiple meiocytes usually numbering from four to nine were involved in cytomixis simultaneously, though in some cases only two PMCs were found involved in cytomixis. It was common to find two to four cytoplasmic channels emerging from a single meiocyte and connecting many cells together. There was unidirectional flow of chromatin material from donor to a recipient cell. The chromatin material was found migrating to different adjoining pollen mother cells from a single meiocyte simultaneously. In case of PMCs simultaneously showing cytomictic channels, the chromosomes passed from one meiocyte to the second and from second to the third and so on. The unidirectional migration of chromatin material has been observed by Gottschalk (1970). However, in this investigation, a single pollen mother cell was found donating its chromatin materials to two PMCs through two independent cytoplasmic channels simultaneously. This type of chromatin migration is unlike to the findings observed by Gottschalk (1970). Some donor cells became empty after passing the entire chromatin materials to the recipient cells. The partial migration of the chromatin material resulted in the formation of aneuploid cells. In the present investigation, in some cases, no visual transfer of chromosomes through cytoplasmic connections could be detected. The formation of cytoplasmic channels between two or more cells at different places in the cells suggests that screening of EMS, gamma rays, MMS and their combination treated population for varied ploidy levels and its utilisation in plant breeding may produce the polyploids.

Some recipient cells were found with additional number of chromosomes which did not show synapses with the chromosomes of recipient cells but showed synapses among themselves and went on divisions separately. This is similar with the findings of Pantulu and Manga (1972). The fate of such additional number of chromosomes is unknown, but they may result in the formation of either micronuclei or micropollen. Sarvella (1958) reported loss of chromosomes in meiotic and somatic cells in *Gossypium.* Thakur (1978) opines that a loss or gain of one or more chromosomes has two possibilities. The gametes with deficient chromosomes will be lethal and get eliminated, and gametes with different number of chromosomes may result in the formation of aneuploids. In the condition where all the chromosomes will move, it results in the formation of aneuploids. Sometimes chromatin material moves without nucleolus, and sometimes nucleolus moves without chromatin material. The movement of chromosomes in between and among the meiocytes ranges from one to six. It signifies that chromosome movement breaks at different intervals at different stages of meiosis, and the PMCs showing cytomixis were also at different stages.

The early stages of meiosis showed more frequent PMCs involved in cytomixis, and gradual decline was observed towards the end of meiosis; same findings were observed earlier by Maheshwari (1950) and Bhat et al. (2006a, b, c, d) but in contrary to the earlier findings of Verma et al. (1984) that all the stages of meiosis are equally susceptible to cytomixis.

During observation, the larger or smaller cells with or without nucleus were also observed than the normal. This is possibly because of additional chromatin material resulting due to transfer and accumulation and by their retransfer to the subsequent cells. That could be the reason for the presence of hexapolar anaphase II, pentads and hexads. The presence of cytomixis may be because of abnormal genetic behaviour as the dose-dependent chromosomal stickiness and cytomixis were observed with the physical, chemical and combined mutagenic treatments. We consistently failed to observe chromosomal stickiness, and cytomixis in controlled plants justifies our view. The decrease in frequency of meiocytes showing deviation in basic chromosome number was observed in the higher concentration-treated plant population, but the cytoplasmic connection between the PMCs was of frequent occurrence which suggests that stickiness inhibits the chromosome migration from one cell to another.

Migration of chromatin material among the adjacent PMCs also occurred through cytoplasmic connections originating from the plasmodesmata formed in the tissues of the anther. The plasmodesmata become closed by the deposition of secondary cell wall depositions, but in some instances, they persisted during microsporogenesis and increase in dimensions, forming evident intermeiocyte connections that permitted the transfer of chromosomes (Falistocco et al. 1995a, b).

11.4 Chromosomal Aberrations at Telophase I and II

11.4.1 Micronuclei

Micronuclei were observed at telophase I/II stages but were more frequent at telophase II. The micronuclei were observed in all the treatments in var. minor with a single exception (0.1 %EMS), but in var. major, the micronuclei were not observed in 0.01 %EMS and at 10 Kr ϒ-rays. The frequency of PMCs with micronuclei ranged from 0.55 to 1.12 % (EMS), 0.20 to 1.70 % (ϒ-rays), 0.36 to 1.50 % (MMS), 0.40 to 1.64 % (ϒ-rays + EMS), 0.54 to 1.85 % (ϒ-rays + MMS) in var. minor and 0.40 to 0.90 % (EMS), 0.54 to 1.10 % (ϒ-rays), 0.18 to 1.28 % (MMS), 0.18 to 1.45 % (ϒ-rays + EMS) and 0.36 to 1.64 % (ϒ-rays + MMS) in var. major. The highest frequency of PMCs with micronuclei was 1.85 % and 1.64 % at 20Kr + 0.04 % MMS in var. minor and var. major, respectively.

Micronuclei generally arise from fragments and laggards which did not reached the poles and got assimilated in the daughter nuclei (Kumar and Dubey 1998b). Suggested that irregular distribution of acentric fragments or laggards result in the formation of micronuclei at telophase resulting in variation in number and size of pollen grains obtained from the pollen mother cell. Micronuclei at dyad or tetrad stage of PMCs in mutagen-treated population might have resulted due to non-orientation of chromosomes and laggards since they were of frequent occurrence. Micronuclei lead to the loss of genetic material. Their presence, therefore, suggests that the resultant product of meiotic division is deficient in one or the other chromosome. This usually leads to the formation of sterile pollen grains. Multinucleate condition may be explained to a particular genotypic change suppressing the organising capacity of nucleolar chromosome and inducing the formation of adventitious nucleoli.

The occurrence of large pollen grains (possibly 2n pollen grains) with a frequency of 1.15 % was found in treated population. An unreduced diploid gamete is a meiotic product which shows the sporophytic rather than the gametophytic chromosome number. Such gametes are the result of abnormalities during microsporogenesis (2n pollen). Unreduced gametes are known to produce individuals with higher ploidy level through sexual hybridisation (Villeux 1985). According to Harlan and deWet (1975), sexual polyploidisation is the main route to the production of naturally occurring polyploids.

Intervarietal differences in the radiosensitivity have been reported by many workers (Kumar and Dubey 1998a, b; Dhamyanthi and Reddy 2000; Zeerak 1990; Singh 2003; Bhat et al. 2006a). These differences were attributed to differences in cell volume, nuclear volume, chromosome volume and DNA amount and the presence of protective or sensitising substances (Sparrow and Evans 1961; Ahmad and Godward 1981; Bhat et al. 2005b).

Most of the workers have in general concluded that gamma rays were more effective than chemical mutagens in causing chromosomal abnormalities. However, in the present study, the frequency of aberrations was greater in the combination treatments than in the individual treatments. It is, therefore, concluded that EMS and MMS too are able to induce a sufficient amount of meiotic aberrations. Singh et al. (1989) and Roy (1989) in lentil reported that EMS is highly efficient in inducing chromosomal aberrations equal to that of physical mutagens. Among different stages of meiosis, the frequency of meiotic aberrations was maximum at metaphase stage in the present study. Similar observations were reported by Mitra and Bhowmik (1996) in *Nigella sativa* and Kumar and Dubey (1998c) and Bhat et al. (2006a, b, c, d) in *Lathyrus sativus* L.

For most of the combination treatments, the coefficient of interaction (K) was less than additive; however, additive or synergistic effect was obtained at 20kR + 0.2 % EMS in var. major in M_1 plants and 20kR + 0.2 % EMS in both var. minor and major in M_2 and M_3 plants. Similar results have been reported by Sharma et al. (2004) in chickpea. Less than additive effects for chromosomal aberrations were also reported by Khalatkar and Bhatia (1975) in barley and by Bhamburkar and Bhalla (1985) in black gram.

11.4.2 Multinucleate Condition

Multinucleate condition was observed at telophase I/II, but their frequency was more at telophase II. The multinucleate condition was observed almost in all treatments except 0.1 % EMS in both varieties. The frequency of PMCs with multinucleate condition ranged from 0.36 to 1.12 %(EMS), 0.55 to 1.50 % (ϒ-rays), 0.92 to 1.70 % (MMS), 0.91 to 1.83 % (ϒ-rays + EMS)) and 1.09 to 2.23 % (ϒ-rays + MMS) in var. minor, whereas the frequency of PMCs with multinucleate condition ranged from 0.20 to 0.90 % (EMS), 0.40 to 1.29 %(ϒ-rays), 0.73 to 1.47 % (MMS), 0.72 to 1.64 % (ϒ-rays + EMS) and 0.92 to 2.01 % (ϒ-rays + MMS) in var. major. The maximum frequency of multinucleate condition was 2.23 % and 2.01 % at 20Kr + 0.04 % MMS in var. minor and var. major, respectively.

11.4.3 Non-disjunction

Non-disjunction of chromosomes was observed in all the mutagenic treatments in var. minor, while in var. major, some lower concentrations/doses of individual mutagenic treatments did not show the non-disjunction (0.1 % EMS,0.2 % EMS, 10 Kr ϒ-rays and 0.01 % MMS). The frequency of non-disjunction ranged from 0.18 to 0.74 % (EMS), 0.20 to 0.92 % (ϒ-rays), 0.20 to 1.12 % (MMS), 0.40 to 1.28 % (ϒ-rays + EMS) and 0.40 to 1.50 % (ϒ-rays + MMS) in var. minor, while the frequency of PMCs with non-disjunction of chromosomes ranged from 0.20 to 0.40 % (EMS), 0.36 to 0.55 % (ϒ-rays), 0.55 to 0.92 % (MMS), 0.18 to 1.09 % (ϒ-rays + EMS) and 0.18 to 1.28 % (ϒ-rays + MMS) in var. major.

11.5 Chiasma Frequency

A number of PMCs from treated as well as untreated (control) plants were studied and chiasmata per cell was calculated (Table). Results pertaining to the frequency of chiasmata per cell in the treated plants as compared to their respective controls clearly indicate that the chiasmata per cell decreased more in combined treatments with the increase of univalents and rod bivalents than the individual mutagenic treatments. The pooled mean of chiasmata per cell showed 19.05 (EMS), 19.18 (ϒ-rays), 18.85 (MMS), 18.83 (ϒ-rays + EMS) and 18.60 (ϒ-rays + MMS) in var. minor, while in var. major the chiasmata per cell was 19.30 (EMS), 18.80 (ϒ-rays), 18.70 (MMS), 17.80 (ϒ-rays + EMS) and 17.40 (ϒ-rays + MMS).

The pooled mean frequency of univalents was 0.55 (EMS), 0.73 (ϒ-rays), 0.90 (MMS), 1.08 (ϒ-rays + EMS) and 1.08 (ϒ-rays + MMS) in var. minor, whereas in var. major the pooled mean frequency of univalents was 0.45 (EMS), 0.73 (ϒ-rays), 0.80 (MMS), 1.20 (ϒ-rays + EMS) and 1.98 (ϒ-rays + MMS). The pooled mean frequency of multivalents was 0.65 (EMS), 0.83 (ϒ-ray), 0.95 (MMS), 1.28 (ϒ-rays + EMS) and 1.30 (ϒ-rays + MMS) in var. minor, while in var. major the pooled mean frequency of multivalents was 0.55 (EMS), 0.88 (ϒ-rays), 0.95 (MMS), 1.18 (ϒ-rays + EMS) and 1.25 (ϒ-rays + MMS). The pooled mean frequency of rod bivalents was 1.0 (EMS), 1.20 (ϒ-rays), 1.03 (MMS), 1.13 (ϒ-rays + EMS) and 1.18 (ϒ-rays + MMS). The pooled mean of ring bivalents was 3.55 (EMS), 3.65 (ϒ-rays), 3.20 (MMS), 3.13 (ϒ-rays + EMS) and 3.51 (ϒ-rays + MMS) in var. minor, while the pooled mean of ring bivalents was 4.08 (EMS) ,3.25 (ϒ-rays), 3.26 (MMS), 2.80 (ϒ-rays + EMS) and 2.63 (ϒ-rays + MMS).

In the present investigation, a dose-dependent decrease in the chiasmata frequency was observed in various treatments. The average frequency of chiasmata per cell decreased with an increase in univalents and rod bivalents in each treatment. The reduction in chiasma frequency can be attributed to an increase in rod bivalents and univalents (Anis et al. 2000). Similar observations were reported by Bennett and Rees (1970) in rye and Sadanandam and Subash (1985) in *Capsicum annuum.*

Chiasma formation characterises the pairing of homologous chromosomes at meiosis and controls the degree of recombination. Chiasma counting is the most straightforward method of scoring the total no. of crossing-over events in the genome (Baptista-Giacomelli et al. 2000b). The decrease in the chiasma frequency denotes the induced heterology due to induced damage or changed loci of genes or intra-/intergenic disturbances following the mutagenic treatments. In the present study, the decrease in chiasma frequency was relatively less in M_2 and M_3 generations in comparison to M_1 generation, indicating that some sort of recovery mechanisms must have been operating in the intervening period.

According to Lawrence (1961), the decrease in chiasma frequency following mutagenic treatments might possibly occur at two stages: (i) during DNA synthesis and (ii) during sensitive period at/or slightly before the stage of chiasmata formation. In the former case, the decrease in frequency of chiasmata may be due to disturbances in chromosome coiling, restricted pairing at pachytene and the delay in DNA synthesis, while in the latter, it may be affecting the process leading to chiasmata formation. Sadanandan and Subhash (1985) attributed it to the nature and potency of mutagens and also the underlying factors, such as complex structural changes or the changes in the nature of gene responsible for chiasmata formation. The alteration of chiasmata in the treated plants might also be due to the failure of complete pairing or rapid terminalisation of chiasmata in the bivalents (Tabassum 2002). The decrease in chiasma frequency may also be attributed to the changes at chromosomal/DNA level, such as deletion, inversion, duplication and translocation (Siddique and Ansari 2005).

Comparative mutagenecity of EMS, gamma rays, MMS and their combinations in the two varieties, viz. minor and major of broad bean reflect the differences in their genome architecture as the former exhibited much induced biological damage than the latter. Thus, it can be inferred that mutagenecity is ultimately determined by genome itself. Khamankar (1984) working with *Lycopersicon esculentum* reported that the rate of mutations was different with mutagens at certain loci. Some of the gene loci affected by one mutagen were not necessarily affected by the other. However, the results of the present study differ from the results of some earlier workers who have found that bold seeded types were more sensitive to mutagens than the small seeded types (Sharma 1986 and Reddy et al. 1992).

There are many reports to demonstrate the effect of physical and chemical mutagens and their combination treatments on different biological parameters such as germination, survival, injury, sterility, etc. (Khan 1999; Mitra and Bhowmik 1999; Sareen and Koul 1999). Reduction in seedling height following treatments with gamma rays and EMS was observed in barley (Sharma 1970). Gupta and Yashvir (1975) reported a radioprotective effect of EMS in *Abelmoschus esculentus*. The combined treatments of gamma rays and EMS showed higher germination percentages than the corresponding EMS treatments. Choudhary (1983) reported a symmetric reduction in germination in different varieties of wheat with higher doses of gamma rays.

11.6 Effect of Mutagens on Biological Parameters

Seed germination, seedling height and plant survival decreased with the increase in mutagenic treatments in the present investigation. However, the extent of decrease differed in both the varieties treated with different concentrations of the mutagens and their combination treatments. In general, the combination treatments were found to be more effective than the individual treatments. Based on the extent of damage, it was found that var. minor was more sensitive than var. major.

Several workers have attempted to explain the cause responsible for inhibition of seed germination. Kleinhofs et al. (1978) in barley reported delay in the inhibition of metabolism following germination which in turn resulted in a uniform delay in mitotic activity, seedling growth and ATP and DNA synthesis. Tarar and Dnyansagar (1983) reported in *Turnera ulmifolia* that inhibitory effect of gamma rays was due to damage of DNA. Reduction in seed germination in mutagenic treatments has been explained due to delay or inhibition in physiological and biological processes necessary for seed germination which include enzyme activity (Kurobane et al. 1979), hormonal imbalance (Chrispeeds and Varner 1976) and inhibition of mitotic process (Ananthaswamy et al. 1971).

Intervarietal differences in the radiosensitivity have been reported by many workers (Kumar and Dubey 1998a, b; Dhamayanthi and Reddy 2000; Zeerak 1990; Singh 2003; Bhat et al. 2006a). These differences were attributed to differences in cell volume, nuclear volume, chromosome volume and DNA amount and presence of protective or sensitising substances (Sparrow and Evans 1961; Ahmad and Godward 1981; Bhat et al. 2005b).

In the present study the frequency of aberrations was greater in the combination treatments than in the individual treatments, it is therefore, concluded that EMS and MMS too are able to induce a sufficient amount of meiotic aberrations. Singh et al. (1989) and Roy (1989) in lentil reported that EMS is highly efficient in inducing chromosomal aberrations equal to that of physical mutagens. Among different stages of meiosis, the frequency of meiotic aberrations was maximum at metaphase stage in the present study. Similar observations were reported by Mitra and Bhowmic (1996) in *Nigella sativa*, Kumar and Dubey (1998c) in *Lathyrus sativus* L. Bhat et al. (2005a, 2007a, b) in *Vicia faba* L.

On the basis of the present investigation and discussion, it has been concluded that the cytomorphological variations observed in the present study are due to the above-mentioned genic disturbances induced by the action of chemical mutagens, along with their interactions with environment. The physiological, biochemical and metabolic changes might have indirectly affected the treated plants due to the disturbances at chromosomal and genic level. But selfing each variant in M_1, followed by selection and selfing in M_2 and M_3 generations, eliminates, up to maximum extent, the possibility and overlapping role of other factors and concentrates to stable genetic changes in the mutants obtained.

It has also been revealed that lower and moderate treatments of mutagens used in the present investigation proved to be efficient in increasing the genetic variability for yield-oriented selection in *Vicia faba* L. The isolated mutants possess desirable plant architecture.

Acknowledgement We are thankful to all those scholars, authors and scientists who provided significant inputs and valuable suggestions during the preparation of this manuscript.

References

Agarwal R, Ansari MYK (2001) The effect of aniline on root tip cells of *Vicia faba* L. J Cytol Genet 2:129–134

Ahmad S (1993) Meiotic studies in two cultivars of *Cicer arietinum* L. after gamma irradiation. Cytologia 58:61–65

Ahmad S, Godward MBE (1981) Comparison of radio-resistant with a radio-sensitive cultivar of *Cicer arietinum* L. II. Differences in the number of chromosome aberrations at the same dose. Environ Exp Bot 21:143–151

Alka AMYK, Bhat TM, Choudhary S, Aslam R (2012) Genotoxic effect of ethylmethane sulphonate and sodium azide in *Linum Usitatissimum* L. Int J Plant Anim Environ Sci 1:1–6

Amer SM, Mikhael E (1986) Cytological effects of pesticides. XVI. Effects of the insecticide rotenone on root mitosis of Vicia faba. Cytologia 51:171–176

Amma CKS, Namboodiri AN, Panikkar AON, Sethuraj MR (1990) Radiation induced male sterility in *Hevea brasiliensis* (Willd ex Adr. De Juss) Muell. Arg. Cytologia 55:547–551

Ananthaswamy HN, Vakil UK, Srinivasan A (1971) Biochemical and physiological changes in gamma irradiated wheat during germination. Radiat Bot 11:1–12

Anis M, Wani AA (1997) Caffeine induced morpho-cytological variability in Fenugreek, *Trigonella foenum-graecum* L. Cytologia 62:343–349

Anis M, Sharma N, Abbasi N (1998) Colchicine induced autoteraploidy in *Capsicum frutescens* var. Suryamukhi. J Cytol Genet 33:183–187

Anis M, Abbasi N, Sharma N (2000) Cytological effect of mutagens in *Capsicum annuum*. J Cytol Genet 1:17–21

Aslam R, Choudhary S, Ansari MYK, Alka HI, Bhat TM (2012) Assessment of genetic variability induced by MMS and 6-AP in a medicinal herb *Cichorium intybus* L. Biosci Int 1:30–39

Bahl SR, Tyagi BR (1988) Cytomixis in pollen mother cells of *Papaver dubium* L. Cytologia 53:771–775

Baptista-Giacomelli FR, Pagliarini MS, De Almeida JL (2000a) Meiotic behaviour in several Brazilian oat cultivars (*Avena sativa* L.). Cytologia 65:371–378

Baptista-Giacomelli FR, Pagliarini MS, De Almeida JL (2000b) Chiasma frequency, distribution and terminalization in hexaploid oats (*Avena sativa* L.). Acta Sci 22:269–273

Bennett MD, Rees H (1970) Induced variation in chiasma frequency in Rye in response to phosphate treatments. Genet Res 16:325–331

Bhamburkar S, Bhalla JK (1985) Effects of differential and combined treatments of gamma rays, hydrazine and mitomycin-C in black gram. J Indian Bot Soc 8:82–87

Bhat TA, Wani AA (2015) Studies on Ethyl Methane Sulphonate induced Desynapsis in *Vicia faba* L. Eur J Acad Essays 2(1):23–28

Bhat TA, Khan AH, Parveen S (2005a) Comparative analysis of meiotic abnormalities induced by gamma rays, EMS and MMS in Vicia faba L. J Indian Bot Soc 84:45–48

Bhat TA, Parveen S, Khan AH (2005b) Clastogenic effect of EMS and MMS in *Vicia faba* L. J Cytol Genet 6:117–122

Bhat TA, Khan AH, Parveen S, Ganai FA (2005c) Clastogenic effects of EMS and MMS in *Vicia faba* L. J Cytol Genet 6:117–122

Bhat TA, Khan AH, Parveen S (2006a) Effect of EMS on certain cytomorphological parameters in two varieties in *Vicia faba* L. Adv Plant Sci 19:227–232

Bhat TA, Khan AH, Parveen S, Ganai FA (2006b) Studies on the effect of EMS on meiosis and pollen fertility in *Vicia faba* L. Adv Plant Sci 19:243–247

Bhat TA, Khan AH, Parveen S, Wani NA (2006c) Radiation Induced pod and seed mutants and MMS induced closed flower mutant in broad been (*Vicia faba* L.). Indian J Genet 66(3):249–250

Bhat TA, Parveen S, Khan AH (2006d) MMS-induced cytomixis in pollen mother cells of Broad Bean (*Vicia faba* L). Turk J Bot 30:273–279

Bhat TA, Parveen S, Khan AH (2007a) Meiotic studies in two varieties of *Vicia faba* L. (Fabaceae) after EMS treatment. Asian J Plant Sci 6:51–55

Bhat TA, Parveen S, Khan AH (2007b) Meiotic studies in two varieties of *Vicia faba* L. (fabaceae) after MMS treatment. Asian J Plant Sci 61(1):51–55

Bhat TA, Sharma M, Anis M (2007c) Comparative analysis of meiotic aberrations induced by Diethylsulphate and Sodium Azide in Broad Bean (*Vicia faba* L.). Asian J Plant Sci 6(7):1051–1055

Bhat TA, Sharma M, Anis M (2007d) Comparative analysis of mitotic aberrations induced by Diethyl Sulphide (DES) and Sodium Azide (SA) in *Vicia faba* L. (Fabaceae). Pak J Biol Sci 10(5):783–787

Bhattacharjee SK (1953) Cytogenetics of *Lens esculenta* Monesh. Caryologia 5:159–166

Bond DA, Lawes DA, Hawtin GC, Saxena MC Stephens JS (1985) Faba bean (*Vicia faba* L.). In: Summerfield RJ, Roberts EH (eds) Grain legume crops. William Collins Sons Co. Ltd, London, pp 199–265

Bose S, Saha A (1970) Cytogenetic effects of X-rays and chemicals on tomato studies in M_1 generation. Nucleus 13:29–40

Bowes BG (1973) Note on apparent case of cytomixis in the root apex of *Allium cepa*. Cytologia 38:125–129

Chaghtai SA, Hasan Z (1979) Effect of EMS, MES and MMS on chromosomal behaviour of *Lens esculenta* during microsporogenesis. Biol Plant 21:280–283

Choudhary RK (1983) A note on effect of gamma irradiation in wheat. WIS 57:21–23

Chrispeeds MJ, Varner JE (1976) Gibberellic acid induced synthesis and release of L-amylase and ribonuclease by isolated barley aleurone layers. Plant Physiol 42:396–406

Cooper DD (1952) The transfer of deoxyribose nucleic acid from the tapetum to the microsporocytes at the onset of meiosis. Am Nat 86:219–229

Cubero JI (1973) Evolutionary trends in *Vicia faba*. Theor Appl Genet 43:59–65

Das SK, Roy SK (1989) Radiocytogenetical studies on *Solanum* I. Meiotic abnormalities. Cytologia 54:477–481

Defani-Scoarize MA, Pagliarini MS, Aguiar CG (1995) Case of partial male sterility in an inbred maize line. Cytologia 60:311–318

Dhamayanthi KPM, Reddy VRK (2000) Cytogenetic effects of gamma rays and ethyl methane sulphonate in chilli pepper (*Capsicum annuum* L.). Cytologia 65:129–133

Duke JA (1981) Handbook of legumes of world economic importance. Plenum Press, New York, pp 199–265

Evans HJ (1962) Chromosome aberrations induced by ionizing radiations. Int Rev Cytol 13:221–321

F.A.O. (1994) Production year book-Rome, Italey

Falistocco E, Tosti T, Falcinelli M (1995a) Cytomixis in pollen mother cells of diploid Dactylis, one of the origins of 2n gametes. Heredity 86:448–453

Falistocco E, Tosti T, Falcinelli M (1995b) Microsporogenesis in desynaptic mutant of diploid Dactylis. Cytologia 59:309–316

Ganai FA, Khan AH, Bhat TA, Parveen S, Wani NA (2005) Cytogenetic effects of Methyl Methane Sulphonate (MMS) on varieties of Chickpea (*Cicer arietinum* L.). J Cytol Genet 6:97–102

Gates RR (1911) Pollen formation in *Oenothera gigas*. Ann Bot 25:909–940

George K, Geethamma S (1985) Role of cytomixis in the speciation of *Jasminum*. Curr Sci 54:586–587

Gottschalk W (1970) Chromosome and nuclear migration during microsporogenesis of *Pisum sativum*. Nucleus 13:1–9

Gottschalk W, Kaul MLH (1980a) Asynapsis and desynapsis in flowering plants. Asynapsis. Nucleus 23:1–15

Gottschalk W, Kaul MLH (1980b) Asynapsis and desynapsis in flowering plants. Desynapsis. Nucleus 23:97–120

Gottschalk W, Kleine HD (1976) The influence of mutated genes on sporogenesis. A survey on the genetic control of meiosis in Pisum sativum. Theor Appl Genet 48:23–34

Gulfishan M, Khan AH, Bhat TA (2010) Studies on cytotoxicity induced by DES and SA in *Vicia faba* L. var. major. Turk J Bot 34:31–37

Gulfishan M, Khan AH, Haneef I, Bhat TA (2011) Genotoxic effect of Diethyl Sulphate in two varieties of *Capsicum annuum* L. Nucleus 54(2):107–111

Gulfishan M, Khan AH, Jafri IF, Bhat TA (2012) Assessment of mutagenicity induced by MMS and DES in *Capsicum annuum* L. Saudi J Biol Sci 19:251–255

Gulfishan M, Jafri IF, Khan AH, Bhat TA (2013) Methyl Methane Sulphonate induced desynaptic mutants in *Capsicum annuum* L. Chromosome Bot 8:59–62

Gupta PK (1975) Induced mutations in foxtail millet (*Setaria italica* Beauv). 1. Chlorophyll mutations induced by gamma rays, EMS and DES. TAG Theor Appl Genet 45:242–249

Gustafsson JP (1983) Cytogenetics of triticale. In: Swaminathan MS, Gupta PK, Sinha U (eds) Cytogenetics of crop plants. Macmillan, New Delhi, pp 225–250

Harlan JR, Dewet JMJ (1975) On *O. winge* and a prayer: the origins of polyploidy. Bot Rev 41:361–390

Haroun SA, Al Shehri AM, Al Wadie HM (2004) Cytomixis in the microsporogenesis of *Vicia faba* L. (Fabaceae). Cytologia 69:7–11

Health MC, Pilbeam CJ, Mckenzie BA, Hebblethwaite PD (1994) Plant architecture, competitive ability and crop productivity in food legumes with particular emphasis on pea (Pisum sativum L.) and faba bean (Vicia faba L.). In: Muehlbauer FJ, Kaiser WJ (eds) Expanding the production and use of cool season Food Legumes. Kluwer Academic Publishers, Dordrecht, pp 771–790

Jabee F, Ansari MYK (2005) Mutagenic effectiveness and efficiency of hydrazine sulphate (HS) in inducing cytomorphological mutations in *Cicer arietin*um L. var. K-850. J Cytol Genet 6:161–166

Jabee F, Ansari MYK, Shahab D (2008) Studies on the effect of maleic hydrazide on root tip cells and pollen fertility in *Trigonella foenum-graecum* L. Turk J Bot 32:337–344

Jafri IF, Khan AH, Sharma M, Gulfishan M (2012) Comparative mutagenicity of EMS and HZ *in Hordeum vulgare* L. var. RD2052. Chromosome Bot 7:79–83

Jambunathan R, Blain HL, Dhindsa KH, Hussain LA, Korgure K, Lijuan L, Youssef MM (1994) Diversifying use of cool season food legumes through processing. In: Muehlbauer FJ, Kaiser WJ (eds) Expanding the production and use of cool season food Legumes. Kluwer Academic Publisher, Dordrecht, pp 98–112

Jayabalan N, Rao GR (1987a) Gamma radiation induced cytological abnormalities in *Lycopersicon esculentum* Mill. Var. Pusa Ruby. Cytologia 52:1–4

Jayabalan N, Rao GR (1987b) Effect of physical and chemical mutagens on seed germination and survival of seedling in *Lycopersicon esculentum* Mill. J Indian Bot Soc 10:133–137

Kallo Z (1972) Chromosomal alterations in mitotic and meiotic system as influenced by gamma rays in Pisum. Cytologia 37:643–651

Kaltsikes PJ (1973) Early seed development in *triticale*. Can J Bot 51:2291–2300

Kamra OP (1960) Chromatin extrusion and cytomoixis in pollen mother cells of *Hordeum*. Heriditas 46:415–421

Katiyar RB (1978) Radiocytogenetical studies on *Capsicum* L. Meiotic anomalies. Cytologia 43:415–421

Kaul KK (1990) Cytomixis in Pollen mother cells of *Alopecurus arundinaceus* poir. Cytologia 55:169–173

Kaul MLH, Nirmala C (1993) Male sterility in Pea II. Male sex specific desynapsis. Cytologia 58:67–76

Kay D (1979) Crop and product digest no. 3-food legumes. Tropical Products Institute, London, pp 26–47

Khalatkar AS, Bhatia CR (1975) Synergistic effect of combined treatments of gamma radiation and ethylmethane sulphonate in barley. Radiat Bot 15:223–229

Khamankar YG (1984) Differential sensitivity to tomato genes to chemical and physical mutagens. Indian J Genet Plant Breed 44:80–83

Khan MN (1999) Mutagenic effectiveness and efficiency of EMS, gamma rays and their combinations in blackgram (Vigna mungo L. Hepper). Ad. Plant Sci 12:203–205

Khan AA, Ansari MYK, Khursheed T, Kalra N (1998) Studies on the effect of 6-benzyl amino purine on meiosis in faba bean (*Vicia faba* L.). Adv Plant Sci 11:121–125

Khan AA, Ansari MYK, Bhat TA, Shahab D (2007a) Clastogenic effect of 8-hydroxy quinoline in *Vicia faba* L. Adv Plant Sci 20:85–87

Khan AA, Ansari MYK, Shahab D, Bhat TA (2007b) Comparative analysis of meiotic abnormalities induced by caffeine and DES in *Vicia faba* L. Adv Plant Sci 20:73–76

Kleinhofs A, Owais WM, Nilan RA (1978) Azide. Mutat Res 55:165–195

Koduru PKR, Rao MK (1981) Cytogenetics of synaptsic mutants in higher plants. Theor Appl Genet 59:197–214

Korniche M (1901) Uber ortsveranderrung von Zellkarnern. S.B. Niedernhein, Ges. Natur-und Heilkunde Boun, pp 14–25

Koul KK (1990) cytomixis in pollen mother cells of *Alopecurus arundinaceus* Poir. Cytologia 55:169–173

Kumar S, Dubey DK (1998a) Effect of separate and simultaneous application of gamma rays and EMS on germination, growth, fertility and yield in cultivars Nirmal and LSD-3 of khesari (*Lathyrus sativus* L.). J Phytol Res 11:165–170

Kumar S, Dubey DK (1998b) Mutagenic efficiency and effectiveness of separate and combined treatments with gamma rays, EMS and DES in khesari (*Lathyrus sativus* L.). J Indian Bot Soc 77:1–4

Kumar S, Dubey DK (1998c) Effect ofgamma rays, EMS and DES on meiosis in Lathyrus sativus. J Cytol Genet 33:139–147

Kumar G, Gupta P (2009) Induced karyomorphological variations in three phenodeviants of *Capsicum annuum* L. Turk J Biol 33:123–128

Kumar G, Rai PK (2007a) EMS induced karyomorphological varation in maize (*Zea mays* L.) inbreds. Turk J Biol 31:187–195

Kumar G, Rai PK (2007b) Comparative genotoxic potential of mercury and cadmium in soybean. Turk J Biol 31:13–18

Kumar G, Rai PK (2007c) Colchiploidy in soybean (*Glycine max*). J Cytol Genet 8:17–23

Kumar G, Sharma V (2002) Induced cytomixis in chickpea (*Cicer arietinum* L.). Nucleus 45:24–26

Kumar G, Singh V (2003) Genetic analysis of induced desynaptic mutant in *Vigna mungo* L. var. PDU.1. Nucleus 46:107–109

Kumar G, Sinha SSN (1991) Genotoxic effects of two pesticides (roger and Bavistin) and antibiotic (Streptomycin) in meiotic cells of grass pea (Lathyrus sativus). Cytologia 56:209–214

Kumar G, Tripathi A (2003) Comparative mitotoxicity and mutagenecity of two heavy metals in *Allium Cepa* L. J Cytol Genet 4:169–173

Kumar G, Tripathi A (2004) Mutagenic response of caffeine in *Capsicum annuum* L. J Indian Bot Soc 83:136–140

Kumar G, Kesarwani S, Sharma V (2003) Clastogenic effect of individual and combined treatment of gamma rays and EMS in *Lens culinaris*. J Cytol Genet 4:149–154

Kurobane I, Yaroaguehi H, Sunder C, Nilan R (1979) The effects of gamma irradiation on the production and secretion of enzymatic activities in barley. Environ Exp Bot 19:75–84

Ladizinsky G (1975) On the origin of the broad bean *Vicia faba* L. Israel J Bot 24:80–88

Lawrence CW (1961) The effects of irradiation of different stages in microsporogenesis on chiasma frequency. Heredity 16:83–89

Laxmi V, Singh RB, Singh BD, Singh RM (1975) Meiotic anomalies induced by gamma rays and ethyl methane sulphonate treatments in *pearl millet*. Indian J Exp Biol 13:465–467

Lea DE (1955) Action of radiations in living cells. Second edn. Cambridge University Press, Cambridge, p 416

Lewis KR, John B (1966) The meiotic consequences of spontaneous chromosomes breakage. Chromosoma 18:287–308

Maheswari P (1950) An introduction to the embryology of angiosperms. McGraw-Hill Book Co. Inc, New York

Malik A, Shrivastava AK (2007) Induced autopolyploidy in *Carthamus*. J Cytol Genet 8:75–80

Mc Clintock B (1929) A cytological and genetical study of triploid maize. Genetics 14:18–222

Mitra PK, Bhowmik G (1996) Cytological abnormalities in Nigel/a sativa induced by gamma rays and EMS. J Cytol Genet 31:205–215

Mitra PK, Bhowmik G (1999) Studies in the frequency and segregation of induced chlorophyll mutations in *Nigella sativa* L. Adv Plant Sci 12:125–129

Myers JR, Gritton ET, Struckmeyer BE (1992) Genetic male sterility in the pea (*Pisum sativum* L.) II. Cytology. Euphytica 63:245–256

Newmann LJ (1966) Bridge and fragment aberrations in *Podophyllum peltatum*. Geneties 53:55–63

Nirmala C, Kaul M (1994) Male sterility in pea IV. Cytologia 59:195–201

Nirmala A, Rao PN (1996) Genetics of chromosome numerical mosaism in higher plants. Nucleus 39:151–175

Oram PA, Agacoili M (1994) Current status and future trends in supply and demand of cool season food legumes. In: Muehlbauer FJ, Kaiser WJ (eds) Expanding the production and use of cool season food legumes. Kluwer Academic Publishers, Dordrecht, pp 3–49

Pagliarini MS (1990) Meiotic behaviour and pollen fertility in *Aptenia cordifolia* (Aizoaceae). Caryologia 43:157–162

Pantulu JV, Manga V (1972) Cytogenetics of 16 chromosome plants in pearl millet. Cytologia 37:389–394

Parveen S, Anis M, Bhat TA, Khan AH (2006) Efficiency and effectiveness of physical and chemical mutagens in M_2 generation of *Trigonella foenum-graecum* L. J Cytol Genet 7:1–4

Permjit K, Grover IS (1985) Cytological effect of some organophosphorus pesticides. II Mitotic effect. Cytologia 50:199–210

Picard JJA, Bond DA, Monti L, Steuckardt R (1988) Production of pea. Faba bean and chickpea in Europe. In: Summerfield RJ (ed) World crops: cool season food legumes. Kluwer Academic Publishers, Dordrecht, pp 1065–1080

Pushpalatha KN, Reddy VRK, Indra M, Nalini R (1992) Mutagenic effects of gamma rays, chemical mutagens and combined treatments in triticale. Adv Plant Sci 5:147–160

Rao PN, Laxmi N (1980) Gamma ray induced meiotic abnormalities in *Capsicum annuum* L. Caryologia 33:221–227

Rao PN, Kanganadhaam P, Nirmala A (1990) Behaviour of sticky desynaptic mutant in pearl millet. Genetica 81:221–227

Reddy VRK, Annadurai M (1992) Cytological effects of different mutagens in lentil (*Lens culinaris* Medik). Cytologia 57:213–216

Reddy VRK, Indra M, Revathi R (1992) Meiotic abnormalities in lentil (*Lens culinaris* Medik). Cytologia 57:213–216

Rees H, Thomson J (1955) Localization of chromosome breakage at meiosis. Heredity 9:399–407

Robertson LD, Singh KB, Erskine W, Ali M, El Moneim A (1996) A cytological study of the Hordeae in the Dakotas. Cytologia 31:213–219

Roy SK (1989) Effects of EMS and SA on dry seeds of Lens esculenta Moench. In: Proceedings of national symposium. On recent advances in plant cell research, June 7–8, Trivandrum, pp 41–42

Roy RP, Sinha BMS, Thakur GK (1971) Irradiation studies in *Cucumis sativus* Linn. J Cytol Genet 6:128–135

Sadanandam A, Subash K (1985) Effect of chemical mutagens on chiasma frequency in Capsicum annuum L. Cytologia 48:415–419

Saha A, Datta AK (2000) Gamma rays induced reciprocal translocation in black cumin (*Nigella sativa* L.). Cytologia 67:389–396

Sareen S, Kaul AK (1999) Mutation breeding in improvement of Plantago ovata Forsk. Indian. J. Genet. 59:337–334

Sarvella P (1958) Cytomixis and the of chromosome in meiotic and somatic cells of *Gossypium*. Cytologia 23:14–24

Saylor LG, Smith BN (1966) Meiotic irregularities in species of inter-specific hybrids in *Pisum*. Am J Bot 53:453–468

Seijo G (1996) spontaneous cytomixis in the microsporogenesis of sweet pea (*Lathyrus odoratus* L.) (Leguminosea). Cytologia 61:189–195

Sen O, Bhattacharya S (1988) Cytomixis in *Vigna Glaberscens* Wild. Cytologia 53:437–440

Sengupta S, Datta AK (2004) Desirable macromutants induced by chemical mutagens in sesame (*Sesamum indicum* L.). Cytologia 69:291–295

Sharma RP (1970) Increased mutation frequency and wider mutation spectrum in barley induced by combining gamma rays with ethyl methane sulphonate. Indian J Genet 30:180–186

Sharma B (1986) Increasing the efficiency of mutagenesis for micromutations by early generation selection. Indian J Genet 46:88–100

Sharma V, Kumar G, Kumar R (2004) Comparative mutagenecity of gamma rays and EMS in *Cicer arietinum* L. J Cytol Genet 5:21–26

Sharma M, Khan AH, Bhat TA (2009) Assessment of mutagenecity of individual and combination treatment of gamma rays and MMS in Broad Bean (*Vicia faba* L). Cytologia 74(2):235–241

Sheidai M, Riahi H, Hakim M (1993) Cytomixis in *Asparagus* L. Nucleus 36:59–62

Sheidai M, Armana M, Zehzada B (2002) Chromosome pairing and B-chromosomes in some *Aegilops* (Poaceae) species and populations from Iran. Caryologia 55:263–273

Siddique I, Ansari MYK (2005) Studies on the genotoxic effect of pollution on brinjal (*Solanum melongena* L.) growing around Harduaganj thermal power plant. Nat Environ Pollut Technol 4:13–17

Siddiqui S, Meghvansi MK, Hasan Z (2007) Cytogenetic changes induced by sodium azide (NaN3) on *Trigonella foenum-graecum* L. seeds. S Afr J Bot 73:632–635

Sidhu MC (2008) Meiotic studies in the hybrids of wheat-rye additional lines and wheat (*Triticum aestivum* L.) varieties. J Indian Bot Soc 87:86–91

Singh AK (2003) Effect of doses of gamma rays and EMS on germination and plant survival in mungbean (*Vigna radiata* L. Wilczek). Int J Mendel 20:99–102

Singh AK, Chaudhary BR (2005) Gamma-rays induced chromosome anomalies in two morphologically distinct varieties of *Capsicum annuum* L. Nucleus 48:80–84

Singh D, Singh RM, Singh J (1989) Investigation on induced macromutations in lentil (*Lens culinaris* Medik). Z Pflanzenzucht 83:315–320

Sinha ARP (1988) Morphological and cytological changes induced in Lindenbergia indica following treatments. Genet Polonica 29:335–339

Sinha SSN, Godward MBF (1972) Radiation studies in *Lens culinaris*. Indian J Genet 32:331–339

Soheir E, Antointee H, Atif H (1989) Cytological effect of herbicide Garlon-4 on root mitosis in *Allium cepa*. Cytologia 54:465–472

Sparrow AH, Evans HJ (1961) Nuclear factors affecting radio sensitivity. I. The influence of nuclear size and structure chromosome complement and DNA content. Brookhaven Symp Biol 12:76–100

Srivastava DP (1979) Effect of physical and chemical mutagens on *Solanum melongena* L. Ph.D. thesis. Banaras Hindu University, Banaras

Sudan AS, Wafai BA (1987) Spontaneous occurrence of cytomixis during microsporogenesis in almond (*Prunus amygdalus* Batsch.) and peach (*P. persica* Batsch.). Cytologia 52:361–364

Tabassum S (2002) Studies on physical and chemical mutagenesis in *Ammi majus* L

Tarar JL, Dnyansagar VR (1980) Comparison of ethyl methane sulphonate and radiation induced meiotic abnormalities in *Turnera ulmifolia* Linn. Var. Angustifolia Wild. Cytologia 45:221–231

Tarar JL, Dnyansagar VR (1983) Indian 1. Bot Soc 6:10–18

Thakur V (1978) Chromosomal and nuclear transmigration during microsporogenesis in natural populations of some plant species. In: Dnyansagar VR et al (eds) Recent trends and contacts between cytogenetics, embryology and morphology. Today and Tomarrow, Publishing Co, New Delhi, pp 123–130

Vandana S, Kumar DDK (1996) Meiotic anomalies induced by EMS and DES in faba bean (*Vicia faba*). J Indian Bot Soc 75:237–240

Verma RC, Sarkar A, Das BC (1984) Cytomixis in mulberry. Curr Sci 53:1258–1260

Villeux R (1985) Diploid and polyploid gametes in crop plants. In: Janick J (ed) Mechanisms of formation and utilization in plant breeding, 442. Plant Breeding rev. 3, AVI publishing Co., Wesport

Yen C, Yang J, Sun GL (1993) Intermeiocyte connections and cytomixis in intergeneric hybrids of *Roegneria ciliaris* (Trin.) Nevski with *Psathyrostachys huashanica* Keng. Cytologia 58:187–193

Zeerak NA (1990) Induced morphological variants in brinjal (*Solanum melongena*). Phytomorphology 40:251–256

Zeerak NA (1992) Cytogenetical effects of gamma rays and ethyl methane sulphonate in tomato (Lycopersicon esculentum var. Cerasiforme). Phytomorphology 42:81–86

Structural Changes in Chromosomes

12

Abdul Rauf Shakoori, Saira Aftab, and Khalid Al-Ghanim

Abstract

Chromosomes can be structurally identified by their sizes, positions of centromeres and nucleolar organizers, and patterns of chromomeres, heterochromatin, and bands. There are four types of aberrations in the chromosomal structure such as deletions, duplications, inversions, and translocations which can be detected cytologically under the microscope. Some changes are however too subtle to be detected cytologically. Deletions represent missing segments of chromosomes. The homozygous deletions can be lethal, whereas heterozygous deletions can be nonlethal or lethal and can express recessive genes uncovered by deletion. Duplications can cause an imbalance in the genetic material, thereby producing phenotypic effect in the organism and leading to increased variety of gene functions. Inversions are caused by 180 degree turn of a segment of a chromosome, which cause little problem for an organism under homozygous condition, but inversion heterozygotes often have pairing difficulties during meiosis, leading to formation of inversion loops. Crossing over within loops usually results in inviable products. The crossover products will be different for pericentric and paracentric inversions. Translocations involve relocation of a chromosomal segment to another position in the genome. In heterozygous state, the translocations produce duplication-deletion meiotic products, which can lead to unbalanced zygotes, and new gene linkages. Chromosome rearrangements can cause ill health in human population – infertility and mental retardation being the dominant effects.

A.R. Shakoori (✉) • S. Aftab
School of Biological Sciences, University of the Punjab,
Quaid-i-Azam, Campus, Lahore 54590, Pakistan
e-mail: arshaksbs@yahoo.com; arshakoori.sbs@pu.edu.pk; saira3aftab@gmail.com

K. Al-Ghanim
Department of Zoology, College of Science, King Saud University, Riyadh, Saudi Arabia
e-mail: kghanim@KSU.EDU.SA

T.A. Bhat, A.A. Wani (eds.), *Chromosome Structure and Aberrations*,
DOI 10.1007/978-81-322-3673-3_12

Keywords
Chromosomal aberrations • Deletion • Duplication • Inversion • Translocation • Genetic imbalance • Heterochromatin • Interstitial deletion • Ring chromosome • Cri-du-chat syndrome • Angelman syndrome • Beckwith-Wiedemann syndrome • Down syndrome • Prader-Willi syndrome

Chromosome contains all the genetic information needed for an organism to live and respond to its environment. Any kind of alteration in this genetic information can have deleterious effect on the organism. These alterations can be the manifestation of variation in the number of chromosome (discussed in another chapter) or alterations in the chromosomal structure. Chromosomes with the change in the gene(s) do not look different from the one bearing the normal allele. Visible changes however do occur regularly which are called chromosomal mutations or chromosomal aberrations. This chapter will focus on chromosomal mutations which are changes in the genome involving parts of chromosomes, whole chromosomes, or whole sets of chromosomes. These changes can be detected cytologically under the microscope or by appropriate genetic tests. Some cytological changes are however too small to be visible under the microscope.

12.1 Changes in Chromosome Structure

Changes in chromosome structure can come about due to deletions or deficiency, duplications, and rearrangements. Loss of a segment of the chromosome is known as deletion, whereas addition of a segment of chromosome on the same or different chromosome is known as duplication. A segment of chromosome may rotate 180 degrees and rejoin the chromosome, which is known as inversion, or parts of two nonhomologous chromosomes may exchange to produce a translocation. Both in inversion and in translocation, the total genetic information and chromatin material remain the same in the cell, and no extra part of chromatin is added or deleted; only the direction or location of chromosome is altered. All these changes are collectively called as chromosome rearrangements.

12.1.1 Deletions (Deficiencies)

In deletion the daughter chromosome loses a part of chromosome compared to parent chromosome, and this deleted part of chromosome is degraded and completely lost, which makes this kind of mutation irreversible (Fig. 12.1a). The chromosomal breakage and rearrangement can be done with the help of ionizing radiations, which cause chromosomal breaks. The way broken ends of the chromosomes rejoin will determine the kind of rearrangement produced. Single break causes a terminal deletion, and two breaks can produce an interstitial deletion (Fig. 12.1b–d). Deletion

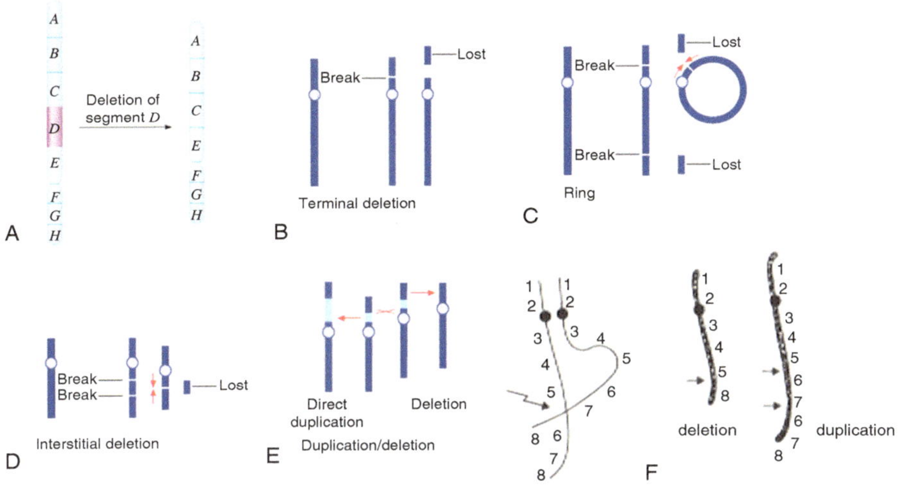

Fig. 12.1 Variety of deletions in the chromosomes: (**a**) A segment of chromosome (D) has been deleted and the resultant chromosome has become shorter in size. (**b**) In terminal deletion the terminal end of the chromosome breaks, which will be lost since it does not have a centromere. (**c**) The two terminal ends of a chromosome break, and the remaining part of the chromosome rounds off to form a ring chromosome. The broken terminal pieces are lost. (**d**) In interstitial deletion the chromosome breaks at two places (between the two ends) leading to joining of the broken ends. The part of chromosome cutout will be lost. (**e**) By recombination between the two chromosomes (in the *center*), the colored segment of the chromosomes has been duplicated in the chromosome on the *left* (direct duplication) and has been deleted in the chromosome on the right (deletion). (*f*) It shows overlapping regions of two chromosomes, which break at the point of overlap and then rejoin leading to deletion in one and overlapping in the other chromosome (http://www.csun.edu/~cmalone/pdf360/Ch08-1%20struct.pdf, http://onlinclibrary.wiley.com/doi/10.1038/npg.els.0001452/abstract;jsessionid=9554ECDCC067053D08F452134C4CEC61.f04t04)

and duplications can be produced by the same event if chromosomal breaks occur simultaneously at different points in two homologues, which can be visualized as occurring when the homologues overlap (Fig. 12.1e, f).

The consequences of deletion can have varying degree of effects which depends on part of the chromosome deleted. In general if the deletion is made homozygous, it is lethal. This is because a greater part of the chromosome is essential for normal viability; the deletion of any part of the chromosome will be deleterious. The heterozygous individuals for deletions may not even survive since deletions upset the specific genetic balance. In some cases, the individuals with relatively large deficiencies may survive if these deletions occur together with normal chromosomes. If the centromere-containing part of chromosome is deleted, the chromosome would become acentric chromosome, and this would fail to attach to microtubules during cell division; hence, whole chromosome would be deleted which would produce lethal effects on organism (Moore and Best 2001). Human disorders caused by deletion of large segment of chromosome are usually the cases in which deletion occurs in heterozygous individuals as the deletion in homozygous would result in lethal condition. Even in heterozygous deletion, the normal copy number of gene in cell is

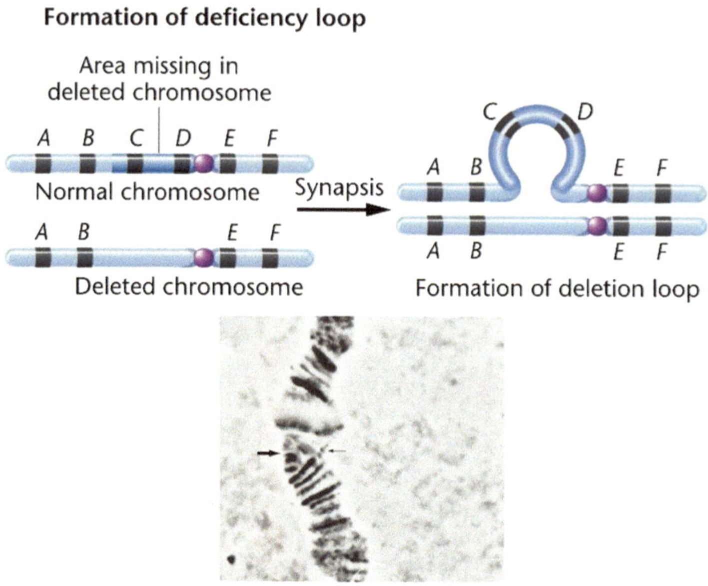

Fig. 12.2 Cytogenetic configurations in a deletion heterozygote. In the *top* figure, the looped-out portion of the meiotic chromosome is because of the deletion of a segment from one of the pair of chromosomes. The genes in this deletion loop have no alleles with which to pair during synapsis. Since the polytene chromosomes in *Drosophila* have specific banding pattern as shown in the *bottom* figure, the missing bands of the deleted chromosome can be observed in the deletion loop (see *arrows*) of the normal chromosome

very important as the deletion of one allele out of many duplicate copies of gene would result in normal phenotype. Normally a phenotypic change in an organism is observed to be due to the deletion of few to several alleles leading to a medical condition called syndrome which is a combination of many disorders due to deletion of many genes at the same time.

1p36 deletion syndrome (also known as monosomy 1p36) is one of the most common (one in every 5000–10,000 births) deletion syndromes caused by a deletion in the outermost band on the short arm (p) of chromosome 1. It is a congenital disorder characterized by delayed growth, malformations, hearing and vision impairment, distinct facial features, moderate to severe intellectual disability, hypotonia, seizures, and limited speech ability.

Deletions can be detected cytologically by deletion loops or by genetic analysis. In heterozygotes the meiotic chromosomes in the region of deletion can form a deletion loop due to failure of pairing with the corresponding segment on the normal chromosome (Fig. 12.2). Deletion loops have also been detected in polytene chromosomes, where the homologue chromosomes are fused. Deletions in some parts of the chromosome produce a unique phenotype such as a dominant notch-wing mutation in *Drosophila*. When heterozygous for the notch deletion, a fly has wings that are notched at the tips and along the edges.

Some of the genetic analyses are based upon the following evidences:

1. Recessive lethality: Mortality of the homozygote. This of course can also be caused by any lethal gene.
2. Suppression of crossing over in the region of deletion, but again, this could occur with other aberrations, and small deletions may have only minor effects on crossing over.
3. Lack of revertability: Chromosomes with deletions can never go back to their normal condition.
4. Phenotypic expression of a recessive gene on a normal chromosome, when the region in which it is located has been deleted from the homologue.

Such pseudodominance, which is expression of a recessive gene when present in a single dose, also allows the cytological localization of the gene when coupled with chromosomal positioning of the deletion loop. This technique permits correlation between the genetic map based on linkage analysis and the cytological map devised by marking the position of deletion loops in specific cases of pseudodominance.

Animals and plants behave differently after genetic deletion. A male animal, which is heterozygous for a deletion chromosome and a normal one, will produce functional sperm with each of the two chromosomes in approximately equal number. In other words, the sperms will function to some extent, regardless of their genetic content. In diploid plants, on the other hand, the pollens produced by a deletion heterozygote are of two types: functional pollen with normal chromosome and nonfunctional pollen carrying the deficient homologue. Thus the pollen cells seem to be sensitive to changes in the amount of chromosomal material. The situation is somewhat different for polyploid plants, which are far more tolerant of pollen deletions. This is because there are several chromosome sets even in the pollen, and the loss of chromosomal segment in one of these sets is less crucial than it would be in a haploid pollen cell. Ovules in either diploid or polyploid plants are also quite tolerant of deletions.

12.1.1.1 Terminal Deletion

A single break at the terminal portion of chromosome causes the loss of genetic material (Fig. 12.1a) and results in many physical abnormalities. A terminal deletion from short arm of chromosome number 4 results in missing genes which account for the anomalies identified by Wolf-Hirschhorn syndrome. This syndrome is characterized by intellectual impairment, microcephaly, developmental delay, abnormal external genitalia, large folded ear, and cleft of the lips and palate (Fig. 12.3). The degree of deletions and the variety of symptoms depend upon the amount of missing genetic material. Example of syndrome caused by the deletion of small arm of chromosome number 5 (5p) in humans is cri-du-chat (cry of the cat) syndrome. It is associated with many congenital anomalies including growth failure, microcephaly, abnormal face, severe mental retardation, and distinctive high-pitched catlike cry in infancy (Fig. 12.4). In 80 % of the cases, the chromosomes carrying deletion comes from the father's sperm.

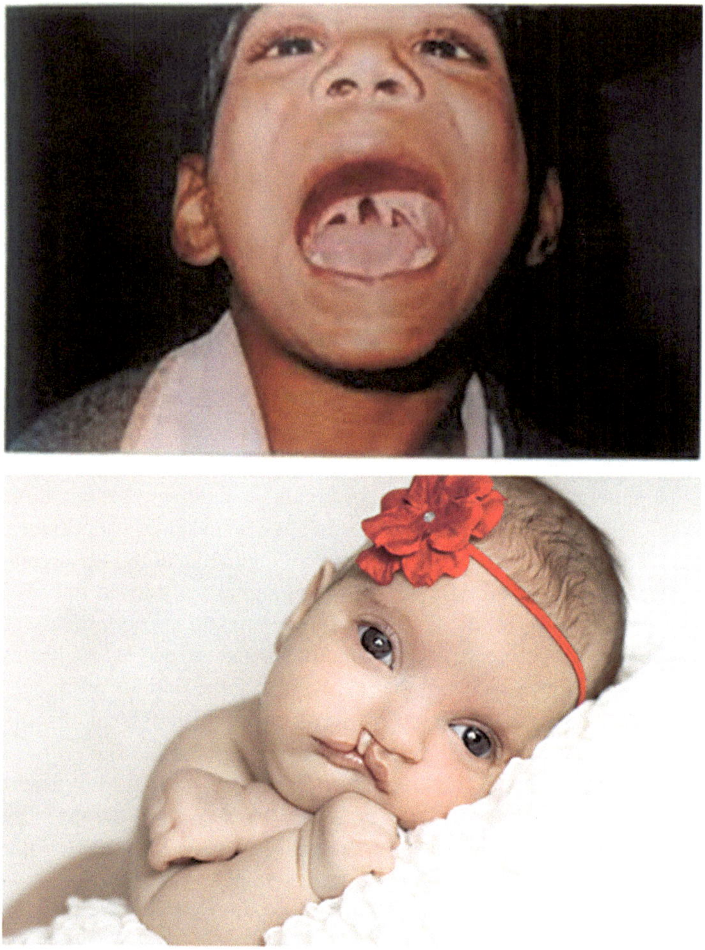

Fig. 12.3 A terminal deletion in the short arm of the chromosome number 4 results in missing genes which cause Wolf-Hirschhorn syndrome. This syndrome is characterized by cleft lips, cleft palate, cleft epiglottis, and intellectual impairment among other disabilities (http://www.wolf-hirschhorn.org/wp-content/uploads/2012/01/img_9365_2.jpg)

Ring chromosomes are the product of terminal deletions from both ends of the same chromosome, which join together to form a ring chromosome (Fig. 12.1b). The ring chromosome is mitotically unstable and is unable to act like normal chromosomes during cell division and causes medical conditions. In humans, ring chromosomes 2, 3, 4, 10, 13, 14, 15 18, 20, and 22 among others have been reported (Fig. 12.5a). Ring chromosomes result in characteristic abnormalities including growth retardation. These ring chromosomes may be very small in size and may be present as an extra chromosome in eukaryotes. Sometimes X chromosome folds back to form a ring shape as in Turner's syndrome, which is characterized by a complete or partial monosomy of X chromosome in a phenotypic female with webbed neck, short stature, and high-curved palate (Fig. 12.5b).

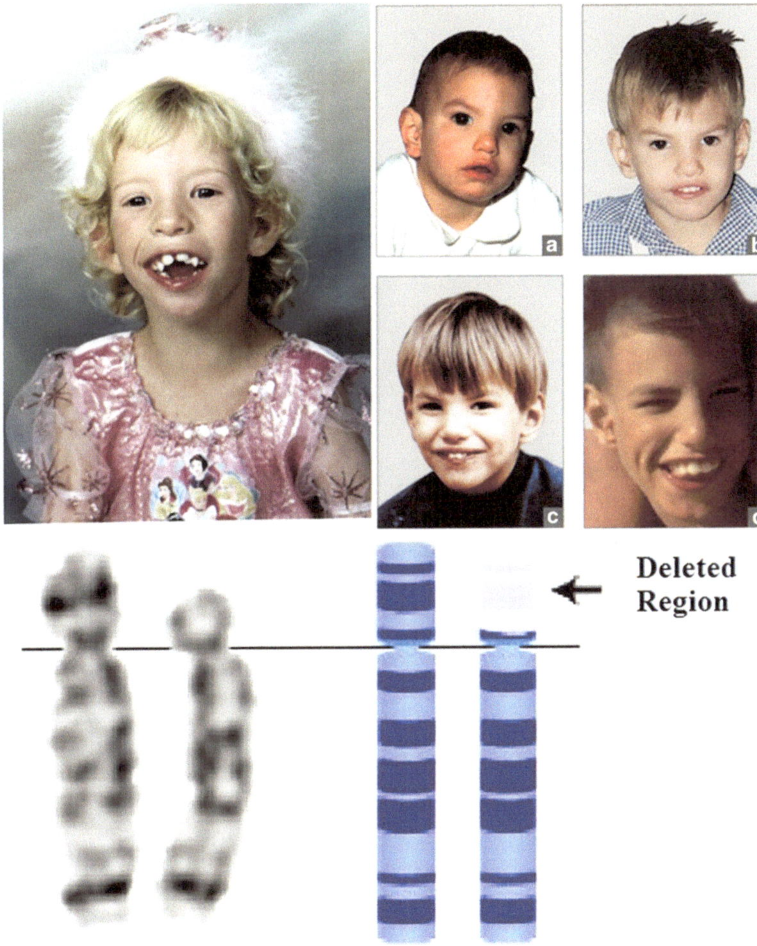

Fig. 12.4 Cri-du-chat syndrome caused by deletion of small arm of chromosome number 5. It is associated with growth failure, abnormal face, and mental retardation (*top left*). The photographs on the *top right* show facial features of patients with cri-du-chat syndrome at age of 8 months (**a**), 2 years (**b**), 4 years (**c**), and 9 years (**d**) (https://www.google.com.pk/imgres?imgurl=https://geneticmutationsperiod6.wikispaces.com/file/view/tessa_1.jpeg/310796006/tessa_1.jpeg&imgrefurl=https://www.emaze.com/@ALOQWQOQ/Cri-du-chat-Syndrome&h=863&w=600&tbnid=hzZo3Vk7eZrOSM:&docid=4j7usLwLqDYZsM&ei=bI2gVs3oD8Oya-WTmuAJ&tbm=isch&ved=0ahUKEwjNzsSvtbrKAhVD2RoKHeWJBpwQMwhkKC4wLg. https://upload.wikimedia.org/wikipedia/commons/c/c9/Criduchat.jpg. http://image.slidesharecdn.com/autosomal-deletion-syndromes3076/95/autosomal-deletion-syndromes-17-728.jpg?cb=1287393140)

12.1.1.2 Interstitial Deletions

Banding pattern analysis of chromosomes at prometaphase stage has led to the discovery of many interstitial deletions causing syndrome in human. Interstitial deletions are caused by the occurrence of two chromosomal breaks at the same time in terminal region, and this leads to the loss of small interstitial chromosomal part (Figs. 12.1c and 12.6a). Very much like terminal deletions, interstitial deletions also

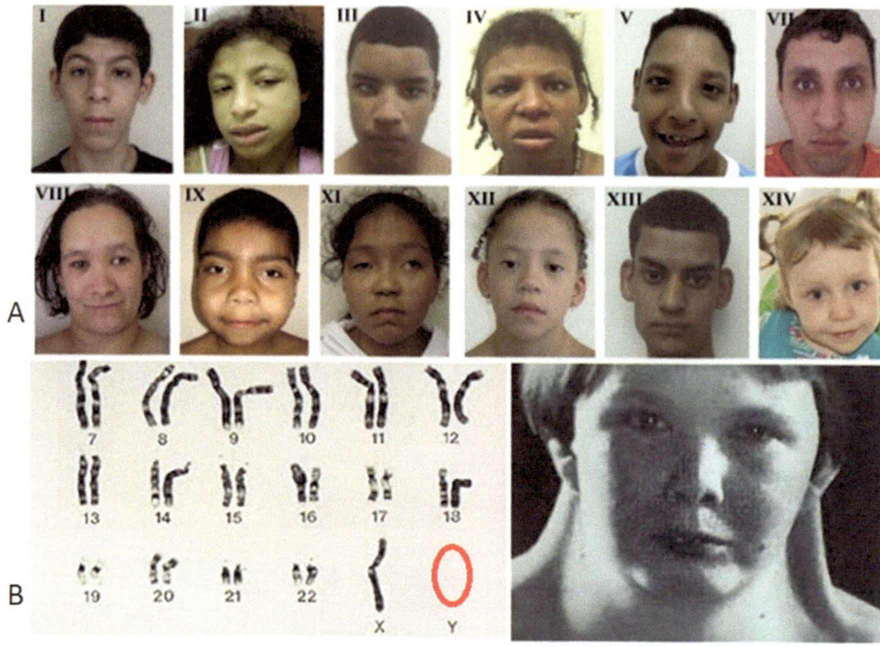

Fig. 12.5 Ring chromosome syndrome: (**a**) Patients I–V, VII–IX, and XI–XIV present ring chromosomes 3, 4, 10,10,13, 14,15, 18,1 8, 22, 22, and 22 at ages 16, 12, 14, 22, 8, 23, 28, 7, 11, 7, 24, and 2, respectively (Guilherme et al. 2011). (**b**) Turner's syndrome: X chromosome usually folds back to form a ring shape (not shown in this figure). It is however characterized by monosomy of X chromosome in a phenotypic female with webbed neck, short stature, and high-curved palate (https://www.google.com.pk/imgres?imgurl=http://static-content.springer.com/image/art%25253A10.1186%25252F1471-2350-12-171/MediaObjects/12881_2011_916_Fig1_HTML.jpg&imgrefurl=http://bmcmedgenet.biomedcentral.com/articles/10.1186/1471-2350-12-171&h=252&w=600&tbnid=o8FKT0l4klAzpM:&docid=1WD7- yO78xP_MM&ei=HrmoVozHKoO1sQGXgJnACw&tbm=isch&ved=0ahUKEwiMztKVgMrKAhWDWiwKHRdABrg49AMQMwhUKFEwUQ. http://sufw.com.au/wp-content/uploads/2013/06/Karyotype-indicating-Downsyndrome-Combined-Screening-Test-for-Down-Syndrome-fig2.png. https://manbironline.files.wordpress.com/2011/05/noonan-syndrome2.jpg?w=300&h=225)

cause severe medical abnormalities including growth defects and mental retardation or even death of embryos. Only those embryos survive which have deletion of very small part of chromosome (Moore and Best 2001). Giordano et al. (2015) have described a rare interstitial deletion of the Xp21.1 chromosome in the long arm of X chromosomes in males, showing intellectual disability, short stature, hearing loss, and dysmorphic facial features (Fig. 12.6b).

Microdeletions are also a type of interstitial deletion in which a very small part is deleted. Deletion of this extremely small portion of chromosome becomes difficult to be detected, and hence in some cases, such small deletions go medically undetected. Usually 1–2 Mb of genetic material is lost from usually the same band or very close bands which make it difficult to detect; hence, these deletions are diagnosed using in situ hybridization techniques. Deletion of 22q11.2 causes heart

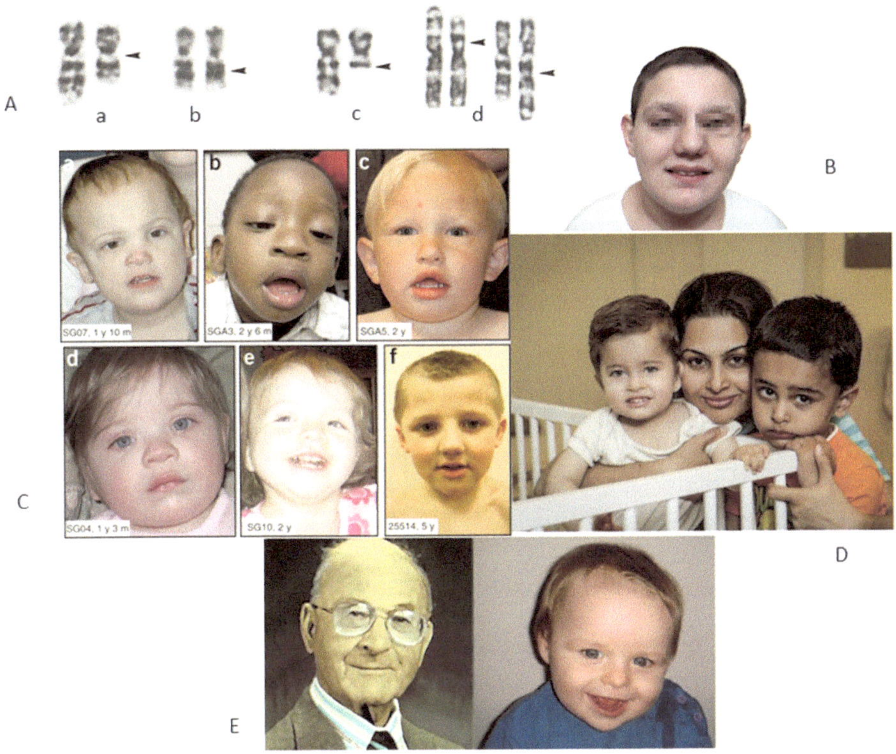

Fig. 12.6 (**a**) Representative partial G-banding karyotypes. (*a*) CML (case no. 1) exhibiting an interstitial deletion del(7)(q11.2q22). (*b*) CML (case no. 2) with a small interstitial deletion del(7)(q22q22). Note that bands 7q21 and 7q31 are closer to each other in the deleted than in the normal chromosome 7 homologue. (*c*) MDS (case no. 3) with a terminal deletion del(7)(q22). (*d*) MDS (case no. 6) exhibiting a reciprocal translocation t(3; 7)(p13; q22) in unstimulated bone marrow and PHA-stimulated blood. (**b**) X-linked intellectual disability, short stature, hearing loss, and dysmorphic facial features of a 15-year-old boy showing an interstitial deletion of the Xq21.1 chromosome. (**c**) 16p12.1 microdeletion associated with childhood developmental delay and neuropsychiatric phenotype (Girirajan et al. 2010). (**d**) A 2-year-old Maheera (*left*) suffers from Prader-Willi syndrome due to interstitial deletion from long arm of paternal chromosome number 15. It has delayed her growth and limited her mental and physical development. (**e**) Harry Angelman identified a structural abnormality in chromosome number 15 with chromosomal deficiency from both parents. The signs of Angelman syndrome were developmental delays, speech impairment, movement and balance issues, and a combination of laughing, smiling, or a constant happy demeanor. This syndrome is not degenerative, and many people who had it were able to overcome some of the symptoms with therapy and or medication (http://www.thestar.com/content/dam/thestar/initiative/santa_claus_fund/2012/11/15/toronto_star_santa_claus_fund_helps_child_with_praderwilli_syndrome/cisantaclausfundsidra.jpeg.size.xxlarge.letterbox.jpeg. http://www.angelmansucks.com/Content/images/TomSmall.jpg. http://ezonpro.com/wp-content/uploads/2015/02/Harry-Angelman.jpg.http://static-content.springer.com/image/art%3A10.1186%2Fs12881-015-0220-z/MediaObjects/12881_2015_220_Fig2_HTML.gif. http://www.nature.com/ng/journal/v42/n3/images/ng.534-F3.jpg. http://www.bloodjournal.org/content/bloodjournal/89/6/2036/F1.large.jpg?width=800&height=600&carousel=1&sso-checked=true)

malfunctions, abnormalities in calcium metabolism, hearing loss, developmental delay, and intellectual impairments. Sometimes microdeletion is linked with gene duplication event (Fig. 12.1d). Girirajan et al. (2010) reported a recurrent 520 bp 16p12.1 microdeletion associated with childhood developmental delay. This microdeletion was detected in 20 of 11,873 cases compared with 2 of 8540 controls. It was found that the 16p12.1 microdeletion not only predisposes to neuropsychiatric phenotypes as a single event but also exacerbates neurodevelopmental phenotypes in association with other larger deletions or duplications (Fig. 12.6c).

Interstitial deletion from long arm of chromosome number 15 from the region just below the centromere causes two characteristically very different syndromes called Prader-Willi syndrome (PWS) and Angelman syndrome (AS), depending on the deletion in the paternal or maternal chromosome. PWS is a congenital syndrome caused by deletion of part of long arm of one paternal chromosome 15 homologue and characterized by obesity, defective muscle tone, hyperphagia, hypoplastic genitalia, delayed growth and development, and mild mental retardation or intellectual abnormalities (Fig. 12.6d). AS with similar genetic deletion from maternal chromosome is a rare congenital disorder, which affects less than one person in 20,000. It was first diagnosed by Harry Angelman (Fig. 12.6e) and was called happy puppet syndrome in 1965. It causes developmental delays; movement and balance issues; seizures; delayed or absence of speech; combination of spontaneous outburst of laughter, smiling, or a consistent happy demeanor; characteristic facial features; smaller than normal head; and severe mental retardation (Fig. 12.6e). It can be hereditary and may need to have copies of the chromosomal deficiency from both parents. Further studies on these syndromes have enhanced the importance of genetic imprinting in chromosome which is different in both diseases (Moore and Best 2001).

12.1.2 Duplications

Repetition of a segment of chromosome containing several genes is called duplication of chromosomal segment resulting in extra genetic material (Fig. 12.7). Depending upon the location and orientation of the duplicated segment, the duplication can be divided into several categories (Fig. 12.7). Tandem sequence duplications in exact order of arrangement and direction are present adjacent to the original segment. The adjacent duplicated segments may occur in tandem sequence with respect to each other (ABCBCD) or in reverse order (ABCCBD). The pairing patterns obtained in these two sequencers are different and shows high affinity of the homologous regions for pairing. Thus chromosomes in meiotic nuclei containing a normal chromosome and a homologue with a duplication are seen to pair in the Fig. 12.7c. Alternatively the duplicated segment may be nonadjacent (Fig. 12.7a), either in the same chromosome or in the separate chromosomes. Reverse tandem duplications are the duplication which are located adjacent to the original strand but are in reverse order as compared to the original segment duplicated. Duplications which are present at the far end or chromosome called telomere are called terminal

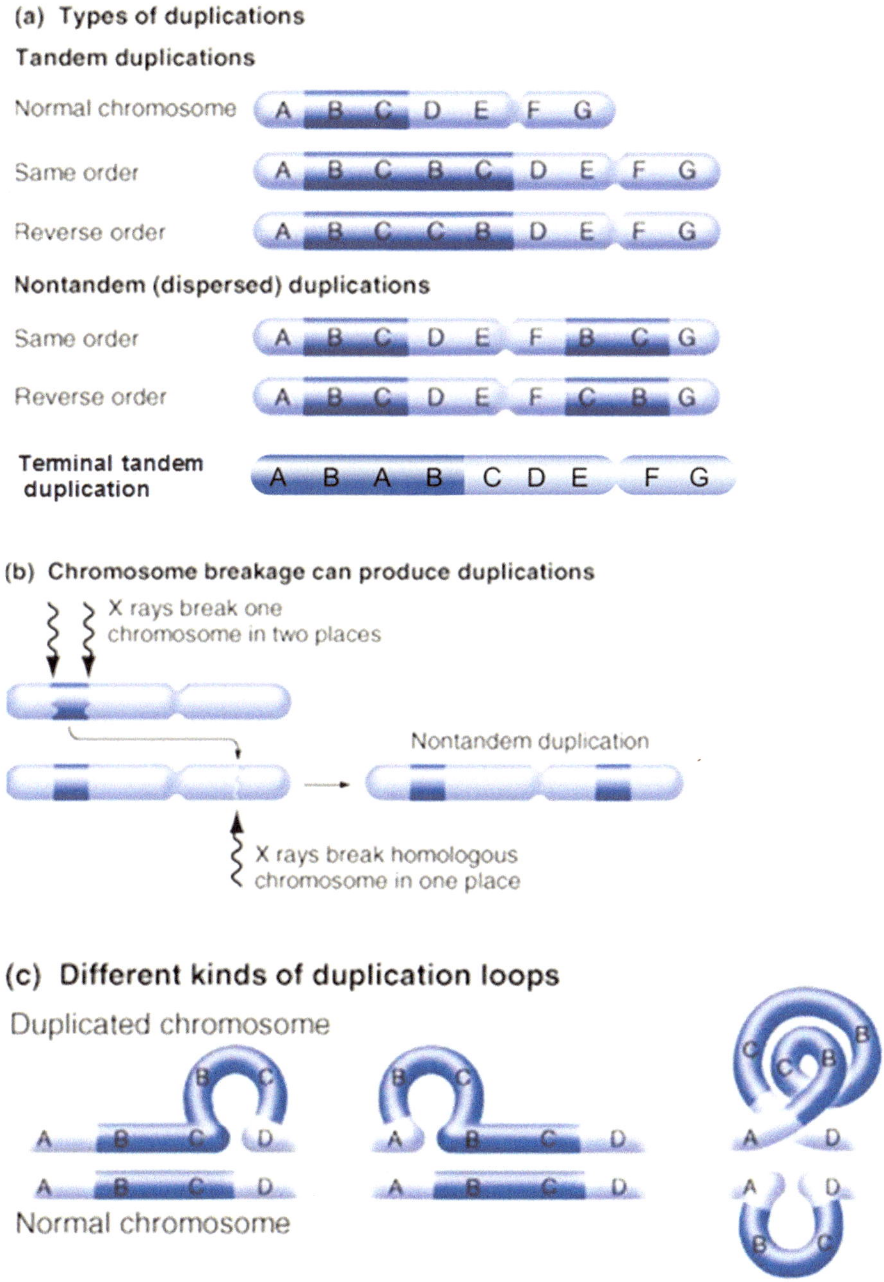

Fig. 12.7 Types of chromosomal duplications: Tandem duplications in the same and reverse order, nontandem (dispersed) duplications in the same and reverse order, terminal tandem duplications, nontandem duplications, and different kinds of duplication loops are shown (http://www.csun.edu/~cmalone/pdf360/Ch08-1%20struct.pdf)

tandem repeats. Nontandem duplications are not adjacent and may be in the same or reverse order as compared to original segment duplicated. Both tandem and nontandem duplications add extra genetic material to chromosome which has effect on the organism's phenotype. Out of many factors, the most important factor inducing duplications is radiation. Radiations induce breaks which act as inducer to repair or to repeat the segment of chromosome damaged (Fig. 12.7c). In Prophase I of meiosis, the duplicated segment makes different loops, and this loop enhances the pairing of related region.

Duplications are example of unbalanced genetic rearrangements like deletions, but unlike deletions, these have milder effects. Duplications are primarily considered to be due to unequal distribution of genetic material during crossing over, which may however also be due to interstitial deletions. These duplications produce a condition of partial trisomy when paired with normal gamete, as also is the case during translocations.

A common example of duplication is found when a segment of chromosome 15 is duplicated in human creating a dicentric chromosome. The symptoms of patients with this dicentric chromosome vary depending upon the size of segment duplicated and parental origin of duplicated segment. Chromosome 15q11.2-13.1 duplication syndrome (dup 15q syndrome) includes extra isodicentric 15 chromosome or an interstitial duplication 15. The physical development abnormalities are not very specific, but the patients may have poor muscle tone, button nose, deep set eyes, and lower-set ears among several other abnormalities (Fig. 12.8a). Duplication of proximal long arm of chromosome number 22 creating a dicentric chromosome leads to mental retardation, coloboma of the eye, and anal atresia (Moore and Best 2001) characteristic of 22q11.2 syndrome (Fig. 12.8b).

Microduplications, just like microdeletions, are hard to detect cytologically, though these are rare compared to microdeletions. Duplication of small portion of chromosome on short arm of chromosome 11 results in Beckwith-Wiedemann syndrome which is characterized by high birth weight and overgrowth of tongue, enlarged kidneys, hearing loss, and musculoskeletal abnormalities (Fig. 12.9). Unlike other genetic disorders, this syndrome lacks any growth defects and mental retardation (Moore and Best 2001).

Duplications of certain genetic segments may produce specific phenotype. An example of chromosomal duplication is the *Bar* allele of *Drosophila* responsible for eye shape. Cytological examination of the *Bar* allele shows that it is a duplicated segment (16A) in X chromosome. *Bar* allele shows incomplete dominance, and heterozygous females for *Bar* allele have a kidney-shaped eye instead of the normal oval having more facets compared to homozygous females. Hemizygous males also have a slit-like eye shape like those of *Bar* females. Cytologically in the polytene chromosomes, Bar was found to be a tandem duplication which probably resulted from an unequal crossover. Evidence for asymmetric pairing and crossing over in *Drosophila* comes from studying homozygous Bar females. Some females occasionally produce offspring with extremely small eyes called double bar, which are because of three doses of the Bar region in tandem (Fig. 12.9).

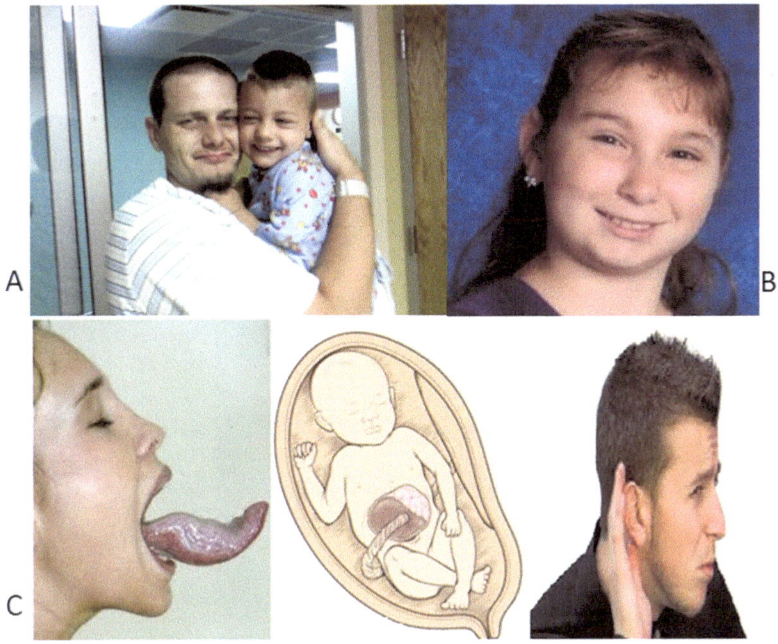

Fig. 12.8 (**a**) The patient (child) with chromosome 15q11.2-13.1 duplication syndrome which includes extra isodicentric 15 chromosome or an interstitial duplication 15. The patient has poor muscle tone, button nose, and low-set ears among other abnormalities. (**b**) A girl with 22q11.2 duplication syndrome. The proximal long arm of chromosome number 22 is duplicated, which creates dicentric chromosome leading to mental retardation and coloboma of the eyes. (**c**) Beckwith-Wiedemann Syndrome caused by duplication of small portion of chromosome on short arm of chromosome 11. It is characterized by high birth weight, overgrowth of tongue, enlarged kidneys, hearing loss, and musculoskeletal abnormalities (Moore and Best 2001) (http://www.skinsheen.com/userfiles/files/Symptoms%20of%20Beckwith%20Weidmann%20Syndrome.jpg. https://www.everribbon.com/img/uploaded/15560/original.jpg. http://www.rarechromo.org/information/Chromosome%2022/22q11.2%20duplications%20FTNW.pdf)

12.1.3 Inversions

Inversion is rearrangement of genetic material in chromosome, and it may be paracentric or pericentric. In pericentric inversion, the inverted genetic material includes the centromere of chromosome, and it may alter the shape of chromosome from metacentric to submetacentric and vice versa. Paracentric inversions do not involve the centromere of chromosome, and genetic material inverted is far away from centromere (Fig. 12.10).

In inversion, no extra genetic material is added or deleted from chromosome; instead total genetic material in chromosome remains the same. Hence it has very little effect on the organism's phenotype. In very rare cases inversion has phenotypic effect and that is only when the inversion is happening in vital genes of organism or a gene is repositioned in a way that it alters the normal phenotype. Almost

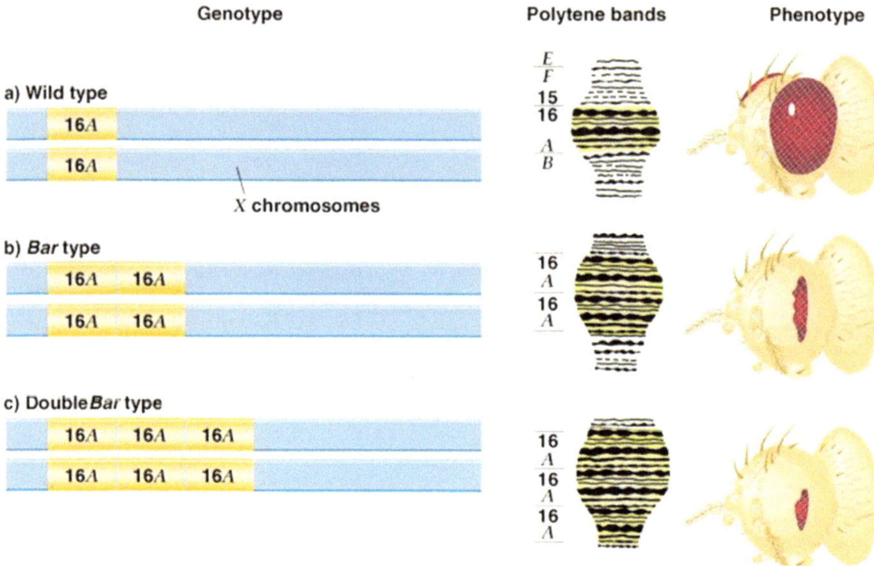

Fig. 12.9 Genotypes and polytene bands of wild type, bar type, and double-bar type eyes of *Drosophila* (http://www.csun.edu/~cmalone/pdf360/Ch08-1%20struct.pdf)

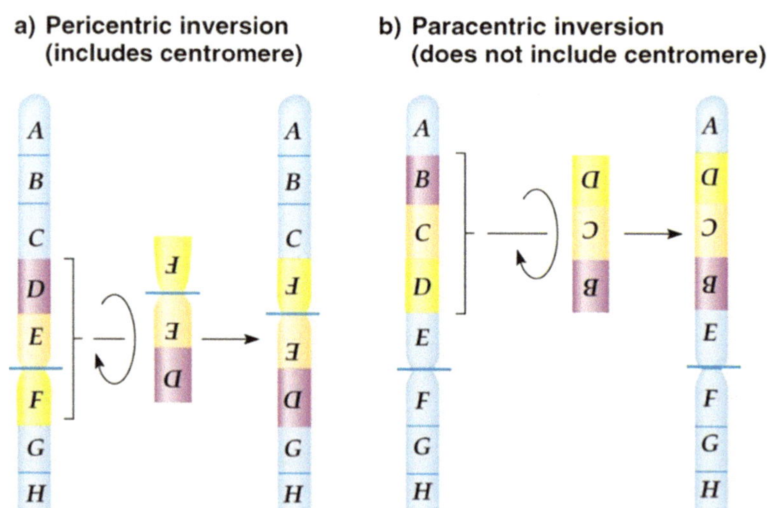

Fig. 12.10 Rearrangement of different parts of chromosomes after pericentric inversion, which involves centromere (*left*), and paracentric inversion, which does not include centromere (*right*) (http://www.csun.edu/~cmalone/pdf360/Ch08-1%20struct.pdf)

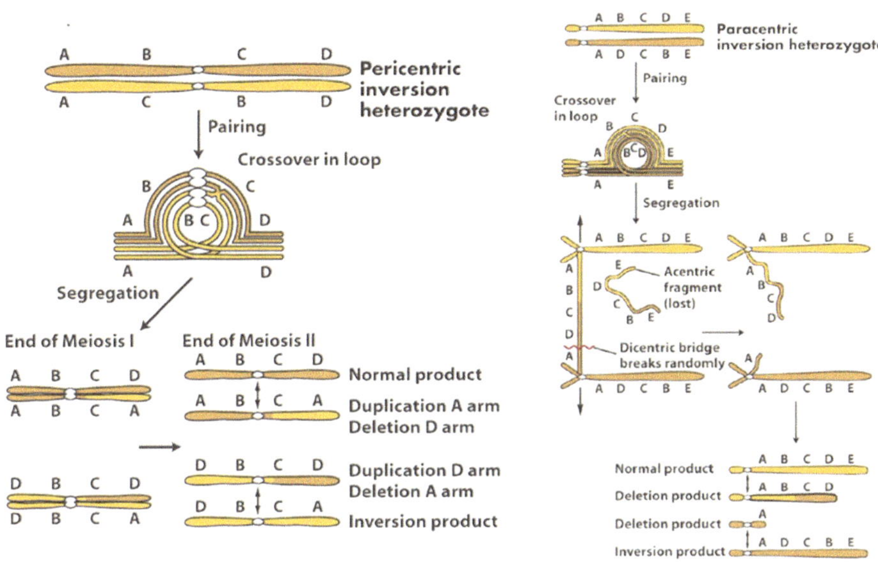

Fig. 12.11 *Right*: Meiotic products resulting from a single crossover within a heterozygous paracentric inversion loop. Crossing over occurs in the four-strand stage, with two identical chromatids connected to each other. *Left*: meiotic products resulting from a meiosis with a single crossover within a heterozygous pericentric inversion loop (http://www.csun.edu/~cmalone/pdf360/Ch08-1%20struct.pdf)

2 % of human population carries inversions which can be detected under microscope through banding technique. All these individuals produce very normal phenotype, but they do produce abnormal offspring due to these inversions.

In a heterozygote for a paracentric inversion, crossing over within the inversion loop has the effect of connecting homologous centromeres in a dicentric bridge, as well as producing an acentric piece of chromosome (Fig. 12.11, right side). Thus, as the chromosome separate during Anaphase I, the disjoining centromeres will remain linked by means of a bridge. This orientates the centromere so that the noncrossover chromatids lie farthest apart. The acentric fragment cannot align itself or move, and consequently it will be lost. In *Drosophila* eggs and in plant megaspores, the dicentric bridge may remain intact long after anaphase I, as the second meiotic division begins, the noncrossover chromatids are directed to the two outermost nuclei (Fig. 12.12). The other two (inner) nuclei either will be linked by the dicentric bridge or will contain fragments of the bridge if it breaks. Fertilization of the nucleus carrying the broken bridge should produce defective zygotes that die because they have an unbalanced set of genes. Consequently in a testcross, the recombinant chromosomes would end up in dead zygotes, and recombinant frequency will be lowered. However in *Drosophila*, the presence of large inversions does not result in a large increase in zygotic mortality. The inner nuclei never participate in fertilization and only one of the outer nuclei can be the egg nucleus. Thus we can see that the

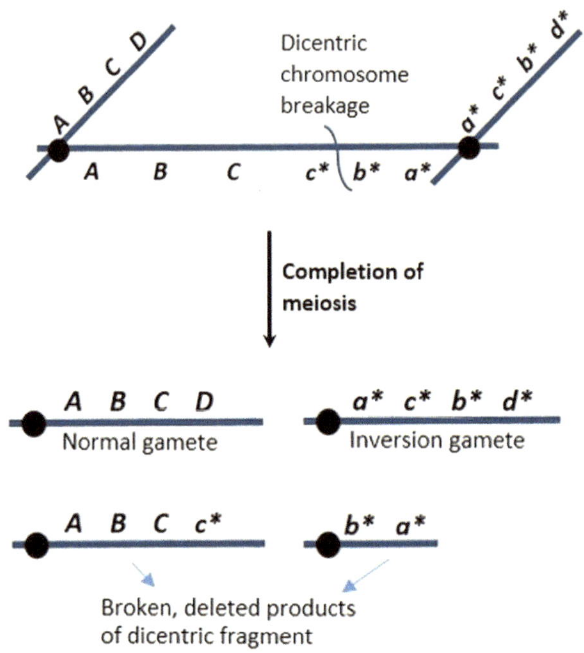

Fig. 12.12 In some organisms such as *Drosophila* and some plants, the dicentric bridge resulting from a single crossover within a heterozygous paracentric inversion loop will not break during anaphase I. As the second meiotic division begins, the noncrossover chromatids are directed to the outer most nuclei. Thus the two inner nuclei either are linked by the dicentric bridge or contain fragments of the bridge, if it breaks (https://en.wikipedia.org/wiki/Dicentric_chromosome#/media/File:Paracentric_inversion_dicentric_chromosome_formation.jpg)

chromatids participating in a crossover event will be selectively retained in the central nuclei, thereby allowing the recovery of noncrossover chromatids in the egg nuclei.

The reason of abnormal offspring lies in the production of imbalance distribution of genetic material during gamete formation. It is observed that if a heterozygous pericentric inversion occurs during gamete formation, then there are different possibilities for each gamete to be normal, with inversion, with duplication or deletion, or with reciprocal duplication or deletion (Fig. 12.11, left side). On the other hand, a heterozygous paracentric inversion during gamete formation would result in one normal gamete, one gamete with inversion, and two with deletion, and some will lose genetic material. All these cases create imbalanced gamete (Fig. 12.11).

The net genetic effect of a pericentric inversion is the same as that of the paracentric one, i.e., crossover products are not recovered, though for different reason. In a pericentric inversion, since centromeres are contained within the inverted region, disjunction of crossover chromosomes is normal. A crossover within the inversion, however, produces chromatids that contain a duplication and deletion for different segments of the chromosome (Fig. 12.11, Left side). In this case

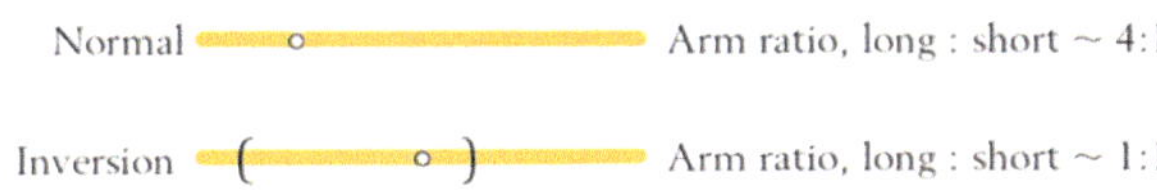

Fig. 12.13 The ratio of the lengths of the left and right arms of a chromosome can be changed by a pericentric inversion. Thus changes in the arm ratio can be used to detect pericentric inversions (Taken from Griffiths AJF, Wessler SR, Lewontin RC, Gelbart WM, Suzuki DT, Miller JH (2014))

fertilization of a nucleus carrying a crossover chromosome generally results in its elimination through zygotic mortality caused by imbalance of genes. The result is again the selective recovery of noncrossover chromosomes as viable progeny.

It has been seen that genetic analysis and meiotic chromosome cytology are both good ways of detecting inversions. As with most rearrangement, there is also possibility of detection through mitotic chromosome analysis. One of the key operational features is to look for new arm ratios. Figure 12.13 shows the ratio of long to short arm changing from 4 to about 1 by the pericentric inversion. Paracentric inversions are more difficult to detect, but they may be detected if the banding pattern or other chromosomal markers are available.

12.1.4 Translocation

Shuffling of genetic material along the length of chromosome is called translocation, and it may be either intrachromosomal in which the translocation occurs in the same chromosome between two different parts of a chromosome, or it may be interchromosomal involving two nonhomologous chromosomes, one donor and the other recipient chromosome (Fig. 12.14). Interchromosomal translocations have gained great medical importance during the past few decades. This shuffling of genetic material between two different chromosomes can either create balanced chromatin or imbalanced chromatin, which are accordingly the result of balanced or unbalanced translocation. In balanced translocation, no genetic material is lost or duplicated. The simple shifting of genetic material does not usually cause a problem for a person who is a carrier, but he/she will have an increased risk of having descendants with an unbalanced translocation. In an unbalanced translocation, the descendant will inherit the affected chromosome with either more or less genetic material compared with the parent's balanced translocation. Since the genetic code is changed as a result, this usually causes problems for the carrier. Two types of translocations are known: reciprocal as described above and Robertsonian or whole arm translocation.

12.1.4.1 Reciprocal Translocation

The most common type of translocation is reciprocal translocation in which a segment from one chromosome is exchanged with a segment from another non homologous chromosome, so that in reality two translocation chromosomes are simultaneously achieved. Reciprocal translocations are induced by the initiation of

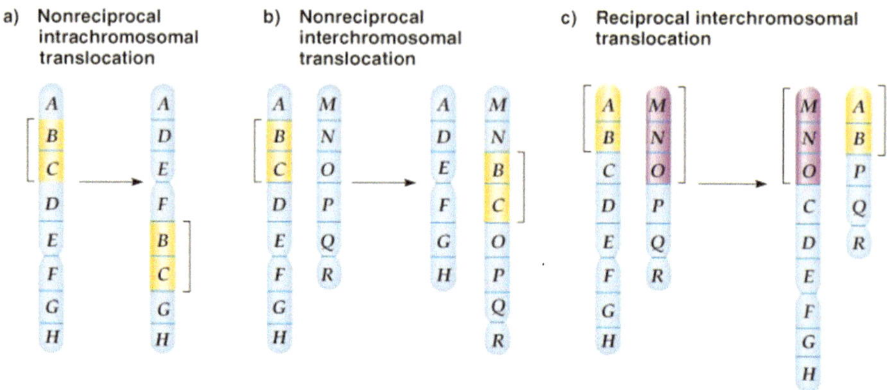

Fig. 12.14 Intrachromosomal and interchromosomal translocations: (**a**) Nonreciprocal intrachromosomal translocation. (**b**) Nonreciprocal interchromosomal translocation. (**c**) Reciprocal interchromosomal translocation (http://www.csun.edu/~cmalone/pdf360/Ch08-1%20struct.pdf)

repair mechanisms after the introduction of breaks in DNA due to mutagens. Reciprocal translocations also occur during abnormal crossovers.

A gamete carrying translocation of chromosome is called translocation carrier gamete. The pairing of homologues in translocation carriers is different compared to normal gametes. Unlike normal pairing and bivalent formation, the derivative chromosomes along with their normal homologues form a quadrivalent during pachytene which is a cross-shaped structure in which each homologous part paired with its counterpart. As pairing and segregation occur after DNA replication, each chromosome consists of two chromatids, and each quadrivalent consists of four chromatids.

Each translocation quadrivalent may segregate in four possible ways. In most of the cases, the chromosomes are distributed equally with two in each daughter cells, but in rare cases, three chromosomes go to one cell and leave on one for another. Daniel (1979) and Jalbert et al. (1980) have listed ways to evaluate a pachytene quadrivalent in order to determine the most likely modes of segregation and viable outcomes (Fig. 12.15). Study of cytogenetic data bases (e.g., Borgaonkar 1994; Schinzel 1994) may help to ascertain whether similar rearrangements have been viable. (i) Alternate segregation: It is an example of balance segregation in which both normal homologues move to one pole and translocation homologues move to another pole. (ii) Adjacent 1 segregation: Adjacent nonhomologous centromeres move to the same pole, thus creating an unbalanced chromosomal complement which will result in a zygote with partial trisomy of one chromosome and partial monosomy of the other when fertilized by a normal haploid gamete. This segregation pattern often is compatible with viability. (iii) Adjacent 2 segregation: In adjacent 2 segregation, the adjacent homologous centromeres move to the same pole. This usually results in large amounts of unbalanced chromatin, which is usually incompatible with embryonic survival. (iv) 3:1 segregation: in this segregation, three of the four chromosomes move to one pole, and only one moves to the opposite pole. This type of segregation often occurs when one of the derivative

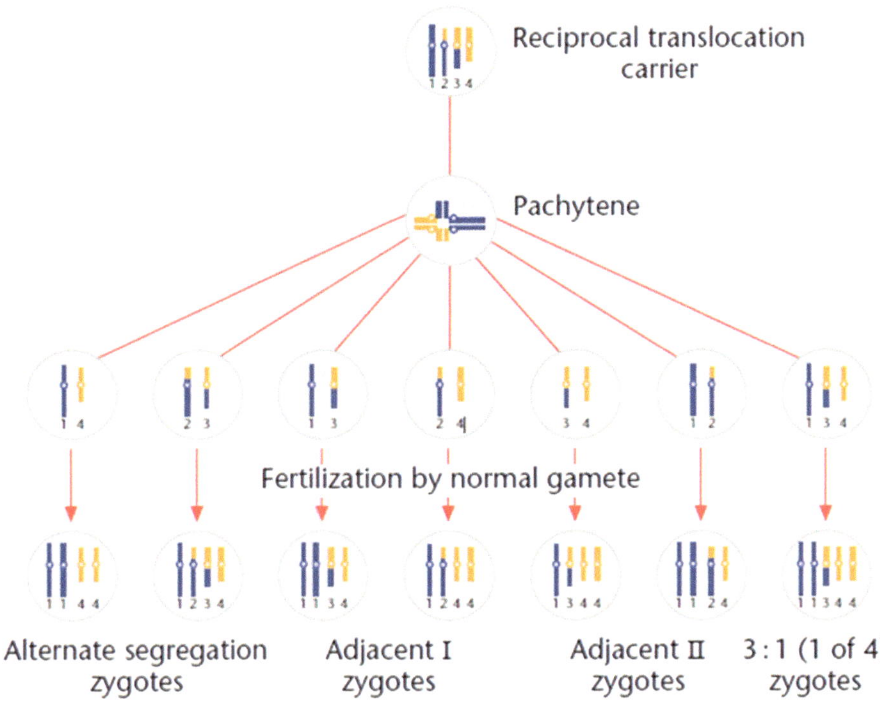

Fig. 12.15 This figure depicts evaluation of reciprocal translocation carrier for segregation of chromosomes and viable outcomes (Hirschhorn 1973; Moore and Best 2001). A pachytene quadrivalent is shown with the results of alternate, adjacent 1, adjacent 2 and 3:1 segregation, and fertilization by a normal gamete. Note that only one of the four possible combinations is represented for 3:1 segregation (http://onlinelibrary.wiley.com/doi/10.1038/npg.els.0001452/abstract;jsessionid=9554ECDCC067053D08F452134C4CEC61.f04t04)

chromosomes is relatively small. Upon fertilization by a normal gamete, the conceptus will have 47 chromosomes. Unless the derivative chromosome is small, the embryo will not be viable.

The exchange of chromosomal parts between nonhomologous establishes new linkage relationships if translocated chromosomes are homozygous or even when heterozygous. The translocations may drastically alter the size of a chromosome as well as the position of its centromere. Figure 12.16 shows large metacentric chromosome which is shortened by one half its length to an acrocentric one, whereas the small chromosome becomes a larger one. There are examples known in natural population, where chromosome numbers have changed by translocation between acrocentric chromosomes and the subsequent loss of the resulting small chromosome elements (Fig. 12.17).

In heterozygotes having translocated and normal chromosome, the genetic and cytological effects are important. The pairing affinities of homologous regions show a characteristic configuration when all chromosomes are synapsed in meiosis. Figure 12.18 shows a cross-shaped configuration of meiosis in a reciprocally

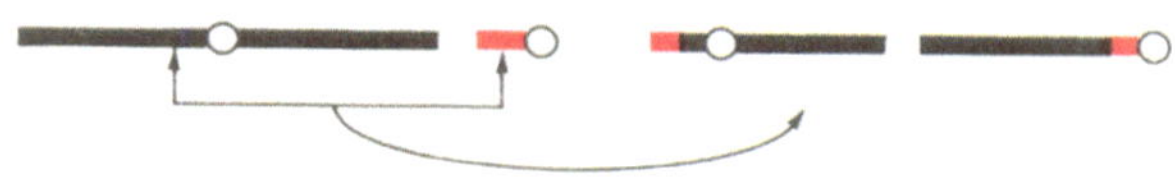

Fig. 12.16 Translocations may drastically alter the size of a chromosome as well as the position of its centromere. Here a large metacentric chromosome is shortened by one half its length to an acrocentric one, whereas the small chromosome becomes a large one (Taken from Griffiths AJF, Wessler SR, Lewontin RC, Gelbart WM, Suzuki DT, Miller JH (2014))

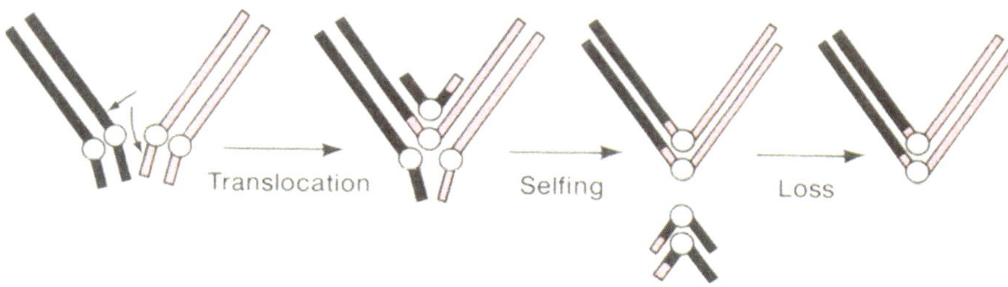

Fig. 12.17 Genome restructuring by translocations. *Small arrows* indicate breakpoints in one homologue of each of two pairs of acrocentric chromosomes. The resulting fusion of the breaks yields one short metacentric and one long metacentric. If, as in plants self-fertilization takes place, an offspring could be formed with only one pair of long metacentrics and one pair of short metacentrics. Under appropriate conditions, the short metacentric may be lost. Thus we see a conversion from two acrocentric pairs of chromosomes to one pair of metacentrics (Taken from Griffiths AJF, Wessler SR, Lewontin RC, Gelbart WM, Suzuki DT, Miller JH (2014))

translocated heterozygote. The homologous paired centromeres disjoin irrespective of translocation. There would be two common patterns of disjunction. (i) Adjacent 1 segregation: in this segregation, each of the structurally normal chromosomes with one of the translocated ones (T1 with N2 and T2 with N1) segregate. Both meiotic products are duplicated and deficient for different regions. (ii) Alternate segregation: In this segregation, the two normal chromosomes may segregate together, as do the reciprocal parts of the translocated ones to produce T1 + T2 and N1 + N2 products. There is another event called adjacent 2 segregation in which homologous centromeres migrate to the same pole, but in general, this is a rare occurrence.

In animals the unbalanced products of adjacent 1 segregation in reciprocal translocation heterozygotes usually produce viable gametes. Thus in *Drosophila*, for example, the gamete population is composed of approximately equal number of alternate and adjacent segregation meiosis. However in diploid plants the T1N2 and T2N1 gametes normally abort, giving a situation normally known as semisterility, often identifiable through shriveled abnormal pollen grains. Even in *Drosophila* and other tolerant animals, the unbalanced gametes, when fertilized by normal gamete, give rise to unbalanced zygotes that tend not to survive unless the imbalance is for very small translocated regions.

Genetically the markers on nonhomologous chromosomes will appear to be linked if these chromosomes are involved in a translocation. Figure 12.19 shows a

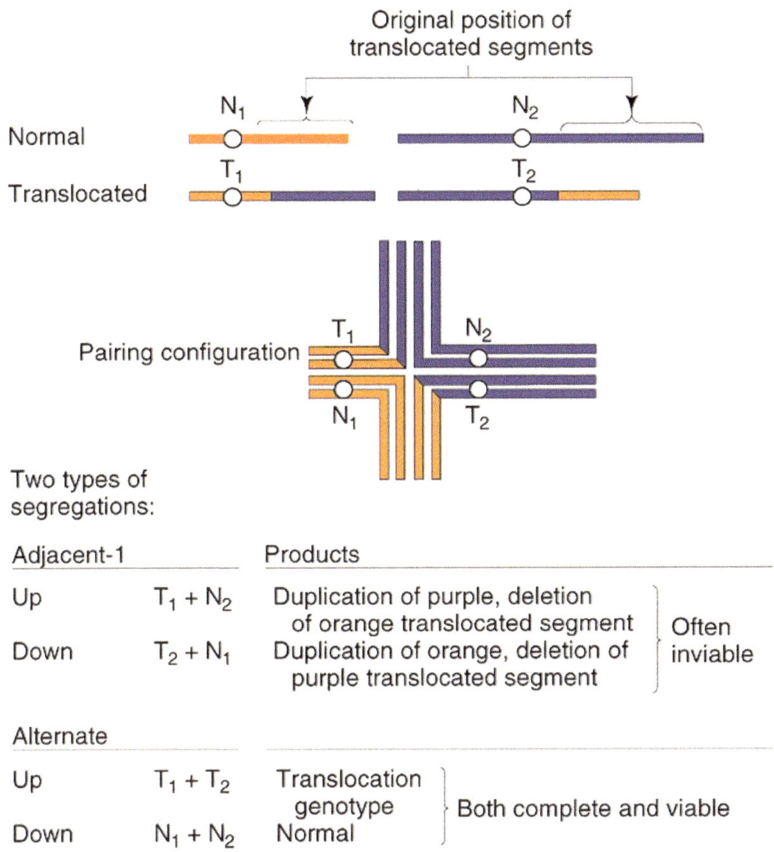

Fig. 12.18 The meiotic products resulting from the two most commonly encountered chromosome segregation patterns in a reciprocal translocation heterozygote (http://images.slideplayer.com/19/5911589/slides/slide_46.jpg)

situation where a translocation heterozygote has been established by crossing an aabb individual with a translocation bearing the wild type genes, assuming that a and b are close to the translocation breakpoint. Upon testcrossing the heterozygotes, the only viable progeny are those bearing the parental genotype, so linkage is seen between loci on different chromosomes. In fact, if all four arms of the meiotic pairing structure are genetically marked, a cross-shaped linkage map will result. Such interlinkage group linkage is often an indication of the presence of a translocation.

In agriculture, translocations can reduce yield because of formation of unbalanced zygotes, though translocations are potentially useful. It has been proposed that the high incidence of inviable zygotes could be used to control insect pests by introduction of translocations into the wild. Thus 50 % of the offspring of crosses between insects carrying the translocation and wild types would die, and 10/16 of the progeny of crosses between translocation-bearing insects would die.

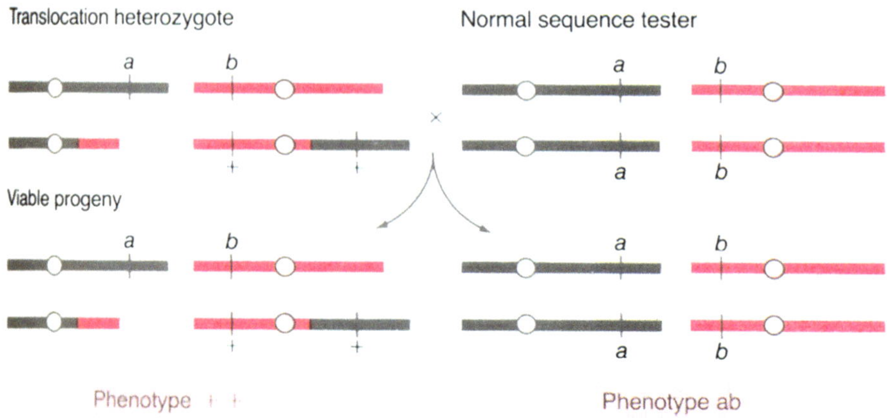

Fig. 12.19 Inviability in some translocation progeny produces apparent genetic linkage because only parental types are recovered in the progeny. This assumes that a and b are close to the breakpoint (Taken from Griffiths AJF, Wessler SR, Lewontin RC, Gelbart WM, Suzuki DT, Miller JH (2014))

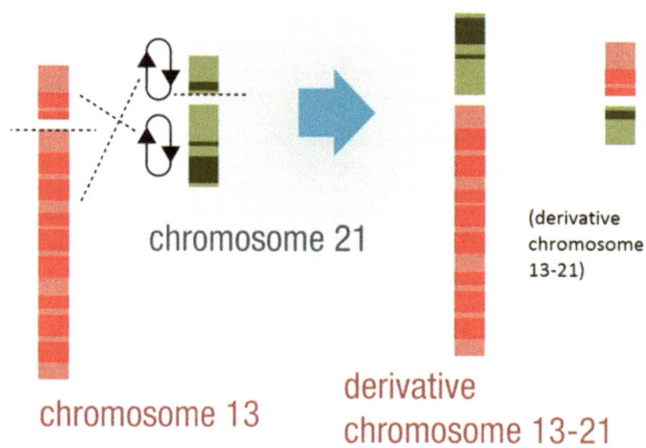

Fig. 12.20 Robertsonian translocation (http://www.brusselsgenetics.be/media/images/Illustraties/ill_eng/ill-10-E_robertsonian_orig.jpg)

12.1.4.2 Robertsonian Translocation

Unbalanced interchromosomal translocations do not exchange genetic material; instead the genetic material is transferred in one direction, i.e., genetic material is lost from chromosome and gained by others. Such translocations are commonly called as Robertsonian translocation (Fig. 12.20), which have great medical importance and are associated with disease and phenotypic abnormalities even to lethality. It occurs in series of steps which starts with the introduction of breaks in nonhomologous acrocentric chromosomes on extreme ends of small arms. The larger parts of chromosome fuse from centromeric region to form a single chromosome (Fig. 12.20). The small acrocentric regions are lost due to lack of centromere. This type of translocation is the most common type of translocation in humans and leads to different syndromes. According to currently known reported cases, it is confined to chromosome number 13, 14, 15, and 21. Robertsonian translocation carriers are

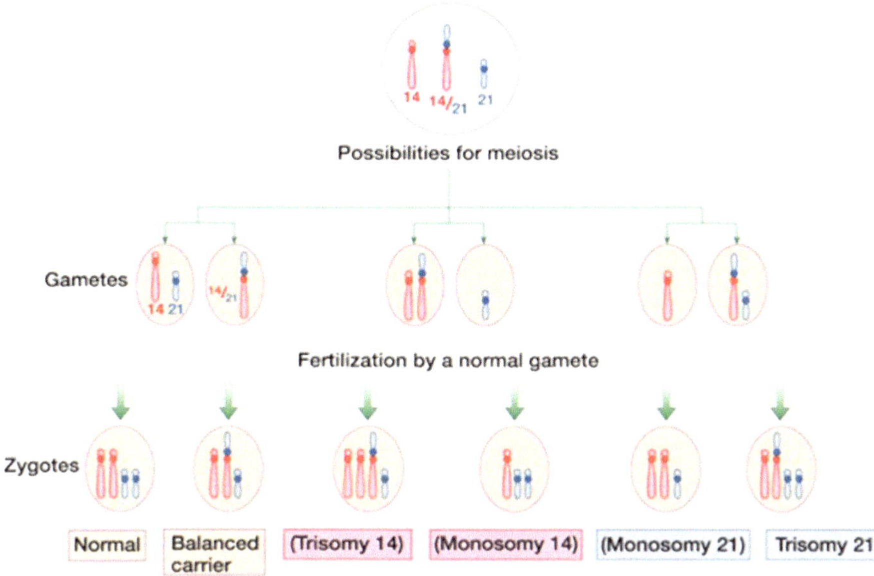

Fig. 12.21 Possibilities of partial and complete trisomy for chromosome number 21 and 14 (http://www.csun.edu/~cmalone/pdf360/Ch08-1%20struct.pdf)

asymptomatic but often produce unbalanced gametes which can result in monosomic or trisomic gametes. Of the possible segregants monosomy 14, trisomy 14 and monosomy 21 are lethal (Fig. 12.21). The remaining possibilities are child with normal chromosomes, a child with balanced translocation and a child with Down's Syndrome due to unbalanced form of translocation causing trisomy 21.

Translocations occur in human, usually in association with a normal chromosome set in a translocation heterozygote. One very common example is familial Down syndrome or trisomy 21 chromosome, and presence of extra 21 chromosome results in structural and physical impairments in patient (Fig. 12.21). Mental retardation may vary from mild to severe. Sometime the whole chromosome is not duplicated; instead a faulty gamete having an extra portion of chromosome 21 present on chromosome number 14 or on normal chromosome number 21. This extra piece of 21st chromosome creates trisomy-like condition, and the new born contains Down syndrome symptoms (Moore and Best 2001). Translocation heterozygote or carrier can repeatedly produce children with Down syndrome. The carriers can do the same in subsequent generations.

12.1.4.3 Translocations and Tumors

There is a variety of chromosomal mutations associated with tumors which may include small chromosomal rearrangement to complex changes in chromosomal structure. Most commonly, translocations are associated with tumor developments. Genetic analysis of tumors shows a variety of mutations, but some are associated to a specific chromosomal abnormality.

Translocation causes tumors in different ways which depend on the part of genome involved in translocation and the site where it integrates in donor chromosome. Usually there is no net gain or loss of genetic material, and the total genetic material remains the same, and patient shows normal phenotype. Sometimes the translocation alters the promoter context, and this causes abnormal gene expression as it happens in Burkitt lymphoma where part of chromosome 8 is translocated on 14th chromosome t(8;14). In another case, the proto-oncogene like *C-myc* is placed in front of highly expressed immunoglobulin gene or promoter which causes the overexpression of proto-oncogene leading to tumor development, e.g., t(14;18) in human follicular lymphoma. Sometimes if translocation occurs within the gene, it alters its structure so that the function of gene may be altered, e.g., chronic myelogenous leukemia with t(9;22), i.e., translocation of part of 9–22 results in a fusion product of gene which is different than the wild type genes.

Glossary

A

Adjacent 1 segregation Adjacent nonhomologous centromeres move to the same pole. This results in an unbalanced chromosomal complement that will result in a zygote with partial trisomy of one chromosome and partial monosomy of the other when fertilized by a normal haploid gamete. This segregation pattern often is compatible with viability.

Adjacent 2 segregation Adjacent homologous centromeres move to the same pole; this usually results in large amounts of unbalanced chromatin, which is usually incompatible with embryonic survival.

Allele One of the possible mutational states of a gene, distinguished from other alleles by phenotypic effects.

Alternate segregation It is an example of balance segregation in which both normal homologues move to one pole and translocation homologues move to another pole.

Anal atresia Anal atresia is an imperforate anus. In anal atresia, the tissue closing the anus may be several centimeters thick or just a thin membrane of skin.

Angelman syndrome A rare congenital disorder characterized by mental disability and a tendency to jerky movement, caused by the absence of certain genes normally present on the copy of chromosome 15 inherited from the mother.

B

Balanced translocation Balanced translocations are the one with an even exchange of material and with no extra or missing genetic information. These translocations have full functionality. The unbalanced translocations are the ones in exchange of chromosome material which is unequal resulting in extra or missing genes.

Beckwith-Wiedemann syndrome Beckwith-Wiedemann syndrome is an overgrowth disorder usually present at birth. It can also cause child behavior problems. Common features are large tongue, above average birth weight and length, midline abdominal wall, ear creases or ear pits, and low blood sugar after birth.

C

Chromosomal aberration Any change resulting in the duplication, deletion, or rearrangement of chromosomal material. Abnormal structure or number of chromosomes includes deficiency, duplication, inversion, translocation, aneuploidy, polyploidy, or any other change from the normal pattern.

Chronic myelogenous leukemia Chronic myeloid leukemia (CML), also known as chronic myelogenous leukemia, is a type of cancer that starts in the blood-forming cells of the bone marrow and invades the blood.

Coloboma A coloboma is a hole in one of the structures of the eye, such as the iris, retina, choroid, or optic disc.

Cri-du-chat syndrome Cri-du-chat (cat's cry) syndrome, also known as 5p- (5p minus) syndrome, is a chromosomal condition that results when a piece of chromosome 5 is missing. Infants with this condition often have a high-pitched cry that sounds like that of a cat. The disorder is characterized by intellectual disability and delayed development, small head size (microcephaly), low birth weight, and weak muscle tone (hypotonia) in infancy. Affected individuals also have distinctive facial features, including widely set eyes (hypertelorism), low-set ears, a small jaw, and a rounded face. Some children with cri-du-chat syndrome are born with a heart defect.

Crossing over The exchange of chromosomal material (parts of chromosomal arms) between homologous chromosomes by breakage and reunion. The exchange of material between nonsister chromatids during meiosis is the bases of genetic recombination.

D

Deficiency (deletion) A chromosomal mutation involving the loss or deletion of chromosomal material. Deficiency heterozygotes are hemizygous for the genes located in the deleted segment; many deficiencies produce genetic effects similar to gene mutations. Deletion also occurs when a block of one or more nucleotide pairs is lost from a DNA molecule.

Dicentric chromosome It is an abnormal chromosome with two centromeres. It is formed through the fusion of two chromosome segments, each with a centromere, resulting in the loss of acentric fragments and the formation of dicentric fragments.

Disjunction An act of disjoining.

Down syndrome An abnormal human phenotype including mental retardation, due to a trisomy of chromosome 21; more common in babies born to older mothers.

Duplication Gene duplication or chromosomal duplication involves duplication of a region of DNA that contains a gene.

F

Follicular lymphoma Follicular lymphoma is a type of non-Hodgkin lymphoma. It develops when the body makes abnormal B lymphocytes – the lymphoma cells. (B lymphocytes are white blood cells that fight infection). The lymphoma cells build up in lymph nodes. The most common symptom is a painless swelling in the neck, armpit, or groin.

G

Genetic imbalance It is to describe situation when the genome of a cell or organism has more copies of some genes than other genes due to chromosomal rearrangements or aneuploidy.

H

Heterozygous A diploid organism is heterozygous at a gene locus when its cells contain two different alleles of a gene. The cell or organism is called a heterozygote specifically for the allele in question; therefore, heterozygosity refers to a specific genotype.

Homologous Chromosomes that occur in pairs and are generally similar in size and shape, one having come from the male parent and the other from the female parent. Such chromosomes contain the same array of genes.

Homologous centromeres During meiosis, homologous chromosomes pair and then segregate from each other at the first meiotic division. Homologous centromeres appear to be aligned when chromosomes are paired.

Hyperphagia An abnormally great desire for food; excessive eating.

Hyperplastic genitalia Underdeveloped or incompletely developed genitalia.

I

Inversion Rotation of a segment of a chromosome by 180° so that the genes in this segment are present in the reverse order; characteristic inversion loops are produced during meiosis in the inversion heterozygotes.

Interstitial deletion Deletion that does not involve the terminal parts of a chromosome.

K

Klinefelter's syndrome Sterile human males with the *XXY* chromosome constitution; other associated symptoms as well.

L

Linkage analysis It is the study aimed at establishing linkage between genes. It is the tendency for genes and other genetic markers to be inherited together because of their location near one another on the same chromosome. Linkage analysis is used for gene hunting and genetic testing.

M

Mental retardation It is a condition diagnosed before age 18, usually in infancy or prior to birth that includes below-average general intellectual function and a lack of the skills necessary for daily living.

Microcephaly Abnormal smallness of the head, a congenital condition associated with incomplete brain development. It is a rare neurological condition in which an infant's head is significantly smaller than the heads of other children of the same age and sex. Sometimes detected at birth, microcephaly usually is the result of the brain developing abnormally in the womb or not growing as it should after birth. Microcephaly can be caused by a variety of genetic and environmental factors.

Microdeletion syndrome Microdeletion syndrome is a syndrome caused by a chromosomal deletion (<5 Mb) spanning several genes that is too small to be detected by conventional cytogenetic methods or high resolution karyotyping (2–5 Mb).[1][2] Detection is done by fluorescence in situ hybridization (FISH). Larger chromosomal deletion syndromes are detectable using karyotyping techniques.

Microduplication syndrome Microduplications, or submicroscopic duplications, are chromosomal duplications that are too small to be detected by light microscopy using conventional cytogenetics methods. Specialized testing is needed to identify these duplications. 1q21.1 duplication syndrome or 1q21.1 (recurrent) microduplication is a rare aberration of chromosome 1.

Mongolism It is a congenital disorder caused by having an extra 21st chromosome; results in a flat face and short stature and mental retardation.

N

Noncrossover chromatids Crossing over, or recombination, is the exchange of chromosome segments between nonsister chromatids in meiosis. Crossing over creates new combinations of genes in the gametes that are not found in either parent, contributing to genetic diversity. The chromatids of two homologous chromosomes are called nonsister chromatids.

P

Paracentric inversion A reversal of the normal order of genes in a chromosome segment involving only the part of a chromosome at one side of the centromere.

Pericentric inversion A reversal of the normal order of genes in a chromosome segment involving parts of a chromosome at both sides of the centromere.

Prader-Willi syndrome It is a rare congenital disorder characterized by learning difficulties, growth abnormalities, and obsessive eating, caused especially by the absence of certain genes normally present on the copy of chromosome 15 inherited from the father.

Proto-oncogene These genes code for proteins that help regulate cell growth. These important genes are called proto-oncogenes. A change in the DNA sequence of the proto-oncogene gives rise to an oncogene, which produces a different protein and interferes with normal cell regulation.

Pseudodominance It is the situation in which the inheritance of a recessive trait mimics a dominant pattern.

R

Reciprocal translocation Reciprocal translocations are usually an exchange of material between nonhomologous chromosomes.

Recombinant frequency When two genes are located on the same chromosome, the chance of a crossover producing recombination between the genes is related to the distance between the two genes. Thus, the use of recombination frequencies has been used to develop linkage maps or genetic maps.

Recombination The process by which offspring derive a combination of genes different from that of either parent; the generation of new allelic combinations. In higher organisms, this can occur by crossing over.

Ring chromosome A ring chromosome is a chromosome whose arms have fused together to form a ring. A ring chromosome is denoted by the symbol r in human genetics or R in *Drosophila* genetics.

Robertsonian translocation Translocation arising from breaks at or near the centromeres of two acrocentric chromosomes. The reciprocal exchange of broken parts generates one large metacentric chromosome and one very small chromosome.

S

Seizure A seizure is a sudden surge of electrical activity in the brain. A seizure usually affects how a person appears or acts for a short time.

Syndrome A group of symptoms that occur together and represent a particular disease.

T

Tandem The occurrence of two identical sequences, one following the other, in a chromosome segment.

Terminal tandem repeats Tandem repeats of nucleotide sequences reaching the extreme terminus of the chromosome.

Translocation Change in position of a segment of a chromosome to another part of the same chromosome or to a different chromosome.

Turner's syndrome In human beings; individuals having *XO* chromosome constitution, being phenotypically female, but having rudimentary sexual organs and mammary glands.

U

Unbalanced genetic rearrangement Chromosome structure abnormalities can be either unbalanced rearrangements or balanced rearrangements. The various unbalanced rearrangements involve deletion, duplication, or both. Deletion of a chromosome segment leads to partial monosomy of that segment. Duplication of a chromosome segment leads to partial trisomy of that segment. Any mutation leading to an imbalance like monosomy, trisomy, or others can result in an abnormal phenotype.

V

Viability It is the ability of a living organism to maintain itself or recover its potentialities.

Z

Zygotic mortality A form of reproductive isolation in which fertilization occurs but development stops soon after preventing much of the warm blood from reaching the surface of the body, where heat could be lost.

References and Additional Reading

Borgaonkar DS (1994) Chromosomal variation in man: a catalog of chromosomal variants and anomalies, 7th edn. Wiley-Liss, New York

Daniel A (1979) Structural differences in reciprocal translocations. Hum Genet 51:171–182

Fischer K, Froehling S, Scherer SW, Brown JM, Scholl C, Stilgenbauer S, Tsiu L-C, Lichter P, Doehner H (1997) Molecular cytogenetic delineation of deletions and translocations involving chromosome band 7q22 in myeloid leukemias. Blood 89(6):2036–2041

Giordano M, Gertosio C, Pagani S, Meazza C, Fusco I, Bozzola E, Bozzola M (2015) A 5.8 Mb interstitial deletion on chromosome Xq21.1 in a boy with intellectual disability, cleft palate, hearing impairment and combined growth hormone deficiency. BMC Med Genet 16:74. doi:10.1186/s12881-015-0220-z

Girirajan S, Rosenfeld JA, Cooper GM, Antonacci F, Siswara P, Itsara A, Vives L, Walsh T, McCarthy SE, Baker C, Mefford HC, Kidd JM, Browning SR, Browning BL, Dickel DE, Levy DL, Ballkif BC, Platky K, Farber DM, Gowans GC, Wetherbee JJ, Asamoah A, Weaver DD, Mark PR, Dickerson J et al (2010) A recurrent 16p12.1 microdeletion supports a two hit model for severe developmental delay. Nat Genet 42:203–209

Griffiths AJF, Wessler SR, Lewontin RC, Gelbart WM, Suzuki DT, Miller JH (2014) An introduction to genetic analysis, 10th edn. W.H. Freeman and Company, New York

Guilherme PS, Meloni VFA, Kim CA, Pellegrino R, Takeno SS, Coilin NK, Christofolini DM, Spinner NB, Kulikowski LD, Melaragno MI (2011) Mechanisms of ring chromosome formation, ring instability and clinical consequences. BMC Med Genet 12:171. doi:10.1186/1471-2350-12-171

Hirschhorn K (1973) Chromosomal abnormalities 1: autosomal defects. In: Mckusick VA, Clairborne R (eds) Medical genetics. HP Publishers, New York, pp 3–14

Jalbert P, Sele B, Jalbert H (1980) Reciprocal translocations: a way to predict the mode of imbalanced segregation by pachytene-diagram drawing. Hum Genet 55:209–222

Moore CM, Best RG (2001) Chromosomal genetic disease: structural aberrations. Encyclopedia of life sciences. Nature Publishing Group, pp 1–8. doi:10.1038/npg.els.0001452

Russell PJ (2010) iGenetics: a molecular approach, 3rd edn. Pearson Education, Inc., San Francisco

Schinzel A (1994) Human cytogenetics database. Oxford University Press, Oxford, pp 397–411

Numerical Changes in Chromosomes 13

Abdul Rauf Shakoori, Saira Aftab, and Farah Rauf Shakoori

Abstract

Every organism has basic specific number of chromosomes, which are constant for a species. Changes in the chromosomal number, however, do occur which reflect high inviability and phenotypic anomalies. Abnormal euploidy will result if whole set of chromosome is involved, and aneuploidy will result if parts of chromosomal set are involved. The most common abnormal euploids are polyploids such as triploids, tetraploids, etc. Allopolyploids can be made by crossing and doubling progeny chromosomes with colchicines. This technique has important applications in crop breeding. Aneuploids result in an unbalanced genotype with an abnormal phenotype. Monosomy ($2n-1$) and trisomy ($2n+1$) are examples of aneuploid. Aneuploid conditions such as Down syndrome, Klinefelter syndrome, and Turner syndrome are well studied in humans. It is believed to be due to chromosomal nondisjunction and constitutes major portion of genetically based ill-health in human population.

Keywords

Euploid • Aneuploid • Polyploid • Allopolyploidy • Monoploid • Chromosomal nondisjunction • Down syndrome • Klinefelter syndrome • Turner syndrome • Trisomy 8 • Trisomy 13 • Trisomy 18

A.R. Shakoori (✉) • S. Aftab
School of Biological Sciences, University of the Punjab, Quaid-i-Azam, Campus, Lahore 54590, Pakistan
e-mail: arshakoori.sbs@pu.edu.pk; saira3aftab@gmail.com

F.R. Shakoori
Department of Zoology, University of the Punjab, Quaid-i-Azam, Campus, Lahore 54590, Pakistan
e-mail: farah.shakoori.zool@pu.edu.pk

T.A. Bhat, A.A. Wani (eds.), *Chromosome Structure and Aberrations*,
DOI 10.1007/978-81-322-3673-3_13

13.1 Chromosome Number

All organisms contain specific number of chromosomes, which is constant for a particular species. The number varies from two (one pair) in jack jumper ant (*Myrmecia pilosula*) and round worm (*Ascaris lumbricoides*) to as many as 308 (154 pairs) in black mulberry (*Morus nigra*) and 268 (134 pairs) in butterfly (*Agrodiaetus shahrami*). A mosquito (*Aedes aegypti*) has 6 (3 pairs), chimpanzee 48 (24 pairs), and man (*Homo sapiens*) 46 (23 pairs) of chromosomes (Table 13.1).

The most commonly present chromosomal number or set of chromosomes is called euploidy number. A human has 46 chromosomes as diploid ($2n$) number in somatic cells, and its gametes (sperms and eggs) have 23 chromosomes which is haploid (n) number. The original set of chromosomes is called monoploid, and its presence in nature as duplicated form is diploidy. Aneuploidy is the condition in which an animal has an extra number of chromosome other than normal euploid chromosomes, e.g., trisomy in humans ($46 + 1 = 47$) and the absence of one ($46-1 = 45$) or more chromosomes from the euploid set of chromosomes are examples of aneuploidy. Aneuploidy generally results from nondisjunction of chromosome and chromosomal loss during the cell division. Nondisjunction occurs when one out of two daughter cells receives an extra chromosome in mitosis ($2n + 1$) or in meiosis ($n + 1$) as a result of nondisjunction of homologous chromosome. The other daughter cell or gamete undergoes loss of chromosome or nullisomy. In some organisms the whole euploid set of chromosomes is repeated several times (polyploidy) making new organism triploid ($3n$) or tetraploid ($4n$) (Fig. 13.1).

13.2 Abnormal Chromosomal Number/Aneuploidy

Changing in chromosome number is usually classified into those changes involving whole chromosome sets and those involving parts of chromosome sets. Any change in chromosomal number leading to disturbance of naturally occurring balance of chromosome is called aneuploidy. In this chapter we will discuss different forms of aneuploidy in plants and animals, and these different forms could be monoploidy, polyploidy, hyperploidy, or hypoploidy.

13.2.1 Monoploidy

Monoploidy or haploidy is a naturally occurring phenomenon in insects like bees, wasps, and aunts which give rise to male population from unfertilized eggs (Fig. 13.2). This process in which the development of an organism occur from an egg without fertilization is known as parthenogenesis. Monoploid cells do not undergo meiosis; hence the gamete production in honeybee and wasp males is produced by mitosis.

Plant breeder uses monoploidy as a selective method to easily grow a plant variety containing desired features by a technique called anther culture (Fig. 13.3). In

Table 13.1 Organisms (plants and animals) listed according to the diploid number of chromosomes (2*n*)

No.	Name of the organisms	2*n*	No.	Name of the organisms	2*n*
1	Ciliate	10	12	Asiatic black bear	74
	(Tetrahymena thermophila)			*(Ursus arctos)*	
	In micronucleus				
2	Black mulberry	308	13	Black nightshade	72
	(Morus nigra)			*(Solanum nigrum)*	
3	Butterfly	268	14	Elk	68
	(Agrodiaetus shahrami)			*(Cervus canadensis)*	
4	Rattlesnake fern	184	15	Chinchilla	64
	(Botrypus virginianus)			*(Chinchilla lanigera)*	
				Horse	
				(Equus ferus caballus)	
5	Northern lamprey	174	16	Donkey	62
				(Equus africanus asinus)	
6	Carp	104	17	American bison	60
				(Bison bison)	
7	Red viscacha rat	102	18	Elephant	56
	(Tympanoctomys barrerae)				
8	Shrimp	86–92	19	Sheep	54
	(Penaeus semisulcatus)			*(Ovis orientalis aries)*	
9	African hedgehog	90	20	Cotton	52
				(Gossypium anatinus)	
10	Pigeon	80	21	Zebra fish	50
	Turkey			*(Danio rerio)*	
11	Chicken	78	22	Chimpanzee	48
	(Gallus gallus domesticus)			*(Pan troglodytes)*	
				Gorilla	
				Potato	
				(Solanum tuberosum)	
				Water buffalo	
				(Bubalus bubalis)	
23	Human	46	32	Cannabis	20
	(Homo sapiens)			*(Cannabis sativa)*	
				Maize	
				(Zea mays)	
24	Dolphin	44	33	Cabbage	18
	Rabbit			*(Brassica oleracea)*	
	(Oryctolagus cuniculus)				

(continued)

Table 13.1 (continued)

No.	Name of the organisms	2*n*	No.	Name of the organisms	2*n*
25	Rat	42	34	Kangaroo	16
	(Rattus norvegicus)				
	Rhesus monkey				
	(Macaca mulatta)				
	Wheat				
	(Triticum aestivum)				
26	Mango	40	35	Pea	14
	(Mangifera indica)			*(Pisum sativum)*	
	Mouse				
	(Mus musculus)				
27	Cat	38	36	Nematode	12/11
	(Felis sivestris catus)			*(Caenorhabditis elegans)*	
	Lion				
	(Panthera leo)				
28	Earthworm	36	37	Fruit fly	8
	(Lumbricus terrestris)			*(Drosophila melanogaster)*	
29	Porcupine	34	38	Mosquito	6
				(Aedes aegypti)	
30	Honeybee	32	39	Australian daisy	2
	(Apis mellifera)			*(Brachyscome dichromosomatica)*	
31	Rice	24		Jack jumper ant	
	(Oryza sativa)			*(Myrmecia pilosula)*	
				Round worm	
				(Ascaris lumbricoides)	

https://en.wikipedia.org/wiki/List_of_organisms_by_chromosome_count

anther culture a pollen grain destined cell may be treated with an alkaloid colchicine which inhibits the spindle fiber development and hence blocks the cell division. These individual cells which have now two sets of chromosomes can be allowed to grow into a mass known as callus in the presence of specific culture media, and then this mass may be seeded to grow into mature plant (Fig. 13.4). This technique has helped to selectively raise insect- and pest-resistant population of plants like soybean and tobacco (Fig. 13.5).

13.2.2 Polyploids

Polyploidy is a condition in which multiple number of natural set of chromosomes is present in the organism. Depending upon the number of chromosomal sets, the organisms are known as triploid (with 3 chromosomes), tetraploid (with 4*n* chromosomes), pentaploid (with 5*n* chromosomes), and so on (Fig. 13.1).

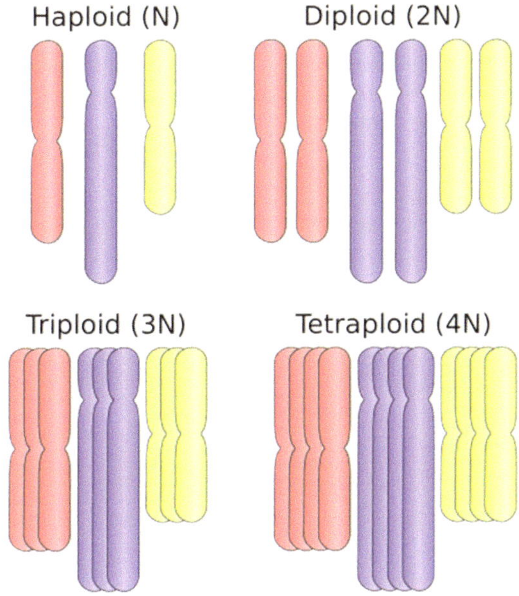

Fig. 13.1 The original set of chromosomes is called monoploid or haploid and its presence in nature in duplicated form is known as diploid. Organisms containing more than two sets of chromosomes are polyploids. Triploid (*3n*) and tetraploids (*4n*) are shown in the figure

13.2.2.1 Polyploidy in Plants

The polyploids evolved from diploids and became established in nature by accidental somatic doubling during mitosis or irregular reductional division in which sets of chromosomes fail to separate completely to the poles in the anaphase of reproductive cell. Once polyploidy is established, intercrossing among plants with different chromosome numbers may give rise to numerous chromosome combinations. Most of these are sterile, but some may be fertile.

Some plant groups have a series of chromosome numbers based on a multiple of a basic number. In the genus *Chrysanthemum*, for example, the basic number is 9 and species are known that have 18, 36, 54, 72, and 90 chromosomes.

Triploids are obtained when a tetraploid ($4n$) and diploid ($2n$) are crossed resulting in population of plants containing triploid gametes ($3n$) which could be further raised as triploid. It could be autopolyploid ($n_1 + n_1$), if the multiple sets of chromosomes are from the same species or allopolyploid ($n_1 + n_2$), if the sets of chromosomes are from different though closely related species. Different varieties of plants in this present day have been developed by this method to accumulate favorite traits of closely related plant varieties. An autotetraploid can be developed by nonsegregation of chromosomes, accidentally, or by treatment with colchicines during cell division.

The first classical allopolyploid was first developed in 1928 by G. Karpechenko in an attempt to make a fertile offspring in a cross between cabbage (*Brassica*) and radish (*Raphanus*). Although these two species are different, they are closely related and both have 18 chromosomes naturally which allows an efficient intercross. He was successful in raising a hybrid, but the resultant plant variety produced was sterile and

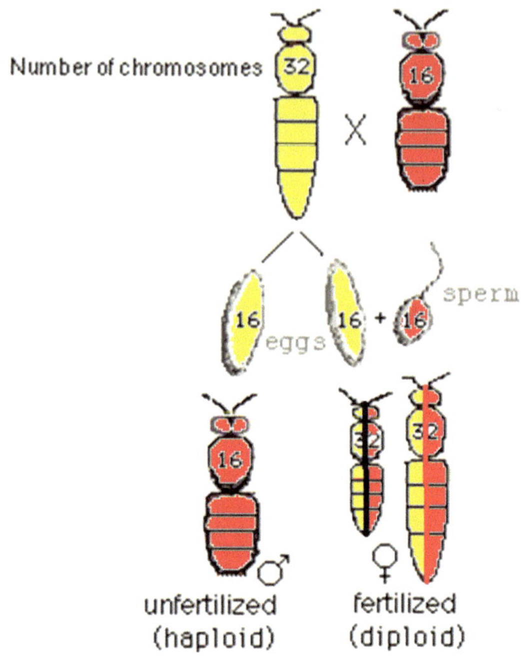

Fig. 13.2 Parthenogenesis in members of Hymenoptera (ants, bees, wasps, and sawflies) which produce haploid male ($n = 16$), while the queen honeybee ($n = 32$) and workers ($n = 32$) are diploid and born from a fertilized egg. This mechanism of sex determination is called haplodiploidy (http://www.glenn-apiaries.com/genetics.html)

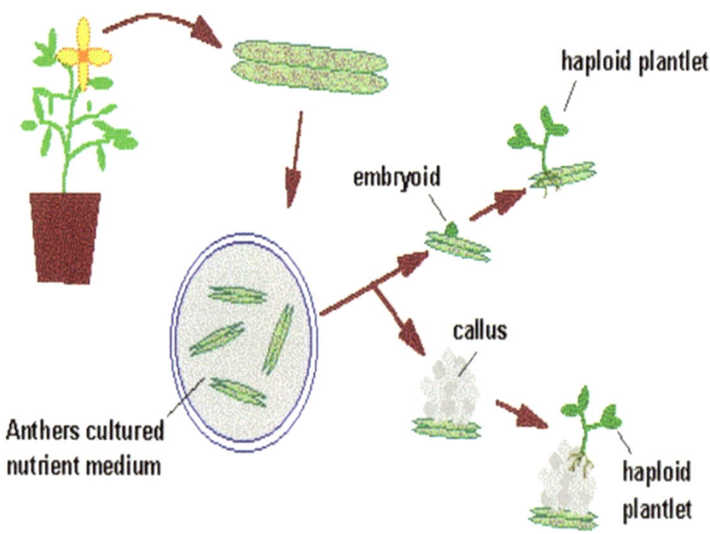

Fig. 13.3 Steps involved in anther culture

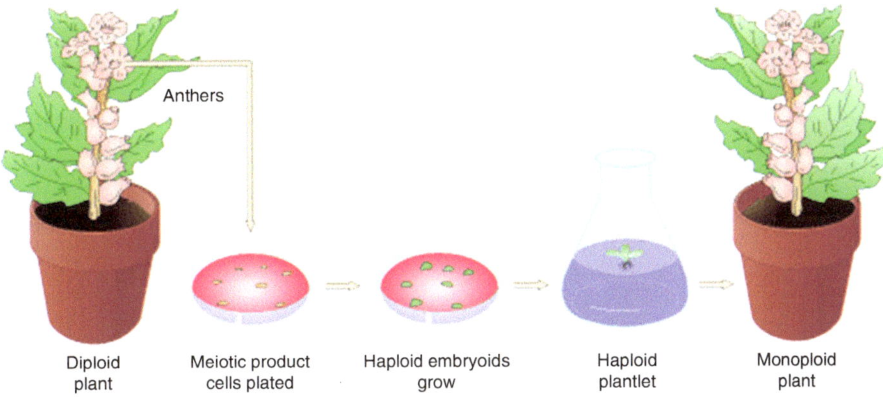

Fig. 13.4 Cells from plant anther are cultured in agar medium and then in liquid plant culture containing specific amino acids which help the development of haploid plantlet, and this plantlet may be grown into full plant with desired traits (http://agridr.in/tnauEAgri/eagri50/GBPR111/lec20.pdf)

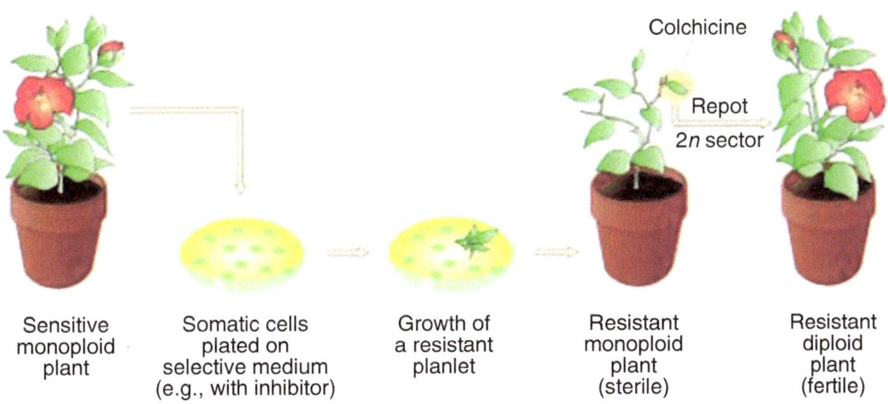

Fig. 13.5 Resistant diploid plant is isolated growing number of somatic cells on agar medium with inhibitors and mutagens to create mutation. Resistant plantlet is selected on basis of desired mutation in agar medium. This plantlet may be treated with colchicine to get a diploid resistant plant which is also fertile (http://agridr.in/tnauEAgri/eagri50/GBPR111/lec20.pdf)

was unable to cross to give rise to new progeny as the nine out of 18 chromosomes of both were too different to pair in normal cell division (Fig. 13.6). Some seeds produced from these plants later on were successful to give rise to new plant species and their chromosomal number was 36. These new variety of plants were fertile and had equal proportion of chromosomes from both species of parent plants. This type of individual is called amphidiploids. Amphidiploids of Karpechenko had roots of cabbage and radish leaves. He called his new plant variety as *Raphanobrassica.*

Using this technique a new variety of wheat (*Triticale*) has been produced by intercrossing wild-type wheat (*Triticum*) and rye (*Secale*). *Triticale* combines the

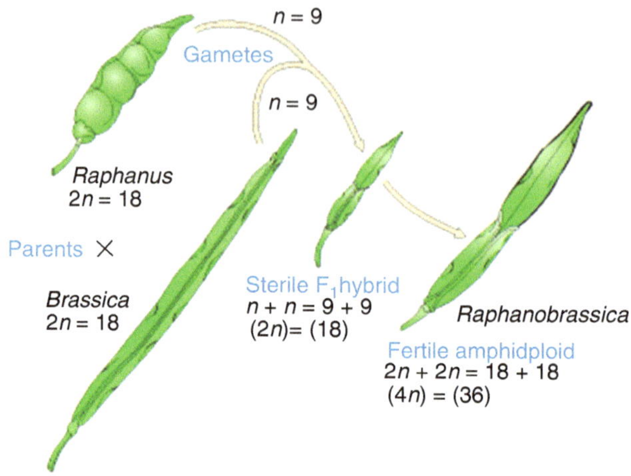

Fig. 13.6 Development of an amphidiploid *Raphanobrassica*, which is a hybrid of radish (*Raphanus*) and cabbage (*Brassica*) by G. Karpechenko in 1928. Both radish and cabbage have 18 chromosomes ($2n$), which on crossing form sterile F hybrid ($n + n = 9 + 9$). Some seeds of this hybrid plant develop into a fertile plant containing 36 chromosomes ($2n + 2n = 18 + 18$; $4n = 36$) http://agridr.in/tnauEAgri/eagri50/GBPR111/lec20.pdf)

high yield of wheat and ruggedness of rye (Fig. 13.7). Bread wheat (*Triticum aestivum*) with 42 chromosomes was produced from a cross between *T. monococcum* (14 chromosomes) and goat grass (*Aegilops*) with 28 chromosomes. Likewise the hybrids of *Gossypium* show a wide range of vigorous fertility. Old world cotton has 13 pairs of large chromosomes. American cotton has 13 pairs of small chromosomes. New World cotton has 26 pairs, 13 large and 13 small. Evidently hybridization and chromosome duplication occurred somewhere in the ancestry of the New World cotton.

13.2.2.2 Polyploidy in Animals

Polyploidy in the animal kingdom is not very common and is found in lower animals like flat worms, leeches, shrimps, and insects. Among insects it is common in ants, wasps, bees, beetles, flies, and butterflies. It is also common in some species of fish and lizards. These animals reproduce by a method called parthenogenesis which lacks normal reduction division of meiosis and cells are divided by mitosis. The offspring are produced by an unfertilized female. A vast majority of insects have polyploidy as a mechanism of sex differentiation. Insects have an advanced type of parthenogenesis in which meiotic division is suppressed and eggs undergo a single maturation division called apomixis. Among insects *Saga pedo* (Pallas) (Fig. 13.8a) which is a grasshopper commonly found in countries in the north of Mediterranean is a tetraploid, and *Physokermes hemicryphus* (Dalman), a hemipteran insect, is triploid. Many members of Lepidoptera, Diptera, and Coleoptera are also polyploid.

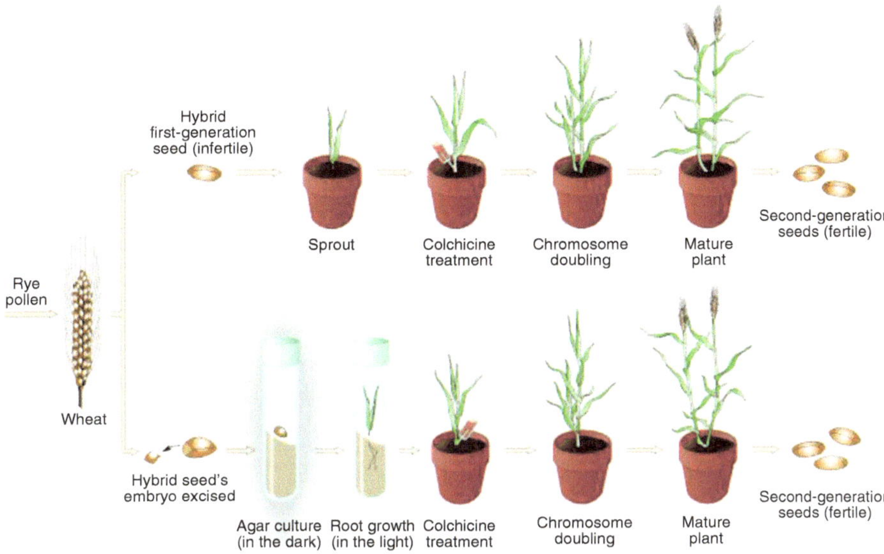

Fig. 13.7 Amphidiploid *Triticale* is developed by intercross between rye pollens and wild type of wheat. The hybrid infertile first generation may sprout on sowing while on treatment with colchicine will develop into a mature plant. The second-generation seed would be fertile. Alternatively the hybrid seed's embryo can grow on culture medium, which on treatment with colchicine will have its chromosomes doubled, and the mature plant so developed will have second-generation fertile seeds (http://agridr.in/tnauEAgri/eagri50/GBPR111/lec20.pdf)

Polyploidy is also found in some species of fish. Initially it was thought that self-fertilization is a basic requirement for polyploidy which is a common phenomenon in plants and is also found in species of fish, but as a matter of fact out of many polyploidy species of fish, only one has self-fertilizing ability, *Kryptolebias (=Rivulus) marmoratus* (Fig. 13.8b). *Catostomids* family (fresh water suckers) is tetraploid. Some very common varieties of fish like goldfish, minnow, and carp are also tetraploid. Figure 13.8c shows a triploid female of sunshine bass, which is bigger in size than the normal diploid female fish.

It was thought that since animals need to have their sex determination system based on chromosome type instead of number so there are less polyploids in the animal kingdom as compared to plants, now new data on chromosomal studies suggests that polyploidy has played an important role in species selection and speciation in the animal kingdom as well.

Among reptiles only some varieties of lizards are known to be naturally polyploids, and they reproduce parthenogenetically (Fig. 13.9). Compared to reptiles, amphibians have a variety of polyploids. Urodeles have triploid females which are supposed to reproduce gynogenetically, a process in which a sperm is donated by male of related bisexual and diploid species and this sperm activates the egg (Fig. 13.9). Members of *Xenopus* may be tetraploid (4x), like *X. vestitus* with 72 chromosomes, and hexaploid (6x), e.g., *X. ruwenzoriensis* with 108 chromosomes. Members

Fig. 13.8 (**a**) A tetraploid grasshopper *Saga pedo* Pallas; (**b**) a triploid hermaphrodite mangrove fish, *Kryptolebias (= Rivulus) marmoratus*; (**c**) the sunshine bass on top is a diploid female which is filled with eggs (gravid). In contrast, the female on the *bottom* is triploid. She is bigger, even compared to a gravid fish. The gigas effect in polyploids is real and effects sport fishing. Everyone wants to catch a bigger fish (https://sites.google.com/site/orthopteraphotos/home/saginae/saga-pedo-pallas-1771, http://www.bbc.com/earth/story/20150519-the-most-extreme-fish-on-earth, https://www.google.com.pk/imgres?imgurl=http://4.bp.blogspot.com/-9-nojIwFZEs/UPRex_ipvxI/AAAAAAAABVY/VtJ9gWCq7kg/s1600/2-%252Bfish.jpg&imgrefurl=http://biologicalexceptions.blogspot.com/2013/01/carp-diem-polyploid-fish-seizeday.html&h=422&w=533&tbnid=kUJbzuLWhzKuaM:&docid=hGhHxxeduAFtMM&ei=xs5_VsG2HtOSuATJ7L2IDg&tbm=isch&ved=0ahUKEwjBsZGH-_vJAhVTCY4KHUl2D-EQMwgZKAAwAA)

of anuran family Leptodactylidae like *Ceratophrys dorsata* and *C. ornata* are octaploids ($8n = 104$, ancestral number is 13).

Meiosis was observed in females of triploid unisexual salamander *Ambystoma tremblayi* and in triploid lizard *Cnemidophorus (=Aspidoscelis) uniparens*. They have three sets of chromosomes instead of normal two. The male's sperm only stimulates egg development and it does not contribute its genetic material. It was found that in initial prophase hexaploid oogonial cells are produced and these oogonial cells turn into triploid eggs after reduction division of meiosis (Lewis 2012).

The fertile hybrids of European water frogs reproduce by hybridogenesis. This is a form of reproduction resembling parthenogenesis, but hemiclonal rather than completely asexual. Half the genome is passed intact to the next generation, while the other half is discarded. It occurs in some animals that are hybrids between species.

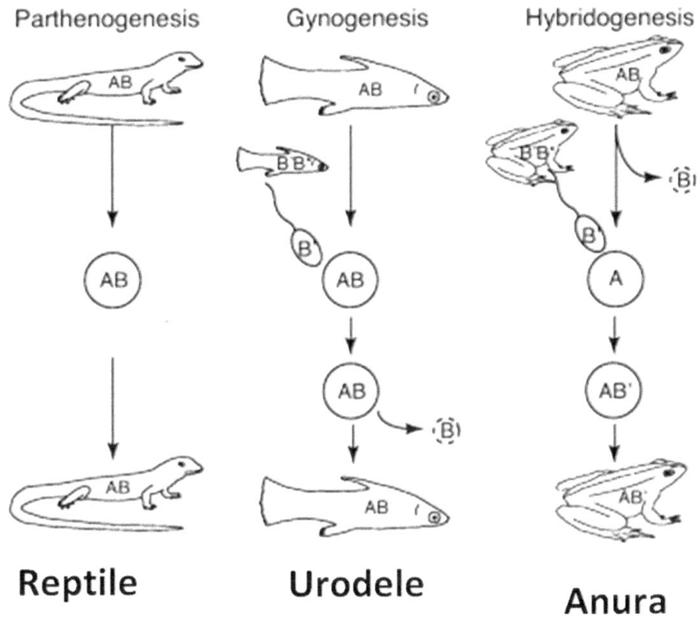

Fig. 13.9 Different modes of unisexual reproduction in animals. Parthenogenesis is a natural form of asexual reproduction in which growth and development of embryos occur without fertilization. In animals, parthenogenesis means development of an embryo from an unfertilized egg cell and is a component process of apomixis. Gynogenesis is a special form of sexual reproduction in which insemination is necessary, but the head of the sperm penetrating into the ovum does not transform into male pronucleus; and the gynogenetic embryo develops at the expense of the ovum nucleus only. Hybridogenesis is a form of reproduction resembling parthenogenesis, but hemiclonal rather than completely asexual: half the genome is passed intact to the next generation, while the other half is discarded. It occurs in some animals that are hybrids (http://alfa-img.com/show/examples-ofparthenogenesis.html)

13.2.3 Aneuploidy in Humans

The exact number of chromosomes in human discovered in 1956 by a group of Swedish scientists is 46 which includes 21 pairs of autosomes and two sex chromosomes (Fig. 13.10). Each human chromosome contains two arms named "p" for petite which means "small" in French and "q" for long arm, and letter "q" is chosen just because it's next to "p" alphabetically. Both long and short arms are connected to each other through a centromere. Both small and long arms have different banding pattern which is important from cytogenetic point of view.

Aneuploidy occurs when spindle fibers fail to attach to any of the chromosome due to nondisjunction which in humans is mostly a lethal condition. According to a report by the American College of Obstetricians and Gynecologists (2005), almost one out of 150 babies is born with chromosomal abnormalities resulting in mild to severe physical defect including mental retardation. Babies born with abnormal chromosomes have either too many or too few chromosomes. Nothing can be done

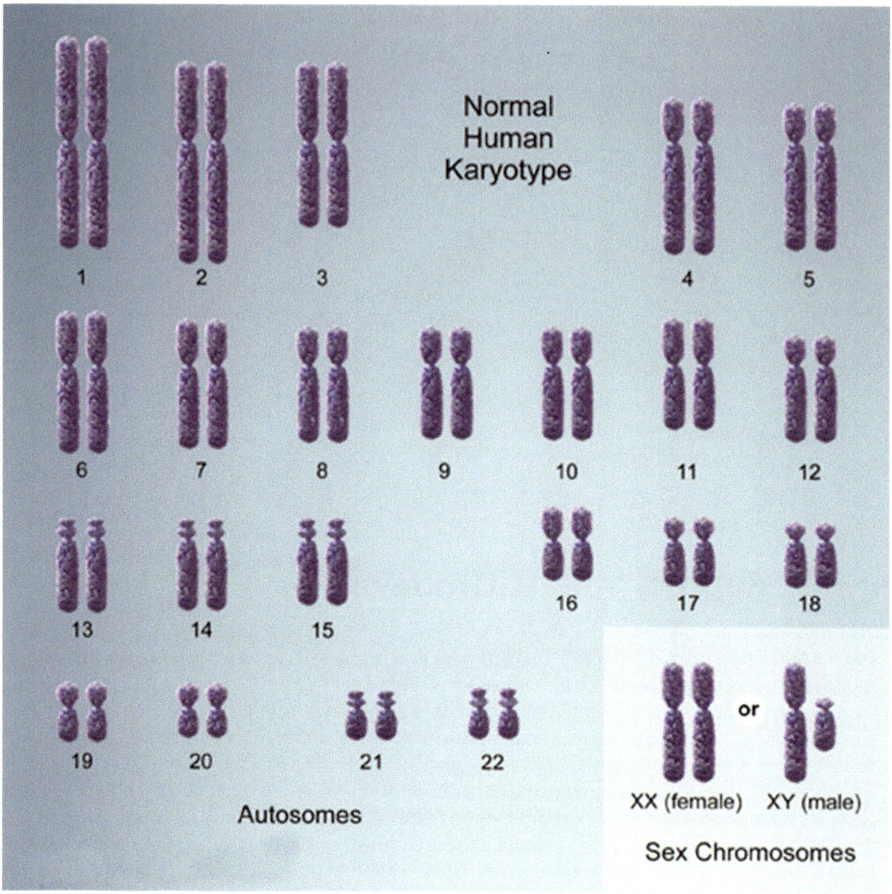

Fig. 13.10 Normal human karyotype with 23 pairs of chromosomes with 22 autosomes and one sex chromosome pair Jerome Lejeune with Down syndrome child (http://ghr.nlm.nih.gov/handbook/illustrations/normalkaryotype)

during pregnancies to prevent chromosomal abnormality as it is caused by the fusion of abnormal gametes either from the mother or from the father. According to this report almost 75 % of abortions are caused by an abnormal number of chromvosomes in the fetus. Most of the changes in chromosomal number or set of chromosomes are lethal and result in abortion or stillbirth. Only some changes in human chromosomes result in live births, though the newborn suffers from serious illnesses and different anomalies designated as syndrome. One of the most common aberrations in human are trisomies ($2n + 1$) in which one chromosome is extra in genome making 47 chromosomes.

13.2.3.1 Trisomy ($2n + 1$)

Trisomy is a condition in which an organism or a cell has one chromosome in excess of the normal somatic complement of the species. One chromosome has three homologous in place of the normal two. It is designated by $2n + 1$. Trisomy in humans results in mental retardation of newborn, physical abnormalities, and some carcinomas as well. Generally trisomy in humans is divided into two main types, trisomy in autosomes and trisomy of sex chromosomes. We will discuss both types one by one.

13.2.3.1.1 Trisomy in Autosomes

Most commonly known autosomal trisomies of chromosome number 21 (Down syndrome), 18 (Edward syndrome), 13 (Patau syndrome), and 8 chromosome are known (Table 13.2).

Trisomy 21 (Down Syndrome)

Down syndrome or trisomy 21 formerly known as Mongolism (due to facial features resembling Mongolian people) is now designated as Down syndrome after Langdon Down who first described the clinical signs in 1866. Jerome Lejeune, a French pediatrician, discovered the link of Down syndrome with chromosomal abnormality in 1958 (Fig. 13.11).

Since Down syndrome causes mental retardation, it was considered as the sole reason of mental retardation in humans for a long time. Other abnormalities associated to Down syndrome are dementia, hypothyroidism, congenital heart diseases, and leukemia (National birth defects prevention network 2000). Children with Down syndrome have more digestive problems compared to normal children like gastroesophageal reflux or celiac disease, swallowing problems, and bowl blockage. These children also experience frequent colds, sinuses, and some infections of the ear and poor muscular tone leading to various physical defects shown in Table 13.3.

The presence of just one extra chromosome (Fig. 13.12) causes several abnormalities related to cardiac, respiratory, and digestive systems. A list of some malformations observed in Down syndrome is given below in Table 13.3.

According to Parker et al. (2010), almost one out of 700 babies born has Down syndrome only in Europe. The remaining reflect fetal loss due to spontaneous abortion. This can be attributed to defect in the maternal and paternal gametes (Fig. 13.13). Recent studies have linked trisomy 21 to mother's age. Chances of women having a Down syndrome child at the age of 30 are 1/1000 and will probably increase as the age increases, e.g., at the age of 35 its 1/400 and at 40 its 1/100. A small number of people have Down syndrome due to translocation of a piece of chromosome 21 to some other chromosome instead of complete trisomy, e.g., "21q21q Robertsonian translocation."

Human chromosome 21 is 33.5 Mb long and only its small portion codes for protein. It contains 127 known genes, 98 predicted genes, and 59 pseudogenes. Recent research has enabled scientist to discover Down syndrome causing loci on chromosome 21, and now different genes are known to cause different abnormalities in Down syndrome (Hattori et al. 2000). Table 13.4 below shows different genes known to cause different abnormalities in Down syndrome.

Recent studies have shown that patients with Down syndrome have reduced chances of solid tumors which could be either due to extra dosage of tumor

Table 13.2 Aneuploidy resulting from nondisjunction in the human population

Chromosome nomenclature	Chromosome formula	Clinical syndrome	Estimated frequency at birth	Main phenotypic characteristics
47,+21	$2n + 1$	Down	1/700	Short broad hands with simian-type palmar crease, short stature, hyperflexibility of joints, mental retardation, broad head with round face, open mouth with large tongue, epicanthal fold
47,+13	$2n + 1$	Trisomy 13	1/20,000	Mental deficiency and deafness, minor muscle seizures, cleft lip and/or palate, polydactyly, cardiac anomalies, posterior heel prominence
47,+18	$2n + 1$	Trisomy 18	1/8000	Multiple congenital malformation of many organs; low-set, malformed ears; receding mandible, small mouth and nose with general elfin appearance; mental deficiency horseshoe or double kidney; short sternum 90 % die in the first 6 months
45,X	$2n-1$	Turner	1/2500 female births	Female with retarded sexual development, usually sterile, short stature, webbing of skin in neck region, cardiovascular abnormalities, hearing impairment
47,XXY	$2n + 1$	Klinefelter	1/500 male births	Male, subfertile with small testes, developed breasts, feminine pitched voice, long limbs, knock knees, rambling talkativeness
48,XXXY	$2n + 2$			
48,XXYY	$2n + 2$			
49,XXXXY	$2n + 3$			
50,XXXXXY	$2n + 4$			
47,XXX	$2n + 1$	Triple X	1/700	Female with usually normal genitalia and limited fertility. Slight mental retardation

suppressor gene Ets2 on chromosome 21 (Sussan et al. 2008) or reduced angiogenesis due to increased dosage of *Erg*, *Jam2*, *Adamts1*, and *Pttglip* (Reynolds et al. 2010).

Fig. 13.11 Jerome Lejeune with Down syndrome child

Table 13.3 Anomalies in Down syndrome

Anomalies in systems	Defects
Cardiovascular system anomalies	Defects in valves
	Hypoplastic ventricles
	Anomalies of aortic arteries
	Anomalies of great veins
	Single umbilical artery
Respiratory system anomalies	Anomalies of bronchus, trachea, and larynx
Anomalies of lungs	
Anomalies of digestive system	Swallowing problems
	Gastroesophageal reflux
	Bowl blockage
Urinary system anomalies	Obstructive defects of ureter, urethra, and bladder
Eyes	Cataracts
Muscular and skeletal defects	Clubfoot
	Polydactyly
	Syndactyly
Nervous system anomalies	Mental retardation
Anomalies of genital organs	Undescended testicles
	Hypospadias and epispadias

Data modified from Torfs and Christianson (1998)

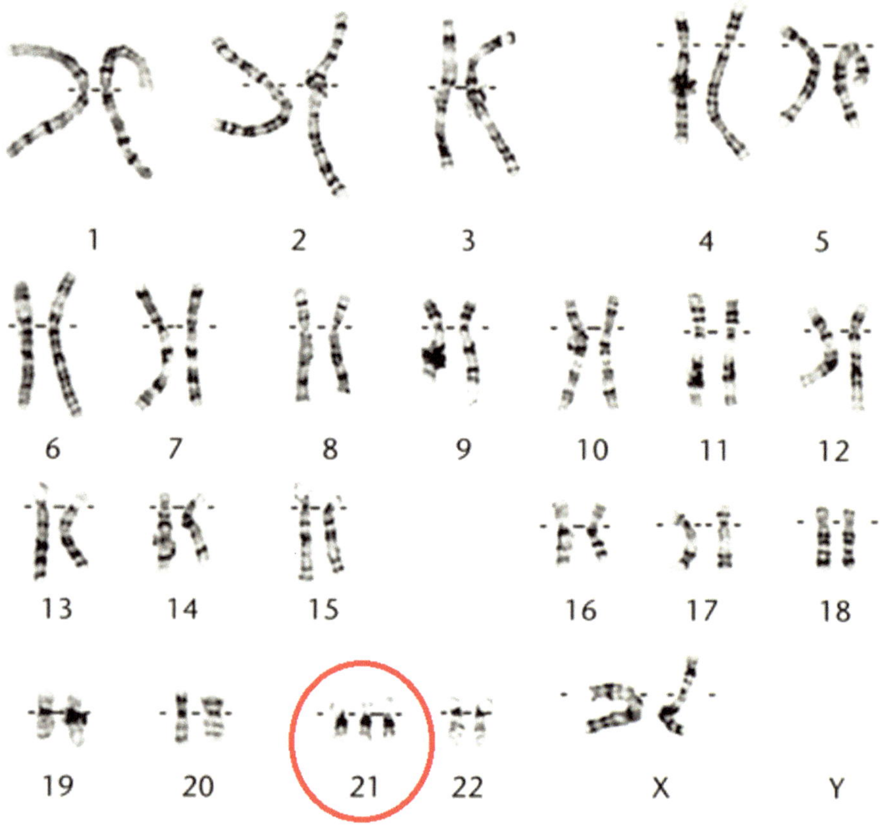

Fig. 13.12 Karyotype of a female patient with Down syndrome (Luthardt and Keitges 2001)

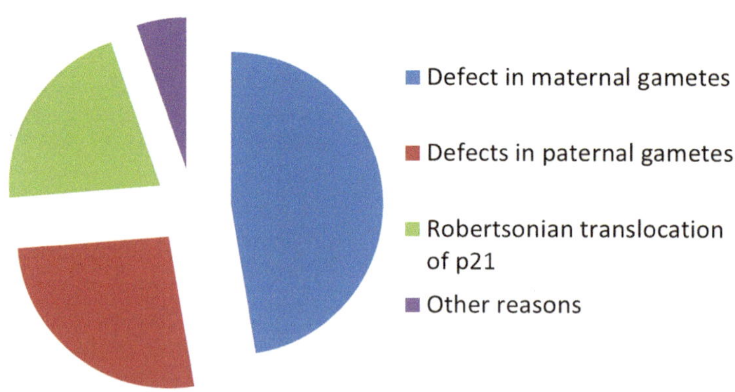

Fig. 13.13 Probable risk factors for Down syndrome (Modified from Gardner et al. 2004. Chromosome abnormalities and genetic counselling

Table 13.4 Genes related to physical defects in Down syndrome

Defects	Genes involved
Defects in learning, memory and brain developmental defects	*dscam,dyrk1a, sim2, app, synj1*
Neurodegenerative disorders	*app, dyrk1a*
Motor control	*app, dyrk1a, synj1*
Cardiac defects	*dscam, slc19a1, col6a1*
Leukemia	*ets2, erg*
Craniofacial alterations	*ets2*

Data modifies from Lana-Elola et al (2011)

Fig. 13.14 Down syndrome patients in sports and society (Various Internet sources)

Prenatal diagnosis plays an important role in early detection of Down syndrome, but nothing can be done to treat it. It is observed that society plays an important role to improve health and learning abilities of the patient. It is observed that life expectancy of a Down syndrome patient has increased from 25 to 50 years over the last few decades due to improved medical facilities and public awareness. According to a survey, kids with Down syndrome are more loveable than normal kids and have same emotional feeling as normal do, and little training has made parents to feel proud of their child (Fig. 13.14).

Trisomy 18 (47,+18; Edwards Syndrome)

Trisomy 18 (Fig. 13.15) is a relatively rare trisomy as compared to Down syndrome, but still it's the second most common autosomal trisomy and has frequency of about 1 in 8000 live births. Trisomy 18 is also known as Edwards syndrome named after John Hilton Edwards who discovered it in 1960 (Fig. 13.16).

Clinical features of Edwards syndrome include defective cardiac system and some congenital abnormalities including kidney problems (Table 13.2). Usually newborn with this condition dies within 6 months (Root and Carey 1994) mostly in 90 % of cases. Females are mostly affected by this condition, usually 80 % which is 3–4 times more than the affected males (20 %). Facial impairments include small head, prominent occiput or back part of the skull, small mouth, small jaw, short neck, low set and malformed ears, unusually shaped chest, short and prominent sternum, wide set nipples, crossed legs, flexed big toe, feet with rounded bottom, clenched hands with overlapping fingers, poorly developed finger nails, cleft or hole in iris, and low birth weight (Luthardt and Keitges 2001). Figure 13.17 shows the abnormalities of Edwards syndrome.

Trisomy 13 (47,+13)

Trisomy 13 (Fig. 13.18) is the third most common trisomy and is caused by an additional chromosome number 13 (Fig. 13.19). It was first identified by Patau and colleagues in 1960 when he discovered an additional chromosome in patient (Smith et al. 1962). Abnormalities common in Patau syndrome were prescribed in 1956 by Thomas Bartholin due to which this syndrome is also called Bartholin-Patau syndrome.

Incident rate of Patau syndrome is 1 in 20,000 live births per year (Delatycki and Gardner 1997). Seventy-five to eighty percent of cases have trisomy. There are different cases of trisomy 13. In one case the whole chromosome 13 is additional and is called classic Patau syndrome or D trisomy 13. Sometimes a part of chromosome 13 is translocated additional through Robertsonian translocation producing symptoms of Patau syndrome. Mosaicism is the case in which some cells of the body have trisomy 13. Usually the life span of a patient is short and a newborn dies in a few years, but some cases have been reported where a patient lives for as long as 11 years (Zoll et al. 1993).

Different anomalies include microcephaly, camptodactyly of both hands, wide gap between first and second toe, and unicuspid heart valves. Several defects in internal genitalia have also been reported like in female patients uterus didelphys (with two vaginas), small ovaries, and multinucleated ova. Male patients had cryptorchidism, small and abnormal scrotum, hypospadias, and many defects in brain development (Moerman et al. 1988).

Trisomy 8

Trisomy 8 was first time reported in 1975 in a 2-month-old boy with aneuploidy and mosaicism for chromosome 8. Like other trisomies it also leads to some psychomotor retardation and bone and joint anomalies. Mild to moderate mental retardation is also prominent in patients. They have lower intellectual abilities and face difficulty

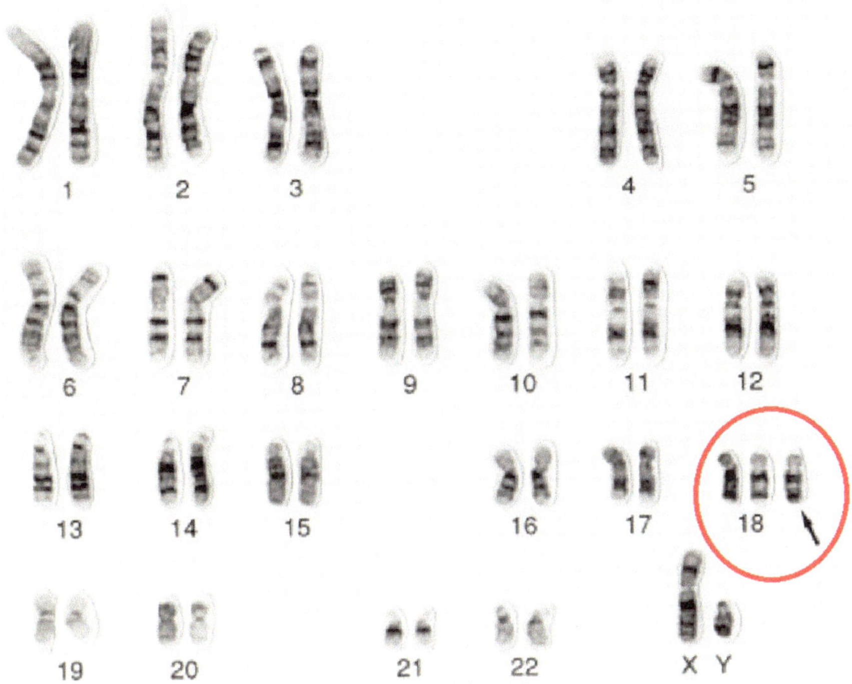

Fig. 13.15 Karyotype of a child with trisomy 18 syndrome, showing three no. 18 chromosomes

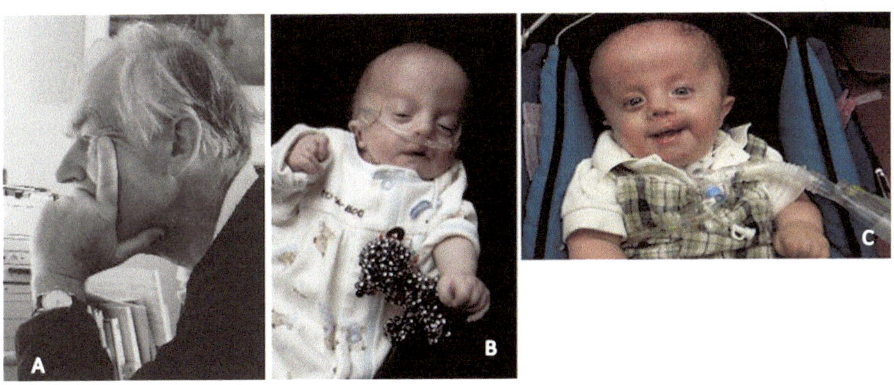

Fig. 13.16 (**a**) John Hilton Edwards; (**b**, **c**) patient with Edwards syndrome having all characteristic features including overriding fingers, tracheostomy, and prominent back skull (Cereda and Carey 2012) (https://www.google.com.pk/imgres?imgurl=https://sites.google.com/site/edwardssyndromejw/_/rsrc/1324046229871/home/edwards_syndrome.jpg%253Fheight%253D320%2526width%253D267&imgrefurl=https://sites.google.com/site/edwardssyndromejw&h=317&w=267&tbnid=fvJAhWDto4KHUeKBW8QMwgvKAAwAA)

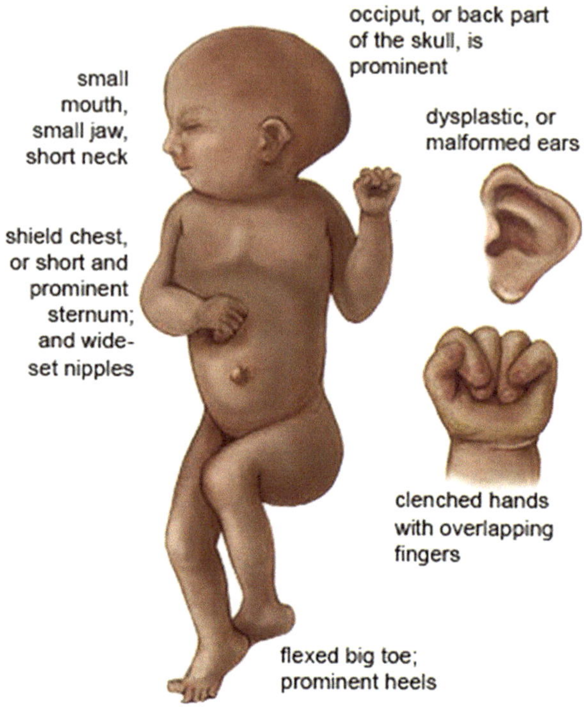

Fig. 13.17 Clinical features of Edwards syndrome

in reasoning and concentration and have poor memory. Trisomy 8 is found to be closely related to leukemia. Almost 9 % of acute myeloid leukemia patients have complete chromosome 8 trisomy; others have either mosaicism or Robertsonian translocation (Mrozek et al. 2004). These patients also have a number of mutations other than complete trisomy of chromosome 8. Almost 90 % of these patients have at least one mutation commonly *RUNX1*, *ASXL1*, and *FLT3* (round about 30 % each). Patients older than 60 years more commonly have these mutations than younger patients (Becker et al. 2014).

13.2.3.1.2 Trisomies in Sex Chromosomes

Normal human male has one X and one Y chromosome while female has two X chromosomes. Any change in normal number of sex chromosomes results in various abnormalities including infertility, mental retardation, structural anomalies, and development-related problems. Trisomies in sex chromosomes are sometimes undetectable, and patients may live normal life without being detected, e.g., XYY in males. We will study the abnormal distribution of sex chromosomes in males and females individually.

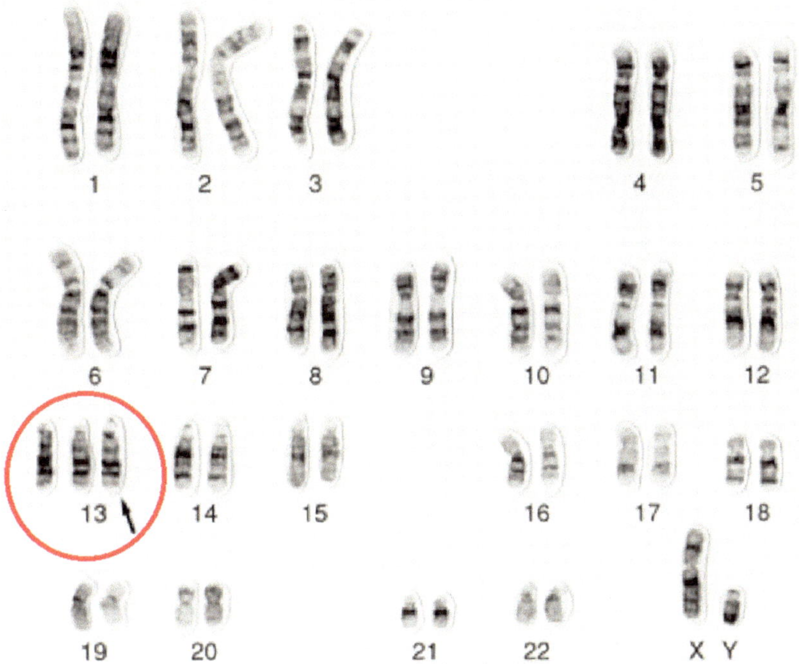

Fig. 13.18 Karyotype of a newborn with trisomy 13 syndrome, showing three no. 13 chromosomes

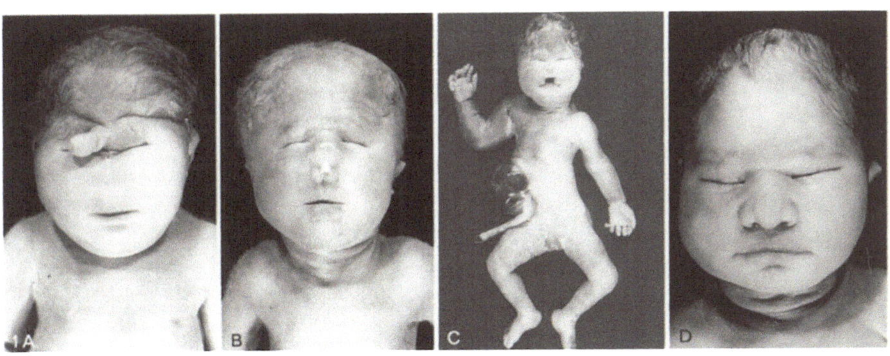

Fig. 13.19 Phynotypic variations in trisomy 13. (**a**) Cyclopia; (**b**) cebocephaly; (**c**) premaxillary agenesis; (**d**) typical holoprosencephaly (Images taken from original study by Moerman et al. 1988)

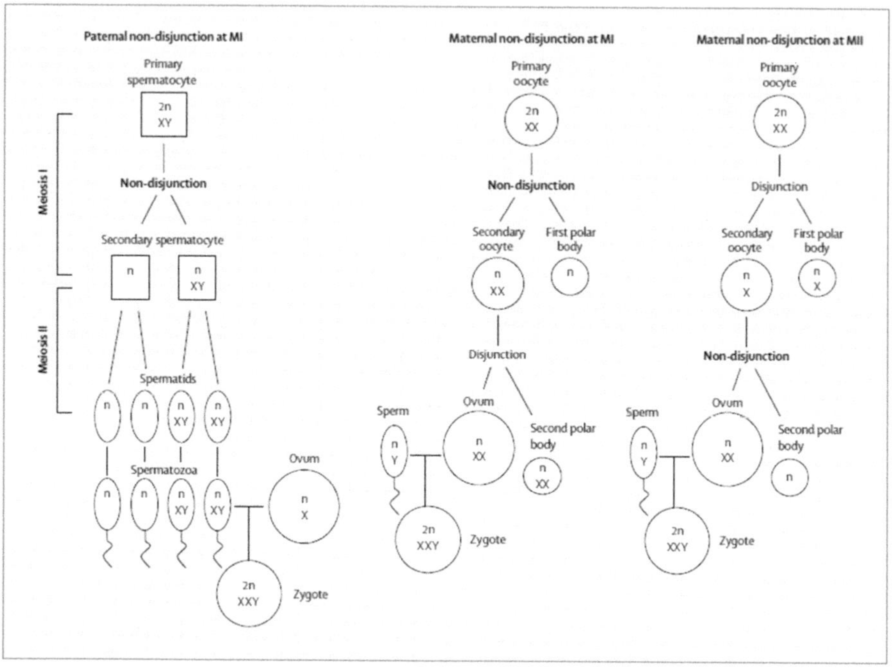

Fig. 13.20 Different forms of nondisjunction leading to the 47, XXY karyotype of Klinefelter syndrome (Lanfranco et al. 2004)

Sex Chromosome Trisomies in Males

47, XXY Klinefelter Syndrome A condition in male with an extra X chromosome is called XXY or Klinefelter syndrome first reported by Klinefelter et al. in 1942 and characterized by hypogonadism. Figure 13.20 shows different forms of nondisjunction leading to XXY karyotype of Klinefelter syndrome. According to a report in 2006 by the National Institute of Child Health and Human Development, the incidence of this syndrome is one in 500–1000 newborn baby boys. The major risk factor is the mother's age. Due to the presence of an extra X chromosome, males tend to have a female-like appearance with large breasts, very less facial and underarm hair growth, and usually small testes. Patients with Klinefelter syndrome usually have height greater than their parents, long legs, and higher chances of gaining weight (Table 13.2).

A complete trisomy results in infertility, but it is commonly observed that almost 50 % of the Klinefelter syndrome patients have sperms which is due to mosaicism of Klinefelter syndrome instead of complete trisomy in which some cells have extra X chromosome while others do not. Patients with Klinefelter syndrome have intellectual disabilities and are often misunderstood as lazy and shy people. Speech delay is very common in Klinefelter syndrome; there are many conditions of Klinefelter syndrome in which sex chromosomes are not in normal proportion like XXYY, XXXY, XXXXY, and XXY/XY. Here it is important to know that XXYY is due to total nondisjunction in paternal chromosomes while XXXXY is due to total nondisjunction maternal chromosomes. As the number of X chromosome increases in Klinefelter syndrome, the condition becomes severe with infertility, severe to complete mental retardation, and prominent physical abnormalities (Fraser et al. 1961).

Tartaglia et al. (2011) have described the medical problems in all these cases of Klinefelter syndrome which include asthma, food allergies, dental problems, cardiac malformation, congenital hip dysplasia, clubfoot, hypothyroidism, type II diabetes, and hypergonadotropic hypergonadism.

Patients with Klinefelter syndrome are also observed to develop various types of cancers like testicular cancer, non-Hodgkin's lymphoma, brain tumor, and various tumors of the digestive system. Breast cancer is not very common in men, but 4 % of all men having breast cancer are Klinefelter syndrome patients (Hasle et al. 1995). Treatment includes hormone therapy to treat hypergonadism, and speech therapy helps young Klinefelter syndrome patients for speech delay. Psychiatric problems are also observed in Klinefelter syndrome patients for which counseling and medication appear helpful.

47, XYY Syndrome.It is the most common trisomy in male population. Almost one out of every 1000 male is born with an extra Y chromosome (Table 13.2). For long it was considered to be related to socially aggressive behavior but no solid evidence has been reported so far. Many studies done on criminals have reported higher number of XYY karyotype persons. The IQ level of XYY is usually below average due to which these patients experience problems in academic life but with extra help it can be overcome. Many of XYY have been known to secure professional degrees. Poor social life is very prominent among XYY people and due to which their marital life has been observed to suffer badly. Usually patients are fertile and most of them are not diagnosed throughout their lives. These people experience delayed motor development or defects in motor coordination in younger age. Speech delay is also observed and patients undergo speech therapy. Usually these patients are taller in height. No specific physical or facial abnormality has been observed so far (Leggett et al. 2010).

Trisomies in Female (XXX, XXXX, XXXXX Meta Female) The presence of an extra X chromosome in female is called X trisomy, and it is a very common sex chromosomal abnormality in female population. One in every 1000 females has this syndrome and risk increases as the maternal age increases. Most of the females with this condition live their life without being ever detected. Usually females with one extra X chromosome have fertility and normal gonadal activity, but these females may have learning disabilities and lower IQ level and also tend to have early menopause as compared to normal people. There are other cases of X chromosomal abnormality where two and three extra X chromosomes are found, but they have severe symptoms and serious mental retardation (Dewhurst 1978). There are some cases where X chromosomal trisomy is found in mosaicism where some cells have an extra X chromosome while others do not. Females with mosaicism have lower fertility as compared to XXX females which are able to conceive and produce offspring, but this offspring sometimes has some abnormalities, while females with mosaicism usually face abortions and stillbirths, and if successful in delivering a child, then a newborn has many genetic abnormalities (Luthardt and Keitges 2001). Females with multiple X chromosomes have also been identified with some psychiatric problems like depression and having suicidal thoughts. These patients also

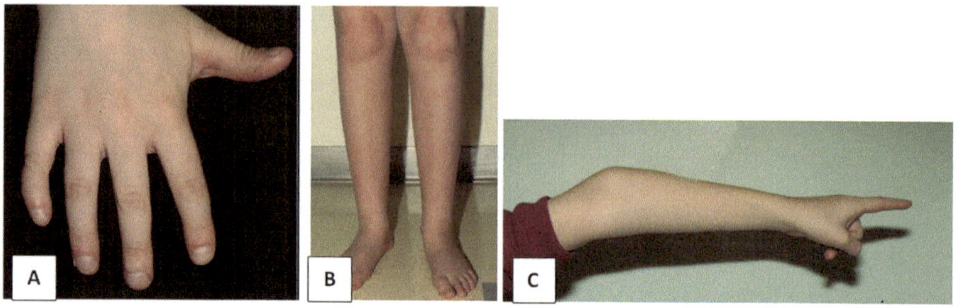

Fig. 13.21 Photograph of physical features in XXYY, XXXY, and XXXXY including (**a**) fifth-digit clinodactyly (and nail biting), (**b**) prominent elbows with hyperextensibility, and (**c**) lower extremities with low muscle bulk in the calves, flat feet, and mild pronation at the ankles (Tartaglia et al. 2011)

face difficulty in accepting their gender as female and desire to be male (Turan et al. 2000). In some cases premature ovarian failure and gonadal abnormality is associated with autoimmune disorder and genitourinary problems (Michalak et al. 1983; Holland 2001). Figure 13.21 shows some physical abnormalities associated with XXYY, XXXY, XXXYY, XXXXY, and XXXXXY genotype.

13.2.4 Nullisomy (2*n*−2)

Nullisomy is the condition in which one pair of chromosome is missing from normal set of chromosomes in cell. Deletion of complete pair of chromosome can occur in autosome as well as in sex chromosomes. This condition is very rare in humans and usually occurs due to defects in cell division.

13.2.5 Monosomy (2*n*−1) (XO or Turner Syndrome)

Monosomy is a condition in which only one chromosome is missing out of 46 chromosomes and karyotype is 45 in human. Turner syndrome is an example. Phenotype of this syndrome was first pointed out by Turner in 1938 as infantilism, cubitus valgus, webbed neck, and short stature in females. Different scientists propvosed different reasons for this phenotype until Charles Ford and his colleagues (1959) proposed XO model with 45 chromosomes and have a loss of one X chromosome (Fig. 13.22). Due to its discovery by Turner, it is also called Turner syndrome. Females with Turner syndrome are short statured; have short neck, mild neck webbing, and low set ears; do not have breast and menstrual cycle like normal females; and are infertile. They also tend to have higher risks of cardiac problems, diabetes, and lower thyroid activity. Life expectancy of these patients is very low due to cardiac problems, and the incident rate is one in 25,000 live births in female population (Elsheikh et al. 2002).

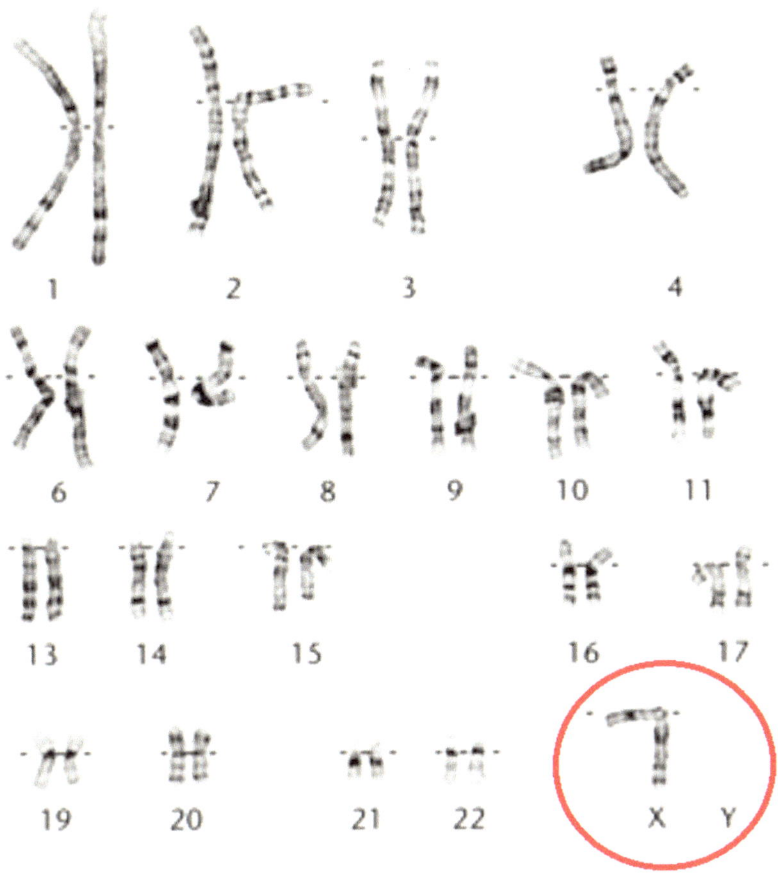

Fig. 13.22 Karyotype of Turner syndrome with only one sex chromosome which is X chromosome. Other counterpart of X chromosome is missing making karyotype 45, XO (Luthardt and Keitges 2001)

Glossary

A

Allele One of the possible mutational states of a gene, distinguished from other alleles by phenotypic effects.

Allopolyploids Polyploidy hybrids in which the chromosome sets coming from two or more distinct, though related, species.

Aneuploid Also known as heteroploid. An individual whose chromosome number is not an exact multiple of the haploid (monoploid) number for the species.

Aneuploidy A condition in which the number of chromosomes is not an exact multiple of the haploid set. It is a condition of a cell or of an organism that has additions or deletions of whole chromosomes.

Anther culture It is the process of formation of haploid plants from microspores (pollen) cultured individually.

Apomixis In flowering plants apomixis is the asexual formation of a seed from the maternal tissues of the ovule, avoiding the processes of meiosis and fertilization, leading to embryo development.

Atavism Reappearance of ancestral trait after several generations because of recessiveness of other masking effects.

Autoimmune disorder An autoimmune disorder occurs when the body's immune system attacks and destroys healthy body tissue by mistake. There are more than 80 types of autoimmune disorders. An autoimmune disorder may result in the destruction of body tissue, abnormal growth of an organ, or changes in organ function.

Autopolyploidy Polyploid condition resulting from the replication of one diploid set of chromosomes.

Autosomes Chromosomes other than the sex chromosomes. In humans, there are 22 pairs of autosomes.

Autosomal set A combination of nonsex chromosomes consisting of one from each homologous pair in a diploid species.

Autotetraploid An autopolyploid condition composed of four similar genomes. In this situation, genes with two alleles (A and a) can have five genotypic classes: AAAA (quadruplex), AAAa (triplex), Aaaa (duplex), Aaaa (simplex), and aaaa (nulliplex).

B

Basic chromosome number (*X*) The number of different chromosomes that make up a single complete set.

Bartholin-Patau syndrome A congenital syndrome of multiple abnormalities produced by trisomy of chromosome number 13. The trisomy is usually due to primary nondisjunction, occasionally to translocation and mosaicism.

C

Camptodactyly It is a medical condition that causes one or more fingers to be permanently bent. It involves fixed flexion deformity of the proximal interphalangeal joints. The fifth finger is always affected. Camptodactyly can be caused by a genetic disorder.

Chromosomal aberration Any change resulting in the duplication, deletion, or rearrangement of chromosomal material. Abnormal structure or number of chromosomes includes deficiency, duplication, inversion, translocation, aneuploidy, polyploidy, or any other change from the normal pattern.

Chromosomes loss A mechanism causing aneuploidy in which a particular chromatid or chromosome fails to become incorporated into either daughter cell during cell division.

Crossing-over The exchange of chromosomal material (parts of chromosomal arms) between homologous chromosomes by breakage and reunion. The exchange of material between nonsister chromatids during meiosis is the bases of genetic recombination.

Cryptorchidism It is the absence of one or both testes from the scrotum. It is the most common birth defect of the male genitalia.

D

Deficiency (deletion) A chromosomal mutation involving the loss or deletion of chromosomal material. Heterozygote deficiency is hemizygous condition for the genes located in the deleted segment; many deficiencies produce genetic effects similar to gene mutations. Deletion also occurs when a block of one or more nucleotide pairs is lost from a DNA molecule.

Diploid A condition in which each chromosome exists in pairs; having two of each chromosome.

Diploidy The state of being diploid, that is, having two sets of the chromosomes (and therefore two copies of genes), especially in somatic cells.

Disjunction The separation of chromosomes at the anaphase stage of cell division.

Down syndrome An abnormal human phenotype including mental retardation, due to a trisomy of chromosome 21; more common in babies born to older mothers.

E

Edwards syndrome Also known as trisomy 18, is a chromosomal abnormality caused by the presence of all, or part of, an extra 18th chromosome. This genetic condition almost always results from nondisjunction during meiosis.

Endopolyploidy The increase in chromosome sets that results from endomitotic replication within somatic nuclei.

Euploid Cells containing only complete sets of chromosomes. An organism or cell having a chromosome number that is an exact multiple of the monoploid (x) or haploid number. Terms used to identify different levels in a euploid series are diploid, triploid, tetraploid, and so on.

Euploidy The condition of a cell or organism that has one or more complete sets of chromosomes.

G

Gynogenesis The development in which the embryo contains only maternal chromosomes due to activation of an egg by a sperm that degenerates without fusing with the egg nucleus.

H

Haploid Also known as monoploid. An organism or cell having only one complete set (n) of chromosomes or one genome. A single set of chromosomes present in the egg and sperm cells of animals and in the egg and pollen cells of plants.

Haplotype Specific combination of linked alleles in a cluster of related genes. A contraction of the phrase "haploid genotype."

Homologous Chromosomes that occur in pairs and are generally similar in size and shape, one having come from the male parent and the other from the female parent. Such chromosomes contain the same array of genes.

Hypergonadism It is a condition where there is a hyperfunction of the gonads. It can manifest as precocious puberty and is caused by abnormally high levels of testosterone or estrogen, crucial hormones for sexual development. Anabolic steroids may also be a major cause of high androgen and/or estrogen functional activity.

Hyperploid Having a chromosome number greater than but not an exact multiple of the normal euploid number.

Hypogonadism Reduction or absence of hormone secretion or other physiological activity of the gonads (testes or ovaries). It describes a diminished functional activity of the gonads. Spermatogenesis and ovulation in males and females, respectively, may be impaired by hypogonadism, which, depending on the degree of severity, may result in partial or complete infertility.

Hypoploidy Any decrease in chromosome number that involves individual chromosomes rather than entire sets, so that smaller than the normal haploid number of chromosome characteristics of the species are present, as in Turner syndrome.

Hypospadias It is a condition in which the opening of the urethra is on the underside of the penis, instead of at the tip.

I

Inversion Rotation of a segment of a chromosome by 180° so that the genes in this segment are present in the reverse order; characteristic inversion loops are produced during meiosis in the inversion heterozygotes.

K

Klinefelter syndrome Sterile human males with the XXY chromosome constitution; other associated symptoms as well.

L

Linkage groups Associations of loci on the same chromosome. In a species, there are as many linkage groups as there are homologous pairs of chromosomes.

Linkage number The number of times one strand of a helix coils about the other.

M

Menopause It is also known as the climacteric. It is the time in most women's lives when menstrual periods stop permanently, and the woman is no longer able to have children. Menopause typically occurs between 45 and 55 years of age. It may also be caused by a decrease in hormone production by the ovaries.

Mental retardation It is a condition diagnosed before age 18, usually in infancy or prior to birth, that includes below-average general intellectual function and a lack of the skills necessary for daily living.

Microcephaly Abnormal smallness of the head, a congenital condition associated with incomplete brain development. It is a rare neurological condition in which an infant's head is significantly smaller than the heads of other children of the same age and sex. Sometimes detected at birth, microcephaly usually is the result of the brain developing abnormally in the womb or not growing as it should after birth. Microcephaly can be caused by a variety of genetic and environmental factors.

Mongolism It is a congenital disorder caused by having an extra 21st chromosome; results in a flat face and short stature and mental retardation.

Mosaicism The condition of being a mosaic.

Mosaics Individuals or tissue having cells of two or more different genotypes. In such cases the cells originate in the same zygote.

N

Nondisjunction Failure of disjunction or separation of homologous chromosomes in mitosis or meiosis, resulting in too many chromosomes in some daughter cells and too few in others. Examples: in meiosis, both members of a pair of chromosomes go to one pole so that the other pole does not receive either of them; in mitosis, both sister chromatids go to the same pole. It is responsible for defects such as trisomy.

Nullisomic An otherwise diploid cell or organism lacking both members of a chromosome pair (chromosome formula $2n-2$).

Nullisomy A type of genome mutation in which a pair of chromosomes that are normally present in the genome is missing. Organisms that exhibit nullisomy are called nullisomes. Nullisomy, especially in higher animals, usually results in death. Viable nullisomes can be found among polyploid plants.

P

Parthenogenesis Reproduction in which offspring are produced by an unfertilized female. Parthenogenesis is common in ants, bees, wasps, and certain species of fish and lizards.

Patau syndrome It is a syndrome caused by a chromosomal abnormality, in which some or all of the cells of the body contain extra genetic material from chromosome 13.

Polyploid An organism with more than two sets of chromosomes ($2n$ diploid) or genomes (e.g., triploid ($3n$), tetraploid ($4n$), pentaploid ($5n$), hexaploid ($6n$), heptaploid ($7n$), octoploid ($8n$).

R

Recombination The process by which offspring derive a combination of genes different from that of either parent; the generation of new allelic combinations. In higher organisms, this can occur by crossing-over.

Robertsonian translocation Translocation arising from breaks at or near the centromeres of two acrocentric chromosomes. The reciprocal exchange of broken parts generates one large metacentric chromosome and one very small chromosome.

S

Syndrome A group of symptoms that occur together and represent a particular disease.

T

Tetraploid Having a chromosome number that is four times the basic or haploid number.

Translocation Change in position of a segment of a chromosome to another part of the same chromosome or to a different chromosome.

Turner syndrome In human beings; individuals having XO chromosome constitution, being phenotypically female, but having rudimentary sexual organs and mammary glands.

X

X linkage The pattern of inheritance resulting from genes located on the X chromosome.

Y

Y linkage Mode of inheritance shown by genes located on the Y chromosome.

References and Additional Reading

American College of Obstetricians and Gynecologists (ed) (2005) Your pregnancy & birth. American College of Obstetricians and Gynecologists Women's Health Care Physicians, Washington, DC

Becak ML (2014) Polyploidy and epigenetic events in the evolution of Anura. Genet Mol Res 13:5995–6014

Becker H, Maharry K, Mrózek K, Volinia S, Eisfeld AK, Radmacher MD, Kohlschmidt J et al (2014) Prognostic gene mutations and distinct gene and microRNA expression signatures in acute myeloid leukemia with a sole trisomy 8. Leukemia 28:1754–1758

Cereda A, Carey JC (2012) The trisomy 18 syndrome. Orphanet J Rare Dis 7:81–95

Delatycki M, Gardner RJM (1997) Three cases of trisomy 13 mosaicism and a review of the literature. Clin Genet 51:403–407

Dewhurst J (1978) Fertility in 47, XXX and 45, X patients. J Med Genet 15:132–135

Elsheikh M, Dunger DB, Conway GS, Wass JAH (2002) Turner's syndrome in adulthood. Endocr Rev 23:120–140

Ford CE, Jones KW, Polani PE, De Almeida JC, Briggs JH (1959) A sex-chromosome anomaly in a case of gonadal dysgenesis (Turner's syndrome). Lancet 1:711–713

Fraser J, Boyd E, Lennox B, Dennison WM (1961) A case of XXXXY Klinefelter's syndrome. Lancet 278:1064–1067

Gardner RM, Sutherland GR, Shaffer LG (2004) Chromosome abnormalities and genetic counseling. Oxford University Press, New York, pp 314–318

Goldstein ML, Morewitz S (2011) Chromosomal abnormalities. In: Chronic disorders in children and adolescents. Springer, New York, pp 31–58

Hasle H, Mellemgaard A, Nielsen J, Hansen J (1995) Cancer incidence in men with Klinefelter syndrome. Br J Cancer 71:416

Hattori M, Fujiyama A, Taylor D, Watanabe H, Yada T, Park HS, Beck A (2000) The DNA sequence of human chromosome 21. Nature 405:311–319

Holland CM (2001) 47, XXX in an adolescent with premature ovarian failure and autoimmune disease. J Pediatr Adolesc Gynecol 14:77–80

Klinefelter HF Jr, Reifenstein EC Jr, Albright F Jr (1942) Syndrome characterized by gynecomastia, aspermatogenesis without A-Leydigism, and increased excretion of follicle-stimulating hormone 1. J Clin Endocrinol Metab 2:615–627

Lana-Elola E, Watson-Scales SD, Fisher EM, Tybulewicz VL (2011) Down syndrome: searching for the genetic culprits. Dis Model Mech 4:586–595

Lanfranco F, Kamischke A, Zitzmann M, Nieschlag E (2004) Klinefelter's syndrome. Lancet 364:273–283

Leggett V, Jacob P, Nation K, Scerif G, Bishop DV (2010) Neurocognitive outcomes of individuals with a sex chromosome trisomy: XXX, XYY, or XXY: a systematic review. Dev Med Child Neurol 52:119–129

Lejeune J, Gautier M, Turpin R (2004) Étude des Chromosomes Somatiques de Neuf Enfants Mongoliens (Study of the somatic chromosomes of nine mongoloid children). Landm Med Genet: Class Pap Comment 51:73

Lewis WH (ed) (2012) Polyploidy: biological relevance, vol 13. Springer Science & Business Media

Luthardt FW, Keitges E (2001) Chromosomal syndromes and genetic disease. eLS. doi:10.1038/npg.els.0001446

Michalak DP, Zacur HA, Rock JA, Woodruff JD (1983) Autoimmunity in a patient with 47, XXX karyotype. Obstet Gynecol 62:667–669

Moerman P, Fryns JP, van der Steen K, Kleczkowska A, Lauweryns J (1988) The pathology of trisomy 13 syndrome. Hum Genet 80:349–356

Mrozek K, Heerema NA, Bloomfield CD (2004) Cytogenetics in acute leukemia. Blood Rev 18:115–136

National Birth Defects Prevention Network (2000) Birth defect surveillance data from selected states, 1989–1996. Teratology 61:86–158

Parker SE, Mai CT, Canfield MA, Rickard R, Wang Y, Meyer RE, Correa A (2010) Updated national birth prevalence estimates for selected birth defects in the United States, 2004–2006. Birth Defects Res A Clin Mol Teratol 88:1008–1016

Reynolds LE, Watson AR, Baker M, Jones TA, D'Amico G, Robinson SD, Hodivala-Dilke KM (2010) Tumour angiogenesis is reduced in the Tc1 mouse model of Down's syndrome. Nature 465:813–817

Roizen NJ, Patterson D (2003) Down's syndrome. Lancet 361:1281–1289

Root S, Carey JC (1994) Survival in trisomy 18. Am J Med Genet 49:170–174

Smith DW, Patau K, Therman E, Inhorn SL (1962) The No. 18 trisomy syndrome. J Pediatr 60:513–527

Sussan TE, Yang A, Li F, Ostrowski MC, Reeves RH (2008) Trisomy represses ApcMin-mediated tumours in mouse models of Down's syndrome. Nature 451:73–75

Tartaglia N, Ayari N, Howell S, D'Epagnier C, Zeitler P (2011) 48, XXYY, 48, XXXY and 49, XXXXY syndromes: not just variants of Klinefelter syndrome. Acta Paediatr 100:851–860

Torfs CP, Christianson RE (1998) Anomalies in Down syndrome individuals in a large population-based registry. Am J Med Genet 77:431–438

Turan MT, Eşel E, Dündar M, Candemir Z, Baştürk M, Sofuoğlu S, Özkul Y (2000) Female-to-male transsexual with 47, XXX karyotype. Biol Psychiatry 48:1116–1117

Zoll B, Wolf J, Lensing-Hebben D, Pruggmayer M, Thorpe B (1993) Trisomy 13 (Patau syndrome) with an 11-year survival. Clin Genet 43:46–50

Techniques of Chromosomal Studies

14

Uzma Qaisar, Asima Tayyeb, and Tariq Ahmad Bhat

Abstract

Chromosomal analysis has been an area of utmost significance for various cytogenetic and medical studies. The techniques to study normal and abnormal chromosomes which initiated as simple observations under the microscope with and without different types of dyes have now developed into more elaborate and invasive techniques. In the early years of cytogenetics, scientists had a difficult time distinguishing individual chromosomes, but over the years, they continued to refine the conditions for preserving and staining chromosomes to the reproducible standard that is now expected in clinical cytogenetics. In today's procedures, metaphase chromosomes are treated with stains that generate distinctive banding patterns, and chromosome pairs are then arranged into a standardized format known as a karyotype. Over the past few decades, versatile methods based on fluorescence in situ hybridization (FISH) have transformed cytogenetics into a molecular science and provided cytogeneticists with powerful new tools. FISH procedures are now routinely employed in clinical cytogenetics. Although chromosomes may appear to be static structures when viewed under a microscope, cytogeneticists know that chromosomes are actually dynamic assemblies made up of a DNA-protein complex called chromatin. This chapter takes stock of all the prevalent techniques, highlighting the principals involved in each method. Karyotyping, genetic mapping, fluorescence in situ hybridization (FISH), multiplex FISH, spectral karyotyping, flow cytometry, and microarray

U. Qaisar, PhD (✉) • A. Tayyeb, PhD (✉)
School of Biological Sciences, University of the Punjab, Lahore, Pakistan
e-mail: uzma.sbs@pu.edu.pk; asima.sbs@pu.edu.pk

T.A. Bhat
Department of Education, Government of Jammu and Kashmir, Srinagar, India
e-mail: bhattariq110@gmail.com

T.A. Bhat, A.A. Wani (eds.), *Chromosome Structure and Aberrations*,
DOI 10.1007/978-81-322-3673-3_14

have been described. New trends in cytogenetics to understand the molecular mechanism have been discussed under new generation sequencing.

Keywords
Karyotyping • Chromosome banding • Genetic mapping • Spectral karyotyping • Flow cytogenetics • Microarray techniques • Next-generation sequencing • Fluorescence in situ hybridization • G band • Q band • R band

14.1 Introduction

Chromosomes are highly organized and discrete subunits of nuclear genomes of humans, animals, and plants. They contain genetic information that guides the development and function of an organism. They are mainly formed by packing DNA into small space with the involvement of histone and nonhistone proteins. The number of chromosome is species specific and any alteration in chromosome, either numerical or structural, is known as chromosomal aberration. People with chromosomal alteration may lead to severe clinical conditions or may have high risk of birth and developmental defects, behavioral glitches, and intellectual incompetence. Moreover, most of these genetic changes are hereditary and can even pass through generation after generation. Therefore, cytogenetic (chromosome) analysis has been an area of utmost interest since the 1960s. Cytogenetic scientists mainly look for variations in the number and structure of the chromosomes and for the loss and gain of genetic material, while molecular cytogeneticists investigate the changes at the DNA or gene level in a chromosome. The techniques through which we can look at chromosomes are getting advanced day by day. This book chapter mainly focuses on the methods to study chromosomes.

14.2 Karyotyping

Karyotyping is a method by which physicians and geneticists take picture of the complete set of chromosomes of a certain species or individual in terms of their number and structure while the cells are undergoing mitosis. Routinely, white blood cells (WBCs) are used to conduct the karyotyping as they do not undergo subsequent cell divisions. Schematic diagram of a simple karyotype protocol using blood is shown in Fig. 14.1. Briefly, WBCs are chemically stimulated with a mitogen phytohemagglutinin (PHA) to enter into mitosis by DNA replication. After 48–72 h, mitosis in the metaphase stage is stopped by adding a mitotic inhibitor colchicine. WBCs are further treated with hypotonic solution, fixed on a slide, stained, and microscopically observed to evaluate chromosomes for abnormalities.

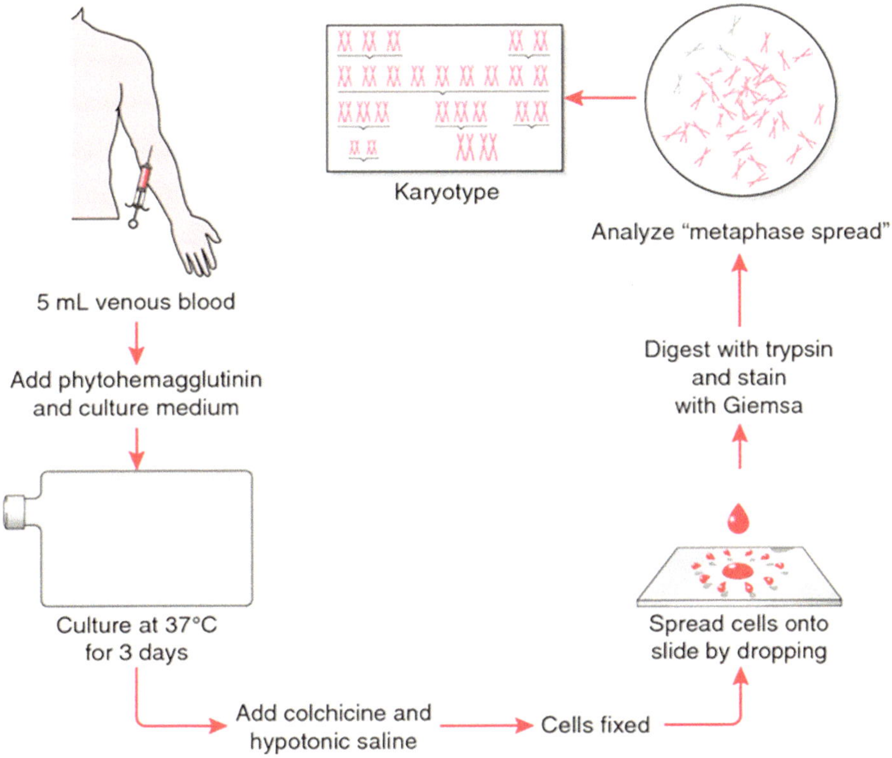

Fig. 14.1 Schematic diagram of karyotyping (Mueller et al. 2001)

14.2.1 Chromosome Banding

Chromosome banding was first introduced in the late 1960s and it is a method to visualize chromosomes with different imaging techniques. Differential staining along the length of a chromosome leads to the production of bands. Stains mainly enhance the contrast between different cellular components giving a pattern of light and dark bands. There are different modes of stain binding to DNA of a chromosome such as through intercalation, minor groove or major groove binding, or external binding (Fig. 14.2). However, direct interaction of the stain with DNA can be either covalent or non-covalent through hydrogen binding, electrostatic interactions, and π-π interactions. There are two types of chromosome staining methods depending on the type of dye being used.

14.2.2 Visible Light Chromosome Staining

Visible light chromosome staining is one of the simplest and classically used chromosome banding methods. It is based on the dyes which are observable under

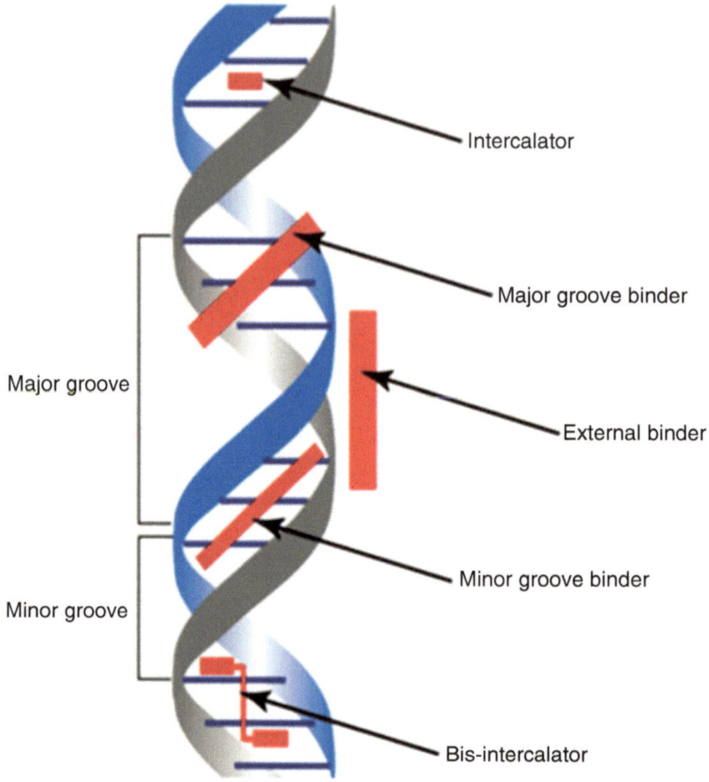

Fig. 14.2 Modes of stain binding to DNA (Estandarte 2012)

visible light. As visible light is being used, major advantage of the method is to produce more stable and clearer bands than fluorochromes. Different types of visible light chromosome staining highlight different regions of the chromosome and mainly differ in the prior denaturation of the chromosomes.

G-banding is the most widely used visible light microscopy-based banding method for cytogenetic analysis. It involves the protease denaturation of chromosomes to extract a subset of proteins. Staining is mainly performed with a dye called Giemsa. Giemsa is a mixture of two components: a cationic thiazine and anionic eosin which interact with both DNA and protein, respectively. The dark G-positive bands and light G-negative bands are obtained as shown in Fig. 14.3. G-positive bands correspond to the AT-rich, hydrophobic, and heterochromatin chromosome regions that favor the formation of the thiazine-eosin precipitate. In contrast, G-negative bands are GC-rich, less hydrophobic, and euchromatin chromosome regions.

R-banding, as the name indicates, shows a reverse banding pattern of G-banding. It mainly highlights the GC-rich G-negative bands. Although same Giemsa stain is being used in R-banding, major difference is the initial denaturation of the chromosomes at a very high temperature 87 °C. This temperature can dissolve

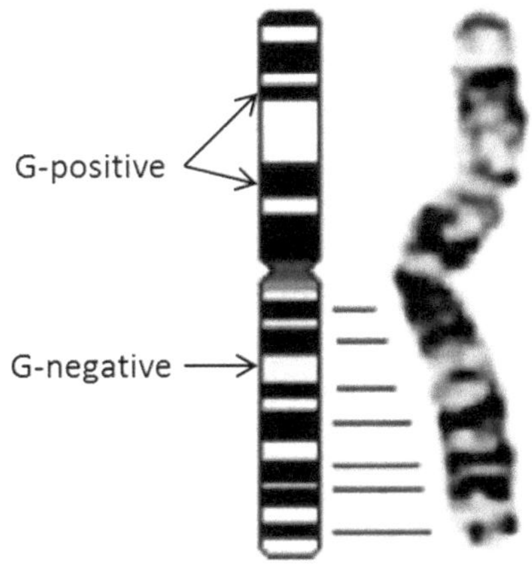

Fig. 14.3 G-banding of a chromosome with corresponding ideogram

low-melting-point AT-rich regions (65 °C) compared to GC-rich regions which have high melting point (105 °C).

C-banding is also a Giemsa-based chromosome staining method which mainly reveals the AT-rich centromeres. Extensive initial denaturation is carried out by either acid treatment, hot saline incubation, or alkali (barium hydroxide) treatment. This hard treatment breaks the whole DNA backbone while saving the centromere region. Prominent C bands are found in chromosomes 1, 9, 16, and the distal Y.

14.2.3 Fluorescent Chromosome Staining

Fluorochromes are organic molecules that are capable of undergoing fluorescence. Chromosome banding with fluorescence compounds under fluorescent microscope is becoming popular. In contrast to visible light dyes, fluorochromes are less stable but do not require any pretreatment or denaturation for protein extraction.

Q-banding is one of the commonly used fluorescence banding which involves staining with guanine alkylating agent, quinacrine, or quinacrine mustard. Quinacrine, like most other chromosome stains, including acridine orange, methylene blue, and DNA staining component of the complex Giemsa stain, is a tricyclic hydrocarbon whose three rings lie in one plane. It intercalates with chromosomal DNA and specifically reacts and fluoresces AT-rich DNA regions brighter than the GC-rich regions. Q-positive bands are comparable with G-positive bands and mainly highlight heterochromatin regions of chromosomes.

Other commonly used fluorochrome stains with a specificity for AT-rich DNA region are 4,6-diamidino-2-phenylindole, DAPI, a bi-benzimidazole derivative, Hoechst 33258, and daunomycin. DAPI and Hoechst bind to the minor groove of DNA, whereas Daunomycin intercalates with DNA. Resistances to fading and excitation-emission spectra compatible with commonly available filters are major advantages of DAPI over other fluorochrome stains. Chromomycin and 7-amino actinomycin D are some other fluorochromes with specificity for GC-rich DNA and give an R-band-like pattern.

14.2.4 Heavy Metal Staining

Heavy metal complexes or compounds containing Eu, Tb, Os, Ur, and DNA metallointercalators are being used especially when chromosomes are being visualized with ultra-magnification with electron microscopy. Platinum complexes, osmium tetroxide, and uranyl acetate are few used DNA metallointercalators.

14.3 Genetic Mapping

Genes are the hereditary units in all living organisms and control the inheritance of traits from one generation to the next. Each chromosome consists of two arms and a centromere in between them. On the basis of the position of the centromere, chromosome has one short and one long arm. Genes reside on the chromosome arms at specific positions and in a peculiar order, while centromere does not have any genes. The localization of genes on a chromosome arm is called gene mapping and these maps are called idiograms.

The International System of Cytogenetic Nomenclature (ISCN) has devised a consistent numbering system for gene maps (idiograms). According to ISCN nomenclature, the shorter arm of the chromosome is named as "p," while the longer arm is designated as "q," and the numbering for each arm starts from the centromere and extends toward the arm endings in ascending order. Conventionally, the "p arm" of a chromosome is always shown at the top in a karyotype. Each arm is divided into specific regions based on staining patterns, e.g., Giemsa-staining bands. Each prominent band is considered as a region and is named as q1, q2, q3, etc., on the long arm and p1, p2, p3, etc., on the short arm. The additional dim bands are numbered by adding a digit after a number. For example, dim bands between region p11 and p12 would be p11.1, p11.5, as described in Fig. 14.4.

Genetic mapping provides information about the location of risk genes causing genetic diseases and enables medical researcher to investigate the root cause and ultimately find the solution of many medical problems. In agriculture sector, genetic maps are utilized for the production of animal breeds and economically important crops with improved traits. Genetic mapping is very useful in crime scene investigation, identification, and paternity tests. Various techniques have been used for genetic mapping of chromosomes in different genomes.

14.3.1 Linkage and Recombination-Based Maps

Genes located on one chromosome are called linked genes and are transferred to the offsprings together (if located close to each other on a chromosome), or recombination occurs due to crossing over event during meiosis. Genes located close to each other have less recombination frequencies as compared to those which are far apart. The genetic distance between two genes is directly proportional to the recombination frequency of these genes.

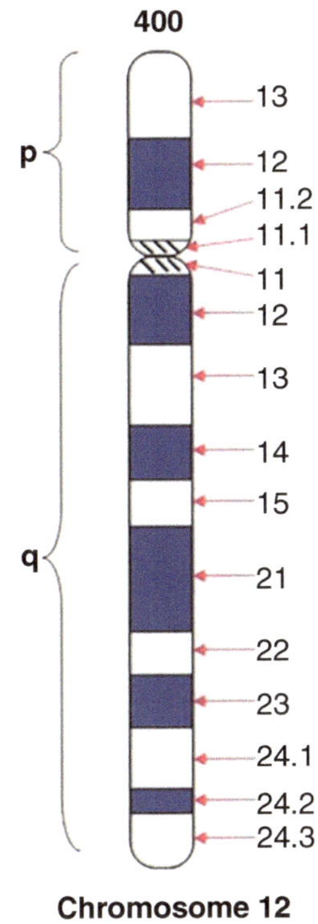

Fig. 14.4 International system of cytogenetics nomenclature

If the recombination frequency between gene A and gene B is 3 %, that means that these genes are at 3 centimorgan (cM) distance from each other. Centimorgan is the map unit and does not correspond to any absolute value but is a relative quantity. On the basis of recombination frequencies of various genes, the distance between them can be calculated. On the basis of this distance, linkage maps are commonly produced to find out the order of genes on a chromosome and the distance between genes. Linkage maps are commonly constructed in the plant breeding programs to study the quantitative trait loci (QTL) associated with specific traits of interest. Linkage maps are very useful to study additions, deletions, and rearrangements of chromosomal regions during the course of evolution. These maps have been extensively utilized to study *Drosophila* genetics. In humans, it is impossible to preselect the genotypes of parents and design crosses for mapping purposes. To study recombination frequencies and linkage in humans, researchers use genotypes of members of successive generations of existing family members. The resolution of linkage map depends on the number of crossovers scored and cannot be efficiently used in humans, so additional techniques were developed to map genes on chromosomes.

14.3.2 Marker-Based Maps

Genetic markers are the tools which can recognize unique sequences in the genomic DNA (packaged in chromosomes) and provide orientation of genetic maps. Restriction fragment length polymorphism (RFLP) markers can identify the polymorphism in a specific restriction endonuclease binding site by analyzing the size and number of restriction fragments observed in a genome. Cleaved amplified polymorphic sequence (CAPS) markers are similar to RFLP markers but require polymerase chain reaction (PCR) to amplify a specific region of genomic DNA. A genomic DNA region is amplified by PCR using region-specific primers, and those amplified fragments are then digested with a restriction endonuclease to reveal the polymorphism. CAPS marker-based genotyping is very feasible due to the requirement of only small amount of template DNA and visualization of bands using gel electrophoresis. RFLP and CAPS markers were widely used to detect the deletions, additions, and other changes in the chromosomes or a part of the chromosome. But these markers have a limitation that they can miss the copy number variations and balanced inversions. Moreover, very few restriction sites are polymorphic in a genome and thus many sites are not mapped by this technique.

Random amplified polymorphic DNA (RAPD) markers are another type of PCR-based DNA markers. These constitute identical 10-mer arbitrary nucleotide sequence primers which are used to PCR-amplify random segments of genomic DNA. These markers do not require the knowledge of nucleotide sequence of target genome. These markers can amplify random fragments from the DNA, and the comparison of the pattern of amplified products of the unknown genome with the reference genome can give information about the chromosomal/DNA changes and mutations.

Simple sequence repeat (SSR)/microsatellite markers target the stretches of repeats of short sequences. These simple sequence repeats are spread all over eukaryotic genomes and hence are very successful markers for studying variations across DNA in the chromosomes. Single-nucleotide polymorphism (SNP) markers took the place of SSR markers in recent years. These markers can detect the changes up to a single base pair and are most widely distributed.

All these markers have some advantages or disadvantages over the others, but none of them can completely map all the mutations, copy number variations, or abrasions across the genome. So there was a need for additional high-throughput techniques to comprehensively study the chromosomes.

14.4 Fluorescence In Situ Hybridization

Fluorescence in situ hybridization (FISH) is known as one of the most substantial development in clinical cytogenetics. Initially, FISH has been extensively used for the localization of genes and specific genomic regions on target chromosomes in metaphase and interphase cells. However, FISH is not only limited to gene mapping or rearrangements of genes in human genetic diseases but is also emerging as a significant advancement to explore the genome organization in various organisms. FISH has been established as an important molecular cytogenetic technique and an integral part of the clinical cytogenetic diagnostics.

The fundamental double helical structure of DNA and its property to bind or hybridize to complementary sequences provide the basis for FISH analysis. FISH mainly works by exploiting the ability of one DNA strand to hybridize specifically to another DNA strand. Small DNA strands called probes are being used by FISH. The probes are complementary to specific parts of a chromosome and are fluorescent labeled. Fluorescent-labeled probes are hybridized with denatured DNA of interest and are detected by the fluorescent emission. FISH probes are of different types depending upon the specific part of chromosome to be labeled such as whole chromosome painting probe, centromeric probes, and locus-specific probes to identify a specific gene or region of chromosome.

One of the major limitations of FISH is that it is probe dependent and probes are only available for the most well-characterized deletion and duplication syndromes. FISH testing does not usually screen all chromosomes for changes, but it will identify only what it is looking for. Moreover, it is less precise in defining genes and breaks which are involved in any chromosomal imbalance. In cases, where FISH is unable to detect the problem, patients are referred to examine all chromosomes for losses and gains of DNA through comparative genomic hybridization. FISH is discussed in more detail in Chap. 9.

14.5 Multiplex FISH

As the name indicates, multiplex FISH uses multiple number of fluorochromes usually five to conduct whole chromosomal analysis in a single FISH experiment. One, two, or three different fluorochromes are used to paint chromosomes in different combinations. Five different narrow microscope filters are used for each fluorochrome, and 24 different pseudo colors are assigned to each unique color combination referring to each chromosome pair.

14.6 Spectral Karyotyping

Spectral karyotyping (SKY) is a novel, hyper-accurate, and hypersensitive cytogenetic technique. Sky can identify chromosomal abnormalities more accurately and without any prior information compared to conventional banding methods and FISH. It mainly uses the same approach of combinational labeling like multiplex FISH. However, unlike multiplex FISH, image is being captured by a combination of fluorescence microscopy, charge-coupled device imaging, and Fourier spectroscopy. It helps to analyze chromosomal changes such as translocations and insertions. Schematic diagram of cytogenetic analysis using spectral karyotyping has been shown in Fig. 14.5.

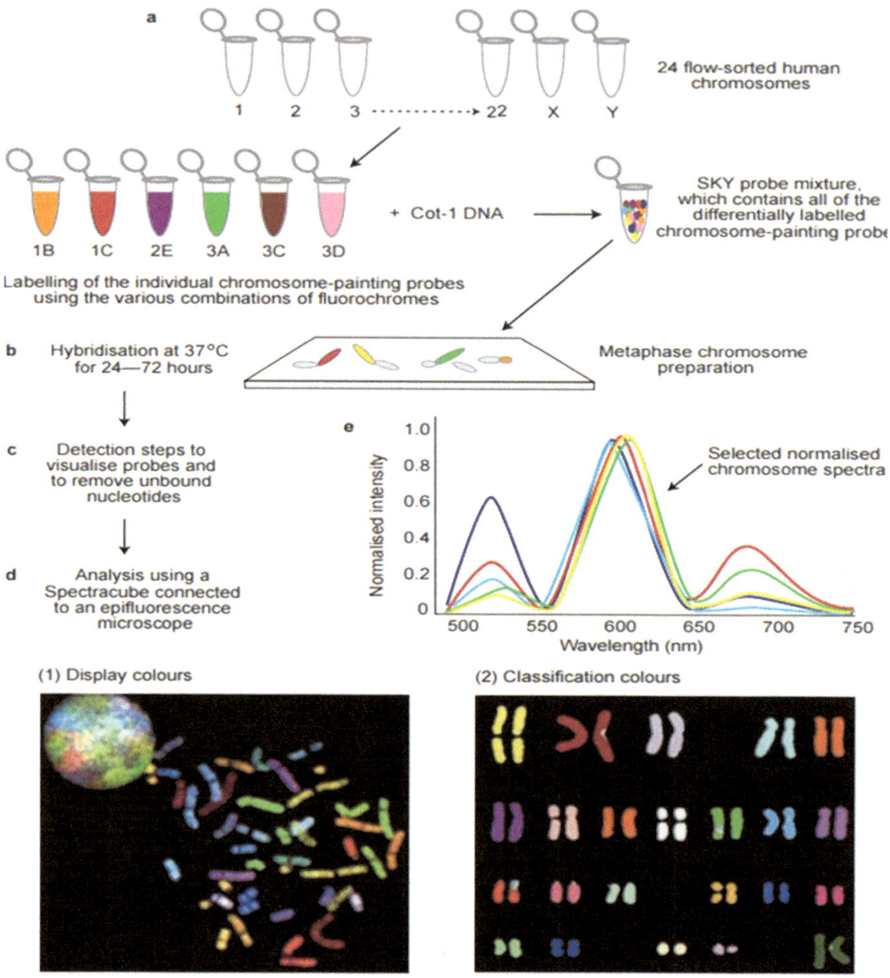

Fig. 14.5 Schematic diagram of cytogenetic analysis using spectral karyotyping (Mcneil and Ried 2000)

14.7 Flow Cytogenetics

Flow cytogenetics is to harness the flow cytometry to perform the cytogenetic analysis. It is the most successful approach to analyze and sort chromosomes from populations of dividing cells into aqueous suspension. Flow cytometry is basically a laser-based technology which sorts the microscopic particles in a fluid based on their optical properties (light scatter, fluorescence). Therefore, a chromosome of interest can be purified in large quantities if it can be resolved from other chromosomes based on its optical properties. Furthermore, as condensed chromosomes are easy to sort and analyze, flow cytogenetics is possible only during the metaphase

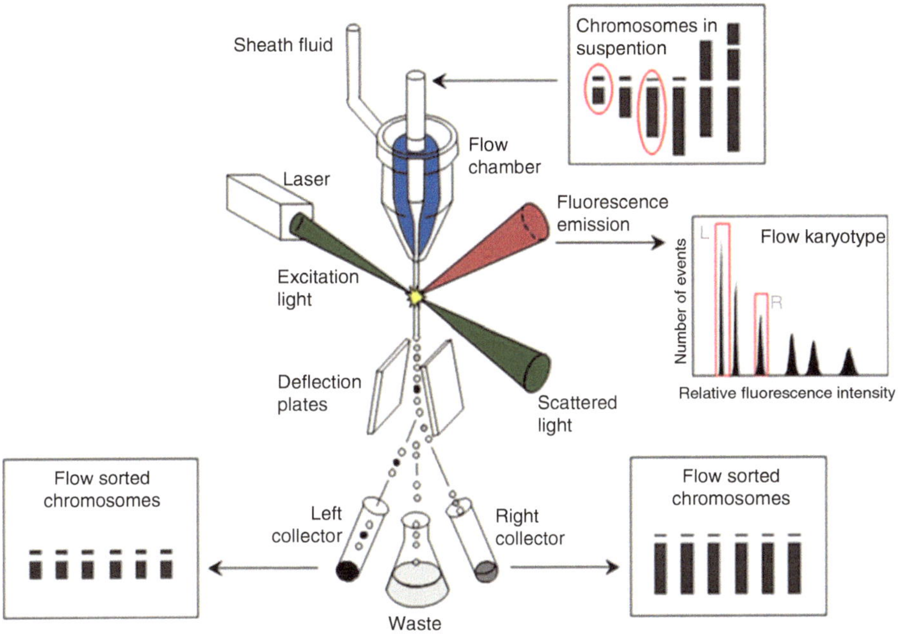

Fig. 14.6 Schematic of flow cytogenetics using a flow cytometer and sorter (Dolezel et al. 2012)

stage of cell division. Characterization of karyotypes, identification of structural chromosome changes, and purification of large quantities of chromosomes for further biochemical and molecular analysis are some of the major applications of flow cytogenetics. A schematic diagram of the flow cytogenetics is provided in the Fig. 14.6. Briefly, a specific DNA dye is used to stain the intact chromosome suspension and introduced to a flow chamber. A narrow stream of liquid carrying chromosomes is being emerged and exposed to laser beam of the flow cytometer. The interaction of the laser with chromosomes scatters light and emits fluorescence in the form of peaks. The quantification of fluorescence intensity and generation of histograms generate flow karyotypes for further analysis.

Each chromosome should form a distinct peak on the flow karyotype, but because of similar size and comparable amount of DNA, peaks of some chromosome types may overlap and chromosomes cannot be separated.

14.8 Microarray

Chromosomes undergo various types of rearrangements in response to cellular stress, incorrect DNA repair, or recombination. Common types of rearrangements are deletion, duplication, inversions, and translocations. These chromosomal changes may involve 50 bp–50 kbp or more of the DNA fragment and are responsible for many genetic disorders in humans. The conventional techniques of chromosomal banding and karyotyping are still successfully being used to detect the major disorders which affect the larger portions of DNA (few megabases) or some time to study the numerical changes in chromosome number (aneuploidy, polyploidy).

At the advent of the twenty-first century, molecular cytogenetic techniques emerged which can study the whole genome of an individual and can detect all types of small-scale and large-scale abrasions. Microarrays are now commonly used to study the chromosomal variations in the whole genome.

14.9 Microarray Technique

DNA microarrays or chips are basically glass chips on which probes of known sequences are printed in an array using specialized robotic DNA spotters. These probes are so tiny that millions of them are printed on a regular-sized glass slide (high-density arrays). Each of these probes is unique and represents a specific sequence of a known portion of a chromosome. In early microarray platforms, cDNAs were printed as probes (100 bp–1 kbp), while later oligos (20–60 bp) took the place of cDNA platforms and are widely being used nowadays. In a standard microarray-based experiment, reference genome DNA/cDNA and test genome DNA/cDNA are prepared, fluorescently labeled (each with a different colored fluorescent dye), fragmented, and hybridized to the microarray platform. Usually, the reference DNA/cDNA is labeled with green dye and test DNA/cDNA is labeled with red dye. After hybridization, the probes representing deleted portions of the chromosomes appear green, while those where copy number has increased appear red and all unaltered parts appear yellow.

This technique was originally developed for gene expression profiling, but now it is commonly used to detect copy number changes (comparative genome hybridization (CGH) microarray), for single-nucleotide polymorphism (SNP) genotyping and detection of miRNAs, and to study protein–DNA interactions (chromatin immunoprecipitation (ChIP) arrays). Main advantages of this technology include specificity, sensitivity, high throughput, genome-wide coverage, and tiny amount of sample requirement.

14.10 Comparative Genomic Hybridization (CGH) Array

In comparative genomic hybridization (CGH) array, whole genomic DNA, a specific chromosome, a small portion of a chromosome, or a BAC/PAC (bacterial/PI-derived artificial chromosome) clone, can be hybridized to the platform as a target. Schematic diagram depicting the methodology of CGH array is given in Fig. 14.7. These arrays are specialized for very efficiently detecting copy number variations (CNVs) across the whole genome. In the last 5 years, genomic microarrays (arrays) have been widely used to diagnose postnatal cases with developmental

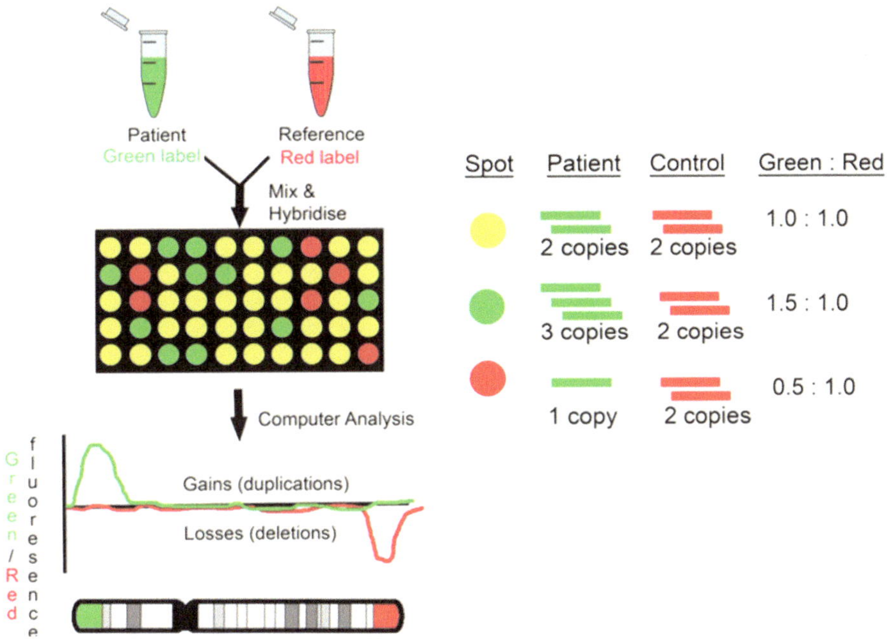

Fig. 14.7 Comparative genome hybridization (CGH) microarray (Karampetsou et al. 2014)

delay (DD), autism, and/or congenital abnormalities. These arrays can efficiently detect the amplifications and deletions on a specific chromosome but cannot detect the inversions and balanced reciprocal rearrangements.

14.11 Tiling Arrays

Tiling array is another modified type of CGH microarray in which the probes are designed in such a way that they completely cover the specific portion of the chromosome under study. Probe sequences overlap some sequence with the previous probe sequence and are tale ending with each other. These arrays can be used to identify all deletions, additions, inversions, rearrangements, and mutations as small as 50 bp.

14.12 Single-Nucleotide Polymorphism (SNP) Arrays

In SNP arrays, each probe represents the discrimination between the two possible SNP alleles (A or B) for a specific position in the genome. SNP arrays along with detecting the addition, deletions, and CNVs can also identify the maternal cell contamination (MCC), polyploidy level, and loss of heterozygosity (LOH). SNP array technique is described in Fig. 14.8.

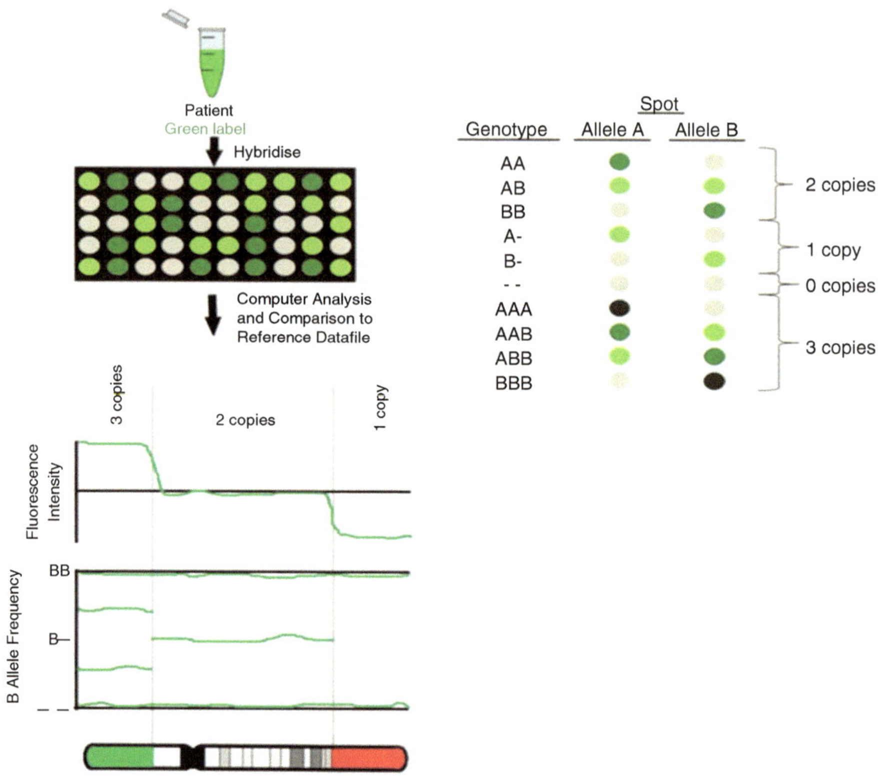

Fig. 14.8 Single-nucleotide polymorphism (SNP) microarray (Karampetsou et al. 2014)

The disadvantage of SNP arrays results from the fact that probe spacing and genome coverage are limited by the nonrandom distribution of SNPs in the genome. Regions of segmental duplications are usually poorly or not at all covered.

14.13 Next-Generation Sequencing

Nucleic acid sequencing is a method for determining the exact order of nucleotides present in a given DNA or RNA molecule. NGS platforms perform massively parallel sequencing, during which millions of fragments of DNA from a single sample are sequenced in unison. Massively parallel sequencing technology facilitates high-throughput sequencing, which allows an entire genome to be sequenced in less than 1 day. Figures 14.9 and 14.10 described the detailed methodology of Next Generation Sequencing. In the past decade, several NGS platforms have been developed that provide low-cost, high-throughput sequencing. The advent of capillary instrumentation and the use of dye-based detection methods, accompanied by automated analysis, have moved traditional DNA sequencing into "next-generation" DNA

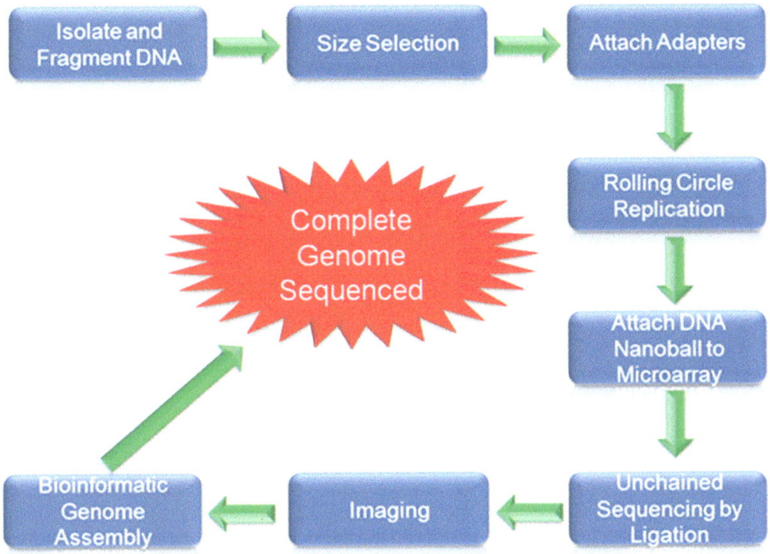

Fig. 14.9 Next-generation sequencing

sequencing. Next-generation sequencing (NGS) is a very high-throughput technique for generating millions of sequences at one time in order to analyze organisms at a genomic level. This allows the researcher to sequence, resequence, and compare data at a rate previously not possible.

Next-generation DNA sequencing (NGS), on the other hand, can detect "all aberration types" including balanced translocations, inversions, and sequence-level variations. CNVs, he says, can be resolved at least 100 times more precisely with NGS than arrays, down to the single-nucleotide level. NGS can detect all kinds of mutations related to diseases up to a single base-pair level.

The limitations of next-generation sequencing (NGS) include that it requires a reference genome for the assembly of short sequences and thus cannot be used for chromosomes of un-sequenced genomes.

There are three main types of NGS which are used according to the requirements of the ongoing study.

14.14 Whole Genome Sequencing

Whole genome sequencing through NGS technique sequences both coding and non-coding regions of the genome and traces all types of mutations, deletions, inversions, and copy number variations in the intronic and exonic regions. Enormous amount of information with comparatively tiny amount of genomic DNA is the major advantage of this technique, while the disadvantages include complexity of data analysis.

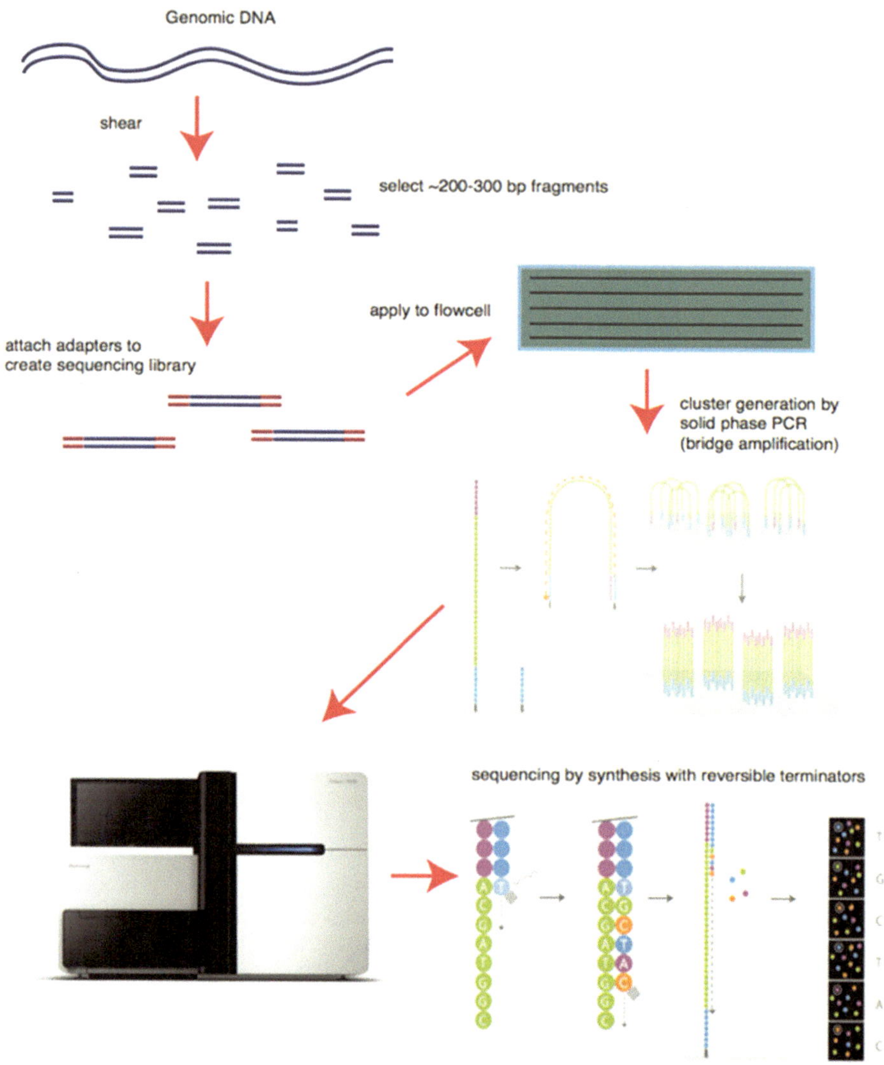

Fig. 14.10 Next-generation sequencing (NGS) (Reference: http://bitesizebio.com/13546/sequencing-by-synthesis-explaining-the-illumina-sequencing-technology/)

14.15 Whole Exome Sequencing

Exome sequencing targets the genome which is transcribed to RNA and subsequently to the proteins. Usually, this is the portion of the genome which includes most of the information. This technique enables us the precise introgression of exons but leaves out the intronic region mutations.

14.16 Targeted Sequencing

The whole genome and whole exome sequencing gives very comprehensive data. But in cases where the regions of interest are known, targeted sequencing of a chromosome or a genomic region is much more affordable and efficient. A specific chromosome or chromosomal region is isolated using combination of other cytogenetic techniques.

Glossary

A

Acridine dyes A class of organic compounds that bind to DNA and intercalate into the double-stranded structure, producing local disruptions of base pairing. The disruptions result in additions or deletions in the next round of replication.

Allele One of the possible mutational states of a gene, distinguished from other alleles by phenotypic effects.

Aneuploidy A condition in which the number of chromosomes is not an exact multiple of the haploid set.

Atavism Reappearance of ancestral trait after several generations because of recessiveness of other masking effects.

Autism A group of complex disorders of brain development. These disorders are characterized, in varying degrees, by difficulties in social interaction, verbal and nonverbal communication, and repetitive behaviors.

Autopolyploidy Polyploid condition resulting from the replication of one diploid set of chromosomes.

B

Bacterial artificial chromosomes (BACs) Artificial chromosomes used for sequencing that are derived from bacterial fertility factors (F plasmids).

C

C-banding It is a technique in which the centromeric regions and other regions containing constitutive heterochromatin of the chromosomes are specifically stained.

Centimorgan A unit of distance between genes on chromosomes. One centimorgan represents a value of 1 % crossing over between two genes. Abbreviated as cM.

Centromere Specialized region of a chromosome to which sister chromatids remain attached after replication and the site to which spindle fibers attach during cell division. Location of the centromere determines the shape of the chromosome during the anaphase part of cell division. Also known as the primary constriction.

ChIP A method for purifying chromatin containing a protein of interest by immunoprecipitating the chromatin with an antibody directed against that protein or against an epitope tag attached to the protein.

Chromosomal aberration Any change resulting in the duplication, deletion, or rearrangement of chromosomal material.

Chromosome banding Technique for the differential staining of mitotic or meiotic chromosomes to produce a characteristic banding pattern or for the selective staining of certain chromosomal regions such as centromeres, the nucleolus organizer regions, and GC- or AT-rich regions. The pattern of banding is highly specific, and all the 23 human chromosomes can be identified. Not to be confused with the banding pattern in polytene chromosomes, which is produced by the alignment of chromomeres.

Chromosome map A diagram showing the location of genes on chromosomes.

Colchicine An alkaloid compound that inhibits spindle formation during cell division. Used in the preparation of karyotypes to collect a large population of cells inhibited at the metaphase stage of mitosis.

Comparative genomic hybridization (CGH) It is a molecular cytogenetic method for analyzing copy number variations (CNVs) relative to ploidy level in the DNA of a test sample compared to a reference sample, without the need for culturing cells.

CGH microarray Through the use of DNA microarray in conjunction with CGH techniques, the more specific form of array CGH (aCGH) has been developed, allowing for a locus-by-locus measure of CNV with increased resolution as low as 100 Kb.

Cleaved amplified polymorphic sequence (CAPS) It is a technique for the analysis of genetic markers. It is an extension to RFLP method, using PCR to more quickly analyze the results. The genetic differences between individuals can create or abolish restriction sites, and these differences can be detected in the resulting DNA fragment length after digestion. The digested PCR products will give readily distinguishable patterns of bands.

Congenital abnormalities Congenital abnormalities are caused by problems during the fetus's development before birth.

Copy number variations (CNV) Large segments of DNA, ranging in size from thousands to millions of DNA bases, can vary in copy number. Genes that were thought to always occur in two copies per genome have now been found to be present in one, three, or more than three copies. In a few rare instances, the genes are missing altogether.

Crossing over Physical exchange between DNAs or chromosomes that occur during recombination.

Cytogenetics A branch of biology in which the techniques of both cytology and genetics are used to study heredity. This is a cytological approach to genetics mainly involving microscopic studies of chromosomes.

Cytological map A diagram showing the location of genes at particular chromosomal sites.

D

Developmental delay A developmental delay is any significant lag in a child's physical, cognitive, behavioral, emotional, or social development, in comparison with norms.

E

Euchromatin Chromatin or chromosomal regions that are lightly stained and are relatively uncoiled during interphase of the cell cycle. Euchromatic regions contain most of the structural genes.

F

Fluorescence in situ hybridization (FISH) Fluorescence in situ hybridization – a technique in which a fluorescent dye is attached to a nucleotide probe that then binds to a specific site on a chromosome and makes itself visible by its fluorescence.

Flow cytometry It is a technology that is used to analyze the physical and chemical characteristics of particles in a fluid as it passes through at least one laser. Cell components are fluorescently labeled and then excited by the laser to emit light at varying wavelengths. It is a laser-based, biophysical technology employed in cell counting, cell sorting, biomarker detection, and protein engineering, by suspending cells in a stream of fluid and passing them by an electronic detection apparatus. It allows simultaneous multiparametric analysis of the physical and chemical characteristics of up to thousands of particles per second.

Fluorochromes or fluorescent dyes Fluorochromes absorb light. The absorption of light by a population of these molecules raises their energy level to a brief excited state. As they decay from this excited state, they emit fluorescent light.

G

Genetic markers Genes identifiable through phenotypic variants that can serve as points of reference in determining whether particular progeny are the result of recombination. These are alleles used as experimental probes to keep track of an individual, a tissue, a cell, a nucleus, a chromosome, or a gene.

Giemsa stain A complex of stains specific for the phosphate groups of DNA.

G-banding or Giemsa banding It is a technique used in cytogenetics to produce a visible karyotype by staining condensed chromosomes. It is useful for identifying genetic diseases through the photographic representation of the entire chromosome complement.

H

Haploid The chromosomal number in the gamete (n).

Haplotype A cluster of alleles on a single chromosome.

Heterochromatin Highly condensed chromosomal regions within which genes are usually transcriptionally inactive. It stains darkly even during interphase, often containing repetitive DNA with few genes.

Histones Small DNA-binding proteins with a preponderance of the basic, positively charged amino acids lysine and arginine. Histones are the fundamental protein components of nucleosomes. They function in the coiling of DNA in chromosomes and in the regulation of gene activity.

I

Idiogram A photograph or diagram of the chromosomes of a cell arranged in an orderly fashion.

Intercalators Class of chemical mutagenic compounds composed of flat, planar molecules that can sandwich themselves between successive base pairs and disrupt the machinery of replication, recombination, or repair.

Inversion Rotation of a segment of a chromosome by 180° so that the genes in this segment are present in the reverse order; characteristic inversion loops are produced during meiosis in the inversion heterozygotes.

K

Karyotype The visual description of a complete set of chromosomes in one cell of an organism; usually presented as a photomicrograph of an individual cell with the chromosomes arranged in a standard format showing the number, size, and shape of each chromosome type, used in low-resolution physical mapping to correlate gross chromosomal abnormalities the characteristics of specific traits of diseases.

L

Linkage The proximity of two or more markers on a chromosome; the closer together the markers are, the lower the probability that they will be separated by recombination, DNA repair, or replication processes, thereby increasing the probability that they will be inherited together. Genes are linked when the frequency of parental type progeny exceeds that of recombinant progeny.

Linkage groups Associations of loci on the same chromosome. In a species, there are as many linkage groups as there are homologous pairs of chromosomes.

Linkage number The number of times one strand of a helix coils about the other.

Loss of heterozygosity (LOH) It occurs when a somatic cell contains only one copy of an allele due to nondisjunction during mitosis, segregation during recombination, or deletion of a chromosome segment. LOH becomes critical when the remaining allele contains a point mutation that renders the gene inactive.

M

Macroarray Microtiter plate-based DNA array.

Marker A gene or mutation that serves as a signpost at a known location in the genome.

Maternal cell contamination (MCC) It is one of the risks associated with prenatal testing, which can occur when a fetal specimen comes into contact with maternal blood or tissue.

Microarray Small glass-slide DNA array. A chip containing many tiny spots of DNA or oligonucleotides used as a dot blot to measure expression of many genes at once.

Micro RNA (miRNA) A short (18–25 nucleotides) RNA produced naturally in cells that can control the expression of cellular genes by causing destruction of specific mRNAs or blocking their translation.

Microsatellite DNA element composed of 15–100 tandem repeats of one, two, or three base-pair sequences, such as CACACACA, dispersed throughout the eukaryotic genome. A given microsatellite is found in varying lengths scattered around a eukaryotic genome. The loci can be studied by polymerase chain reaction amplification.

Minisatellite DNA element composed of 10–40 bp tandem repeating units of identical sequences.

Microsatellite DNA Repeats of very short sequences of DNA.

Multiplex FISH M-FISH is a filter-based multicolor karyotyping technology, which provides the simplest way to label probes in a multiplex fashion as each probe fluorochrome is either completely absent or present. At least five distinguishable fluorochromes are needed for combinatorial labeling to uniquely identify all 24 chromosome types in the human genome using chromosome painting probe sets.

N

Nonhistone proteins The proteins remaining in chromatin after the histones are removed. They are mainly acidic proteins.

P

Phage artificial chromosome or P1-derived artificial chromosome (PAC) It is a form of chromosome derived through biological manipulation and it originates from a "phage" instead of a "plasmid."

p arm of a chromosome The short arm of a chromosome. The "p" comes from the French "petit" meaning small. All human chromosomes have 2 arms – the p (short) arm and the q (long) arm – that are separated from each other only by a centromere.

Physical map Chromosomal map in which distances are in physical units of base pairs. These maps can be of microsatellite markers or of sequence-tagged sites.

Physical markers Cytologically visible abnormalities that make it possible to keep track of specific chromosomes parts.

Polyploidy level Polyploidy level describes cells with three or more sets of chromosomes.

Q

q arm of a chromosome All human chromosomes have 2 arms – the p (short) arm and the q (long) arm – that are separated from each other only by a primary constriction. The symbol "q" was chosen for the long arm simply because it followed "p" in the alphabet.

Quantitative trait loci (QTLs) Loci that control the expression of continuous traits.

Q-banding In this technique, chromosomes are treated with quinacrine mustard solution, a fluorescent stain, to identify specific chromosomes and structural rearrangements. It is especially useful for distinguishing the Y chromosome and various polymorphisms involving satellites and centromeres of specific chromosomes.

R

R-banding or Giemsa reverse banding It is a technique in which the resulting chromosome pattern shows darkly stained R bands and pale G bands. This method is useful for analyzing deletions or translocations that involve the telomeres of chromosomes.

Random amplification of polymorphic DNA (RAPD) A protocol designed to detect single-base changes at polymorphic loci throughout a genome.

Recombination The production of gene combinations not found in the parents by the assortment of nonhomologous chromosomes and crossing over between homologous chromosomes during meiosis. For linked genes, the frequency of recombination can be used to estimate the genetic map distance; however, high frequencies (approaching 50 %) do not yield accurate estimates.

Restriction enzymes Proteins that recognize specific, short nucleotide sequences, and cut DNA at those sites. Bacteria contain over 400 such enzymes that recognize and cut over 100 different DNA sequences.

Restriction fragment length polymorphism (RFLP) Variation between individuals in DNA fragment size cut by specific restriction enzymes; polymorphic sequences that result in RFLPs are used as markers on both physical maps and genetic linkage maps.

S

Satellite DNAs Blocks of repetitive, simple noncoding sequences, usually around centromeres. These blocks have a different chromatin structure and different higher order packaging than other chromosomal regions.

Sequencing (high throughput or next generation) DNA sequencing that uses very rapid automated methods, such as pyrosequencing to produce relatively short reads of DNA in a massively parallel manner that yields a great deal of sequences in a short time.

Simple sequence repeats (SSR) A small segment of DNA, usually 2 to 5 bp in length that repeats itself a number of times.

SINE Sequences of DNA interspersed in eukaryotic chromosomes in many copies. Alu, a 300 base-pair sequence, is found about half a million times in human DNA.

Single-nucleotide polymorphisms (SNPs) Differences between individuals involving single base pairs that are located about every 1.000 bases along the human genome. SNPs are useful for mapping disease genes.

T

Tiling array Tiling arrays are a subtype of microarray chips. Like traditional microarrays, they function by hybridizing labeled DNA or RNA target molecules to probes fixed onto a solid surface. Tiling arrays differ from traditional microarrays in the nature of the probes.

Translocation A chromosomal rearrangement in which a segment of genetic material from one chromosome becomes heritably linked to another chromosome.

V

Variable-number-of-tandem-repeat (VNTR) loci Loci that are hypervariable because of tandem repeats. Presumably, variability is generated by unequal crossing over.

W

Whole genome sequencing (WGS) Also known as full genome sequencing, complete genome sequencing, or entire genome sequencing is a laboratory process that determines the complete DNA sequence of an organism's genome at a single time.

Y

Yeast artificial chromosome (YAC) A vector used to clone DNA fragments up to 400 kb; it is constructed from telomeric, centromeric, and replication origin sequences needed for replication in yeast cells.

References

Dolezel J, Vrana J, Safar J, Bartos J, Kubalakova M, Simkova H (2012) Chromosomes in the flow to simplify genome analysis. Funct Integr Genomics 12:397–416

Estandarte AKC (2012) A review of the different staining techniques for human metaphase chromosomes. Department of Chemistry, University College London, University of London

Karampetsou E, Morrogh D, Chitty L (2014) Microarray technology for the diagnosis of fetal chromosomal aberrations: which platform should we use? J Clin Med 3:663–667

McNeil N, Ried T (2000) Expert reviews in molecular medicine. Cambridge University Press, Cambridge, ISSN 1462–3994

Mueller RF, Young ID, Emery AEH (2001) Emery's elements of medical genetics. Churchill Livingstone, Philadelphia

Cytomixis: Causes and Consequences as a Case Study in *Vicia faba* L.

15

Tariq Ahmad Bhat, Mohd Gulfishan, and Aijaz Ahmad Wani

Abstract

Broad bean (*Vicia faba* L.) is a member of the family Fabaceae which showed cytomixis during microsporogenesis. Cytomixis is proposed to be caused by genes, abnormal cell wall formation, action of chemicals, changes in the biochemical process, imbalanced and sterile genetic systems, persistence of male-sterile mutant genes, and frequent alterations in environmental factors. The cytomixis was found to happen through different methods, i.e., cytoplasmic channels and direct fusion method in the meiocytes. The former was more abundant than the latter, and both the methods were found in various phases of microsporogenesis. The transfer of chromatin material involved the whole diploid set of chromosomes or some chromosomes in the diploid set from donor to recipient cell/cells. The PMCs which showed deviation from the basic diploid chromosome number by cytomixis may lead to the formation of aneuploid and polyploid gametes. Stickiness was found in all the mutagen-treated population and showed dose-dependent increase with the mutagens. There was found positive correlation between the pollen fertility, cytomixis, and chromosomal stickiness. Genetic factors might have been also contributed toward decrease in pollen fertility. The cytomixis was reported in both the varieties of *Vicia faba* L., but the var. minor showed more frequency than var. major along with stickiness and reduction in pollen fertility indicating its greater sensitivity to the mutagens.

T.A. Bhat (✉)
Department of Education, Goverment of Jammu and Kashmir, Srinagar, India
e-mail: bhattariq110@gmail.com

M. Gulfishan
School of Life and Allied Health Sciences, Glocal University, Saharanpur, India

A.A. Wani
Department of Botany, University of Kashmir, Srinagar, Jammu and Kashmir, India

T.A. Bhat, A.A. Wani (eds.), *Chromosome Structure and Aberrations*,
DOI 10.1007/978-81-322-3673-3_15

Keywords
Cytomixis • Microsporogenesis • Stikiness • Chromosomal aberrations • *Vicia faba*

15.1 Description of *Vicia faba* L.

15.1.1 Origin

Faba bean is believed to be originated from Central Asian, the Mediterranean, and South American centers. According to Cubero (1973) the faba bean originated from four centers of eastern origin: (1) Europe, (2) North African coast to Spain, (3) Nile to Ethiopia, and (4) Mesopotamia to India. Secondary centers of diversity are postulated in Afghanistan and Ethiopia. The *Vicia narbonensis* L. and *Vicia galilaea* Plitmann and Zohary are taxonomically closely related to the cultivated crop having $2n = 12$ chromosomes. Ladizinsky (1975) believed its origin to be Central Asia. The China is the biggest center of diversity of *Vicia faba* L. which produces 60 % of total world production (FAO 1994).

15.1.2 Taxonomy, Morphology, and Floral Biology

Vicia faba L. is an unbranched annual herb with upright shoot system 0.3–2 m tall with one or more hollow stems from the base (Bond et al. 1985; Duke 1981; Health et al. 1994). The leaves are compound having 2–6 leaflets without tendrils or with rudimentary tendrils (Kay 1979; Bond et al. 1985). Flowers are large with white and dark purple markings, produced from short pedicels with 1–5 in number per cluster on each axillary raceme between the 5th and 10th node; the cross-pollination is predominant and bumble bees act as main pollinator. The plant shows taproot system with profusely branched secondary roots (Bond et al. 1985).

The systematic position of *Vicia faba* L. under Bentham and Hooker's system of classification is as follows:

Division – Phanerogams
  Class – Dicotyledons
    Subclass – Polypetalae
      Series – Calyciflorae
        Order – Rosales
          Family – Leguminosae
            Subfamily – Papilionaceae (Fabaceae)
              Genus – *Vicia*
                Species – *Faba*

15.1.3 Ecology

Faba bean develops well in cool season. It grows in warm temperate and subtropical areas as a winter annual herb. It tolerates the temperature up to –15 °C (Robertson et al. 1996). It can be grown anywhere particularly in wetter portions of cereal-growing areas. It grows in a range of soil but grows best on rich loams with moderate moisture (Duke 1981). They are least drought resistant among the legumes, but cultivars with high water use efficiency have been developed at ICARDA (Robertson et al. 1996). Faba bean is more tolerant to acidic soils than most of the legumes. Its production is best in optimum temperatures which range from 18 to 27 °C (Duke 1981). The ideal rainfall which favors its best production is 650–1000 mm per annum (Kay 1979). The life cycle ranges from 90 to 220 days depending upon the cultivars and climatic conditions (Bond et al. 1985).

15.1.4 Yield and Economics

The attack of pathogens such as *Botrytis fabae*, *Ascochyta fabae*, *Uromyces fabae*, and *Orobanche crenata* and insect such as *Aphis fabae*, flower drop, seed abortion, lack of adequate pollination, and reduced seed setting can be major constraints to yield. The yield is also influenced by drought, high temperature, deficiency of nutrients, salinity, and water logging. China was the largest producer with estimated annual production of 2.4–2.6 million MT (1161–1447 kg/ha) from 1979 to 1994 (FAO 1994). Argentina reported the production of more than 9000 kg/ha, followed by Switzerland (3350–4375 kg/ha), France (3000–3900 kg/ha), and Belgium (3350–3750 kg/ha) from 1992 to 1994 (FAO 1994). In Southern United States and US Pacific Coast, the faba been has been in use for human consumption, feed for cattle particularly to horse (popularly called as horse beans), and green manuring. The immature seeds of faba been are used as green vegetable.

The major agroecological regions of faba been are Northern Europe, the Mediterranean, the Nile Valley, Ethiopia, Central Asia, East Asia, Oceana, Latin America, and North America (Bond et al. 1985). The production of faba has been increased manyfold in Australia, EEC, West Asia, and North Africa, but there is 25 % decrease in area meant for sowing of faba been in China and decrease in production in Africa from 1982 to 1992 (Oram and Agacoili 1994). Faba been is the second abundant food legume in Europe (Picard et al. 1988).

15.1.5 Uses

The faba been is a source of protein and carbohydrate from the very ancient times. It plays a vital role in crop rotation, improving physical conditions of soil, disease control, and weed population. It is cultivated as a vegetable and used green or dried, fresh, or canned and for stock feed and is considered superior to other legumes in food value. Faba bean is used a human food as well as animal feed both in developing countries and in industrialized countries. It is used as breakfast in the Middle East, the Mediterranean region, China, and Ethiopia (Bond et al. 1985). The most popular dishes of faba bean are medamis (stewed beans), falafel (deep dried cotyledon paste with some vegetables and spices), bissara (cotyledon paste poured on to plates), and nabet soup (boiled germinated beans) (Jambunathan et al. 1994). Roasted seeds are eaten in India (Duke 1981). Straw from faba bean harvest fetches a premium in Egypt and Sudan and is considered as a cash crop (Bond et al. 1985). The straw can also be used for brickmaking and as a fuel in parts of Sudan and Ethiopia.

Broad bean has been considered as a meat and milk substitute. Large-seeded cultivars are grown in gardens and are used as vegetable and are one of the most important winter crops for human consumption in the Middle East. The green seeds are consumed as a cooked vegetable. The dry beans are used as a food and feed to livestock. Broad beans are very nutritious, containing 23 % proteins. The whole dry seeds contain (per 100 g): 344 calories, 10.1 % moisture, 26.2 g protein, 1.3 g fat, 59.4 g total carbohydrate, 6.8 g fiber, 3.0 g ash, 104 mg ca, 301 mg p, 6.7 mg Fe, 8 mg riboflavin, 2.1 mg niacin, 162 tryptophane, and 16 mg ascorbic acid. The faba bean is said to be used for diuretic, expectorant, and tonic.

Broad bean proteins consist of globulins, albumins, and glutelins. They also contain two minor proteoses and two globulins, i.e., legumin and vicilin. The favism which is a paralytic condition has been reported in certain tropical countries, so the faba bean should be used with care and cautious manner particularly when it is part of regular diet.

15.2 Cytology

Vicia faba L. is best cytogenetical material. It has six pairs of relatively large chromosomes; one pair is bigger than the other five pairs and is used as an excellent experimental material for assessing chromosomal aberrations. Its root tips are used for mitotic studies, and pollen mother cells have been the most frequently used plant material for meiotic studies.

15.2.1 Cytological Aberrations

Estimation of cytological abnormalities and their magnitude during mitosis or meiosis is the most important index for evaluating the effect of a mutagen. It gives handsome knowledge to researchers to know the radiosensitivity of plant species to both physical, chemical, and combined doses of mutagens. The chromosomal aberrations induced by mutagens have been reported by many researchers in several plants such as in pea (Kallo 1972), triticale (Pushpalatha et al. 1992), lentil (Reddy and Annadurai 1992), fenugreek (Anis and Wani 1997; Parveen 2006), *Cicer* (Ganai et al. 2005), *Capsicum annuum* (Anis et al. 2000; Gulfishan et al. 2010, 2011, 2012), and broad bean (Bhat et al. 2005a, b, 2006a, b, c, d, 2007a, b).

15.2.2 Cytomixis

Cytomixis was first observed by Korniche (1901) and then by Gates (1911) in pollen mother cells (PMCs) of *Crocus sativus* and *Oenothera gigas* and explained it as transfer of chromatin material from one cell to the other cell. The term cytomixis was coined by Gates (1911). The cytomixis has been reported since that time in several plants during microsporogenesis (Mc Clintock 1929; Bahl and Tyagi 1988; Koul 1990; Yen et al. 1993; Kumar and Sharma 2002; Haroun et al. 2004; Bhat et al. 2006a, b). The cytomixis has been reported in somatic cells also (Bowes 1973; George and Geethamma 1985) and in the interphase between mitotic and meiotic cells (Cooper 1952). In legumes, transmigration of chromatin has been reported in many genera, e.g., *Vigna* (Sen and Bhattacharya 1988), *Lathyrus* (Kumar and Sinha 1991; Seijo 1996), and *Pisum*, (Gottschalk 1970; Nirmala and Kaul 1994) and *Vicia faba* L. (Bhat et al. 2006b).

15.2.3 Origin of Cytomixis

Cytomixis has been observed in several plants, but its origin is yet unknown, though it has been viewed as a process limited to genetically unbalanced types like haploids, hybrids, and apomictics (Gottschalk 1970; Basavaiah and Murthy 1987; Bahl and Tyagi 1988).

15.2.4 Factors Affecting Cytomixis

The factors believed to cause the cytomixis are (i) influence of genes (Kaul and Nirmala 1993), (ii) abnormal cell wall formation (Kamra 1960), (iii) action of colchicines and ethyl methane sulfonate (Sinha 1988), (iv) action of chemical mutagens (Bhat et al. 2006c), (v) rotenone (Amer and Mikhael 1986), (vi) changes in the biochemical process that involve microsporogenesis modifying the microenvironment of affected anthers (Koul 1990), (vii) gamma radiation-induced imbalanced and sterile genetic system (Amma et al. 1990), (viii) presence of a male-sterile mutant gene (Nirmala and Kaul 1994), and (ix) environmental stress and pollution (Haroun et al. 2004).

15.2.5 Frequency and Intensity of Cytomixis

The connection between the cells adjacent to each other is determined by the frequency and intensity of cytomictic flow. The pollen mother cells showed two types of connections, i.e., direct fusion and cytoplasmic channels. The PMCs showed cytomixis at various stages of cell division, and mostly the cells showed single cytoplasmic channel; the multiple cytoplasmic channels were also observed though rarely. The first phase of meiosis showed less abundant cytoplasmic channels than the second phase, but the cells at different phases of cell division were also evident. Some PMCs showed cytomixis to other PMCs both by cytoplasmic channel and direct fusion simultaneously. Similar observations have been reported by many workers (Bahl and Tyagi 1988; Seijo 1996 and Haroun et al. 2004).

The breadth of cytomictic channels was of varying dimensions showing transfer of chromosomes from one cell to the other cell/cells at different stages of meiosis. The multiple meiocytes usually numbering from 4 to 9 were involved in cytomixis simultaneously, though in some cases only two PMCs were found involved in cytomixis. It was common to find 2–4 cytoplasmic channels emerging from a single meiocyte and connecting many cells together. There was unidirectional flow of chromatin material from donor to a recipient cell. The chromatin material was found migrating to different adjoining pollen mother cells from a single meiocyte simultaneously. In case of PMCs simultaneously showing cytomictic channels, the chromosomes passed from one meiocyte to the second and from second to the third and so on. The unidirectional migration of chromatin material has been observed by Gottschalk (1970). However, in this investigation a single pollen mother cell was found donating its chromatin materials to two PMCs through two independent cytoplasmic channels simultaneously. This type of chromatin migration is unlike the findings observed by Gottschalk (1970). Some donor cells became empty after passing the entire chromatin materials to the recipient cells. The partial migration of the chromatin material resulted in the formation of aneuploid cells. In the present investigation, in some cases no visual transfer of chromosomes through cytoplasmic connections could be detected. The formation of cytoplasmic channels between two or more cells at different places in the cells suggests that screening of EMS, gamma rays, MMS, and their combination treated population for varied ploidy level, and its utilization in plant breeding may produce the polyploids.

15.2.6 Consequences of Cytomixis

Some recipient cells were found with additional number of chromosomes which did not show synapses with the chromosomes of recipient cells but showed synapses among themselves and went on divisions separately. This is similar with the findings of Pantulu and Manga (1972). The fate of such additional number of chromosomes is unknown, but they may result in the formation of either micronuclei or micropollen. Sarvella (1958) reported loss of chromosomes in meiotic and somatic cells in *Gossypium*. Thakur (1978) opines that a loss or gain of one or more chromosomes has two possibilities. The gametes with deficient chromosomes will be lethal and get eliminated, and gametes with different number of chromosomes may result in the formation of aneuploids. In the condition where all the chromosomes will move, it results in the formation of aneuploids. Sometimes chromatin material moves without nucleolus, and sometimes nucleolus moves without chromatin material. The movement of chromosomes in between and among the meiocytes ranges from 1 to 6. It signifies that chromosome movement breaks at different intervals at different stages of meiosis and the PMCs showing cytomixis were also at different stages.

The early stages of meiosis showed more frequent PMCs involved in cytomixis, and gradual decline was observed toward the end of meiosis; same findings were observed earlier by Maheswari (1950) and Bhat et al. (2006a, b, c, d) but in contrary to the earlier findings of Verma et al. (1984) that all the stages of meiosis are equally susceptible to cytomixis.

During observation the larger or smaller cells with or without nucleus were also observed than the normal. This is possibly because of additional chromatin material resulting due to transfer and accumulation and by their retransfer to the subsequent cells. That could be the reason for the presence of hexapolar anaphase II, pentads, and hexads. The presence of cytomixis may be because of abnormal genetic behavior as the dose-dependent chromosomal stickiness and cytomixis were observed with the physical, chemical, and combined mutagenic treatments. We consistently failed to observe chromosomal stickiness, and cytomixis in controlled plants justifies our view. The decrease in frequency of meiocytes showing deviation in basic chromosome number was observed in the higher concentration treated plant population, but the cytoplasmic connection between the PMCs was of frequent occurrence which suggests that stickiness inhibits the chromosome migration from one cell to another.

Migration of chromatin material among the adjacent PMCs also occurred through cytoplasmic connections originating from the plasmodesmata formed in the tissues of the anther. The plasmodesmata become closed by the deposition of secondary cell wall depositions, but in some instances they persisted during microsporogenesis and increase in dimensions, forming evident intermeiocyte connections that permitted the transfer of chromosomes (Falistocco et al. 1995).

15.2.7 Comparative Mutagenicity and Cytomixis

Comparative mutagenicity of EMS, gamma rays, MMS, and their combinations in the two varieties, viz., minor and major of broad bean, reflects the differences in their genome architecture as the former exhibited much induced biological damage than the latter. Thus, it can be inferred that mutagenicity is ultimately determined by the genome itself. Khamankar (1984) working with *Lycopersicon esculentum* observed that the frequency of mutations was different with different mutagens at various loci. Some gene loci mutated by one mutagen were not necessarily affected by the other. However, the results of the present study differ from the results of some earlier workers who have found that bold seeded types were more sensitive to mutagens than the small seeded types (Sharma 1986 and Reddy et al. 1992).

The presence of large-sized pollen grains (possibly 2n pollen grains) was found in treated populations. A numerically unreduced diploid, or 2n gamete, is a meiotic product that bears the sporophytic rather than the gametophytic chromosome number. Such gametes are due to abnormal microsporogenesis. Unreduced gametes produce individuals with higher ploidy level through a process known as sexual hybridization (Villeux 1985). Sexual polyploidization is the primary cause to the production of naturally occurring polyploids (Harlan and Dewat 1975).

Intervarietal differences in the radiosensitivity have been reported by many workers (Kumar and Dubey 1998a, b; Dhamayanthi and Reddy 2000; Zeerak 1992; Singh 2003; Bhat et al. 2006a). These differences were attributed to differences in cell volume, nuclear volume, chromosome volume, and DNA amount and the presence of protective or sensitizing substances (Sparrow and Evans 1961; Ahmad and Godward 1981; Bhat et al. 2005a, b; Bhat 2007).

Most of the workers have in general concluded that gamma rays were more effective than chemical mutagens in causing cytomixis. However, in the present study, the cytomixis was greater in the combination treatments than in the individual treatments; it is, therefore, concluded that EMS and MMS too are able to induce a sufficient amount of meiotic aberrations. Singh et al. (1989) and Roy (1989) in lentil reported that EMS is highly efficient in inducing chromosomal aberrations equal to that of physical mutagens. Among different stages of meiosis, the frequency of cytomixis was maximum at prophase-1 stage in the present study. Similar observations were reported by Mitra and Bhowmik (1999) in *Nigella sativa* and Kumar and Dubey (1998a, b) in *Lathyrus sativus* L. Bhat et al. (2006a).

For most of the combination treatments, the coefficient of interaction (K) was less than additive; however, additive or synergistic effect was obtained at different combination doses in both var. minor and major. Similar result has been reported by Sharma et al. (2004) in chickpea. Less than additive effects for chromosomal aberrations were also reported by Khalatkar and Bhatia (1975) in barley and by Bhamburkar and Bhalla (1985) in black gram.

15.2.8 Conclusions

The mutagens used in present investigation were both physical and chemical mutagens. Among the physical mutagens, gamma rays were used, and among the chemical mutagens, EMS and MMS were used. The combined doses of gamma rays with MMS and EMS were also tested. The combined doses were more effective than individual doses of gamma rays, EMS, and MMS. Among the individual mutagens, MMS is more effective followed by gamma rays than EMS in inducing cytomixis, stickiness, and reduction in pollen fertility in both the *Vicia faba* var. minor and *Vicia faba* var. major. In all the mutagen-treated populations, the cytomixis was reported, and there was dose-dependent increase in the PMCs showing cytomixis. The pattern of effectiveness is as follows: gamma rays + MMS > gamma rays + EMS > MMS > gamma rays > EMS. Among the two varieties tested, var. minor is more sensitive than var. major of *Vicia faba* L. For most of the combination treatments, the coefficient of interaction (K) was less than additive; however, additive or synergistic effect was obtained at 20kR + 0.2 % EMS and 20kR + 0.02 % MMS. The cytomixis in the present case may be attributed to abnormal genetic behavior due to treatment with combined and individual physical and chemical mutagens. The cytomixis leads to formation of aneuploids and polyploids which is asset in the hands of cytogeneticist.

References

Ahmad S, Godward MBE (1981) Comparison of radio-resistant with a radio-sensitive cultivar of *Cicer arietinum* L. II Differences in the number of chromosome aberrations at the same dose. Environ Exp Bot 21:143–151

Amer SM, Mikhael E (1986) Cytological effects of pesticides. XVI Effects of the insecticide rotenone on root mitosis of Vicia faba. Cytologia 51:171–176

Amma CKS, Namboodiri AN, Panikkar AON, Sethuraj MR (1990) Radiation induced male sterility in *Hevea brasiliensis* (Willd ex Adr. De Juss) Muell. Arg. Cytologia 55:547–551

Anis M, Wani AA (1997) Caffeine induced morpho-cytological variability in fenugreek, Trigonella foenum-graecum L. Cytologia 62:343–349

Anis M, Abbasi N, Sharma N (2000) Cytological effect of mutagens in *Capsicum annuum*. J Cytol Genet 1:17–21

Bahl SR, Tyagi BR (1988) Cytomixis in pollen mother cells of *Papaver dubium* L. Cytologia 53:771–775

Basavaiah D, Murthy TCS (1987) Cytomixis in pollen mother cells of *Urochloa panicoides*. (Poaceae). Cytologia 52:69–74

Bhamburkar S, Bhalla JK (1985) Effects of differential and combined treatments of gamma rays, hydrazine and mitomycin-C in blackgram. J Indian Bot Soc 8:82–87

Bhat TA (2007) studies on the effect of induced mutagenesis on the cytomorphological parameters of Broad Bean (*Vicia faba* L) with special reference to improvement in economic characters. PhD thesis. Aligarh Muslim University Aligarh

Bhat TA, Khan AH, Parveen S (2005a) Comparative analysis of meiotic abnormalities induced by gamma rays. EMS and MMS in Vicia faba L. J Indian Bot Soc 84:45–48

Bhat TA, Parveen S, Khan AH (2005b) Clastogenic effect of EMS and MMS in *Vicia faba* L. J Cytol Genet 6:117–122

Bhat TA, Khan AH, Parveen S (2006a) Effect of EMS on certain Cytomorphological parameters in two varieties in *Vicia faba* L. Ad Plant Sci 19:227–232

Bhat TA, Khan AH, Parveen S, Ganai FA (2006b) Studies on the effect of EMS on meiosis and pollen fertility in *Vicia faba* L. Ad Plant Sci 19:243–247

Bhat TA, Khan AH, Parveen S, Wani NA (2006c) Radiation Induced pod and seed mutants and MMS induced closed flower mutant in broad been (*Vicia faba* L.). Indian J Genet 66:249–250

Bhat TA, Parveen S, Khan AH (2006d) MMS-induced Cytomixis in pollen mother cells of Broad Bean (*Vicia faba* L). Turk J Bot 30:273–279

Bhat TA, Parveen S, Khan AH (2007a) Meiotic studies in two varieties of *Vicia faba* L. (Fabaceae) after EMS treatment. Asian J Plant Sci 6:51–55

Bhat TA, Sharma M, Anis M (2007b) Comparative analysis of mitotic aberrations induced by Diethyl sulphate (DES) and Sodium Azide (SA) in *Vicia faba* L. (Fabaceae). Pak J Biol Sci 10:783–787

Bond DA, Lawes DA, Hawtin GC, Saxena MC Stephens JS (1985) Faba bean (*Vicia faba* L.). In: Summerfield RJ, Roberts EH (eds) Grain legume crops. William Collins Sons Co. Ltd, London, pp 199–265. WIX 3LA.UK

Bowes BG (1973) Note on apparent case of cytomixis in the root apex of *Allium cepa*. Cytologia 38:125–129

Cooper DD (1952) The transfer of deoxyribose nucleic acid from the tapetum to the microsporocytes at the onset of meiosis. Am Nat 86:219–229

Cubero JI (1973) Evolutionary trends in Vida faba. Theor Appl Genet 43:59–65

Dhamayanthi KPM, Reddy VRK (2000) Cytogenetic effects of gamma rays and ethyl methane sulphonate in chilli pepper (*Capsicum annuum* L.). Cytologia 65:129–133

Duke JA (1981) Handbook of legumes of world economic importance. Plenum Press, New York, pp 199–265

F.A.O (1994) Production year Book-Rome. F.A.O, Italey

Falistocco E, Tosti T, Falcinelli M (1995) Cytomixis in pollen mother cells of diploid Dactylis, one of the origins of 2n gametes. Heredity 86:448–453
Gates RR (1911) Pollen formation in *Oenothera gigas*. Ann Bot 25:909–940
Ganai FA, Khan AH, Bhat TA, Parveen S, Wani NA (2005) Cytogenetic effects of Methyl Methane Sulphonate (MMS) on varieties of Chickpea (Cicer arietinum L.). J Cytol Genet 6:97–102
George K, Geethamma S (1985) Role of cytomixis in the speciation of *Jasminum*. Curr Sci 54:586–587
Gottschalk W (1970) Chromosome and nuclear migration during microsporogenesis of *Pisum sativum*. Nucleus 13:1–9
Gulfishan M, Khan AH, Bhat TA (2010) Studies on cytotoxicity induced by DES and SA in Vicia faba L. var. Major. Turk J Bot 34:31–37
Gulfishan M, Khan AH, Haneef I, Bhat TA (2011) Genotoxic effect of Diethyl Sulphate in two varieties of Capsicum annum L. Nucleus 54(2):107–111
Gulfishan M, Khan AH, Gefferi IF, Bhat TA (2012) Assessment of mutagencity induced by MMS and DES in Capsicum annum L. Saudi J Biol Sci 19:251–255
Harlan JR, Dewat JMJ (1975) On *O. winge* and a prayer: the origins of polyploidy. Bot Rev 41:361–390
Haroun SA, Al Shehri AM, Al Wadie HM (2004) Cytomixis in the microsporogenesis of *Vicia faba* L. (Fabaceae). Cytologia 69:7–11
Health MC, Pilbeam CJ, Mckenzie BA, Hebblethwaite PD (1994) Plant architecture, competitive ability and crop productivity in food legumes with particular emphasis on pea (Pisum sativum L.) and faba bean (Vicia faba L.). In: Muehlbauer FJ, Kaiser WJ (eds) Expanding the production and use of cool season food legumes. Kluvser Academic Publishers, Dordrecht, pp 771–790
Jambunathan R, Blain HL, Dhindsa KH, Hussain LA, Korgure K, Lijuan L, Youssef MM (1994) Diversifying use of cool season food legumes through processing. In: Muehlbauer FJ, Kaiser WJ (eds) Expanding the production and use of cool season Food Legumes. Kluwer Academic Publisher, Dordrecht, pp 98–112
Kallo Z (1972) Chromosomal alterations in mitotic and meiotic system as influenced by gamma rays in Pisum. Cytologia 37:643–651
Kamra OP (1960) Chromatin extrusion and cytomixis in pollen mother cells of *Hordeum*. Heriditas 46:415–421
Kaul MLH, Nirmala C (1993) Male sterility in Pea II. Male sex specific desynapsis. Cytologia 58:67–76
Kay D (1979) Crop and product digest no. 3-food legumes. Tropical products Institute, London, pp 26–47
Khalatkar AS, Bhatia CR (1975) Synergistic effect of combined treatments of gamma radiation and ethylmethane sulphonate in barley. Rad Bot 15:223–229
Khamankar YG (1984) Differential sensitivity to tomato genes to chemical and physical mutagens. Indian J Genet Plant Breed 44:80–83
Korniche M (1901) Uber ortsveranderrung von Zellkarnern. S.B. Niedernhein, Ges. Natur-und Heilkunde Boun. pp 14–25
Koul KK (1990) cytomixis in pollen mother cells of *Alopecurus arundinaceus* Poir. Cytologia 55:169–173
Kumar S, Dubey DK (1998a) Effect of separate and simultaneous application of gamma rays and EMS on germination, growth, fertility and yield in cultivars Nirmal and LSD-3 of khesari (*Lathyrus sativus* L.). J Phytol Res 11:165–170
Kumar S, Dubey DK (1998b) Mutagenic efficiency and effectiveness of separate and combined treatments with gamma rays, EMS and DES in khesari (*Lathyrus sativus* L.). J Indian Bot Soc 77:1–4
Kumar G, Sharma V (2002) Induced cytomixis in chickpea (*Cicer arietinum* L.). Nucleus 45:24–26

Kumar G, Sinha SSN (1991) Genotoxic effects of two pesticides (roger and Bavistin) and antibiotic (Streptomycin) in meiotic cells of grasspea (Lathyrus sativus). Cytologia 56:209–214

Ladizinsky G (1975) On the origin of the broad bean *Vicia faba* L. Israel J Bot 24:80–88

Maheswari P (1950) An introduction to the embryology of Angiosperms. McGraw-Hill Book Co.Inc, New York

Mc Clintock B (1929) A cytological and genetical study of triploid maize. Genetics 14:18–222

Mitra PK, Bhowmik G (1999) Studies in the frequency and segregation of induced chlorophyll mutations in *Nigella sativa* L. Adv Plant Sci 12:125–129

Nirmala C, Kaul M (1994) Male sterility in pea IV. Cytologia 59:195–201

Oram PA, Agacoili M (1994) Current status and future trends in supply and demand of cool season food legumes. In: Muehlbauer FJ, Kaiser WJ (eds) Expanding the production and use of cool season food legumes. Kluwer academic Publishers, Dordrecht, pp 3–49

Pantulu JV, Manga V (1972) Cytogenetics of 16 chromosome plants in pearl millet. Cytologia 37:389–394

Parveen S, Anis M, Bhat TA, Khan AH (2006) Efficiency and effectiveness of physical and chemical mutagens in M2 generation of Trigonella foenum-graceum L. J Cytol Genet 7:1–4

Picard JJA, Bond DA, Monti L, Steuckardt R (1988) Production of pea. Faba bean and chickpea in Europe. P. In: Summerfield RJ (ed) world crops: Cool season food legumes. kluwer Academic publishers, Dordrecht, pp 1065–1080

Pushpalatha KN, Reddy VRK, Indra M, Nalini R (1992) Mutagenic effects of gamma rays, chemical mutagens and combined treatments in triticale. Adv Plant Sci 5:147–160

Reddy VRK, Annadurai M (1992) Cytological effects of different mutagens in Lentil (Lens culinaris Medik). Cytologia 57:213–216

Reddy VRK, Indra M, Revathi R (1992) Meiotic abnormalities in lentil (*Lens culinaris* Medik). Cytologia 57:213–216

Robertson LD, Singh KB, Erskine W, Ali M, El Moneim A (1996) A cytological study of the Hordeae in the Dakotas. Cytologia 31:213–219

Roy SK (1989) Effects of EMS and SA on dry seeds of Lens esculenta Moench. Proc. Natl. Symp. On Recent Advances in Plant cell Research, June 7–8, Trivandrum:41–42

Sarvella P (1958) Cytomixis and the of chromosome in meiotic and somatic cells of *Gossypium*. Cytologia 23:14–24

Seijo G (1996) spontaneous cytomixis in the microsporogenesis of sweet pea (*Lathyrus odoratus* L.) (Leguminosae). Cytologia 61:189–195

Sen O, Bhattacharya S (1988) Cytomixis in *Vigna Glaberscens* Wild. Cytologia 53:437–440

Sharma B (1986) Increasing the efficiency of mutagenesis for micromutations by early generation selection. Indian J Genet 46:88–100

Sharma V, Kumar G, Kumar R (2004) Comparative mutagenicity of gamma rays and EMS in *Cicer arietinum* L. J Cytol Genet 5:21–26

Singh AK (2003) Effect of doses of gamma rays and EMS on germination and plant survival in mungbean (*Vigna radiata* L. Wilczek). Int J Mendel 20:99–102

Singh D, Singh RM, Singh J (1989) Investigations on induced macromutants in lentil/Lens culinaris Medik). Z. Pjlanzenzucht 83

Sinha ARP (1988) Morphological and cytological changes induced in Lindenbergia indica following treatments. Genet Polonica 29:335–339

Sparrow AH, Evans HJ (1961) Nuclear factors affecting radio sensitivity. The influence of Nuclear size and structure chromosome complement and DNA content. Brookhaven Symp Biol 12:76–100

Thakur V (1978) Chromosomal and Nuclear transmigration during microsporogenesis in natural populations of some plant species. In: Dnyansagar VR et al (eds) Recent trends and contacts between cytogenetics, embryology and morphology. Today and Tomarrow, publishing Co, New Delhi, pp 123–130

Verma RC, Sarkar A, Das BC (1984) Cytomixis in mulberry. Curr Sci 53:1258–1260

Villeux R (1985) Diploid and polyploid gametes in crop plants. In: Janick J (ed) Mechanisms of formation and utilization in plant breeding, 442. Plant Breeding rev. 3. AVI publishing Co.,: Wesport.

Yen C, Yang J, Sun GL (1993) Intermeiocyte connections and cytomixis in intergeneric hybrids of *Roegneria ciliaris* (Trin.) Nevski with *Psathyrostachys huashanica* Keng. Cytologia 58:187–193

Zeerak NA (1992) Cytogenetical effects of gamma rays and ethyl methane sulphonate in tomato (*Lycopersicon esculentum* var. Cerasiforme). Phytomorphology 42:81–86; Koduru PKR, Rao MK (1981) Cytogenetics of synaptic mutants in higher plants. Theor Appl Genet 59:197–214

Fluorescence In Situ Hybridization (FISH) and Its Applications

16

Abdul Rauf Shakoori

Abstract

Fluorescence in situ hybridization (FISH) is the most convincing technique for locating the specific DNA sequences, diagnosis of genetic diseases, gene mapping, and identification of novel oncogenes or genetic aberrations contributing to various types of cancers. FISH involves annealing of DNA or RNA probes attached to a fluorescent reporter molecule with specific target sequence of sample DNA, which can be followed under fluorescence microscopy. The technique has lately been expanded to enable screening of the whole genome simultaneously through multicolor whole chromosome probe techniques such as multiplex FISH or spectral karyotyping or through an array-based method using comparative genomic hybridization. FISH has completely revolutionized the field of cytogenetics and has now been recognized as a reliable diagnostic and discovery tool in the fight against genetic diseases.

Keywords

FISH • Fluorescence microscopy • Chromosomal aberrations • Diversification of FISH • Principle of FISH • FISH probes

16.1 Introduction

The classical cytogenetics used trypsin-Giemsa or fluorescent banding pattern for identification and characterization of different chromosomal abnormalities such as polycentric chromosomes, ring chromosomes, or chromatid interchanges.

A.R. Shakoori (✉)
School of Biological Sciences, University of the Punjab,
Quaid-i-Azam Campus, Lahore 54590, Pakistan
e-mail: arshaksbs@yahoo.com; arshakoori.sbs@pu.edu.pk

T.A. Bhat, A.A. Wani (eds.), *Chromosome Structure and Aberrations*,
DOI 10.1007/978-81-322-3673-3_16

Though chromosome banding techniques based on Giemsa staining revolutionized cytogenetic analysis, they did not become popular because of limited resolution involving only >3 Mb of DNA. Certain chromosomal aberrations such as reciprocal translocations and inversions were not easily recognizable with Giemsa stain. Besides that these techniques are very time consuming, and interpretation of karyotype is very cumbersome and uncertain.

In situ hybridization techniques initially developed by Joseph Gall and Mary Lou Pardue in 1960s (Pardue and Gall 1969) and John et al. (1969) have proved to be powerful tools for determining the chromosomal location of hybridized nucleic acid. Soon after that fluorescent labels quickly replaced radioactive labels in hybridization probes because of their greater safety, stability, and ease of detection.

Early in situ studies used radioactive RNA or DNA probes that were labeled with ^{3}H or ^{135}I, and the sites of hybridization were detected by autoradiography. These techniques have been successfully applied to both animals and plants. RNA probes can be designed for any gene or any sequence within a gene for visualization of mRNA, long noncoding RNA and miRNA in tissues and cells. These probes, often derived from the fragments of DNA that were isolated, purified, and amplified for use in Human Genome Project, consist of about 20 oligonucleotide pairs and cover a space of 40–50 bp of target RNA. In 1982, a new method was described to localize DNA sequences hybridized in situ to chromosome. This method utilized a biotin-labeled analogue of thymidine (TTP) which could be incorporated enzymatically into DNA probes by nick translation. The sites of hybridization were detected either cytochemically by using avidin conjugated to horseradish peroxidase, or fluorometrically by using fluorescein-labeled antibodies. Compared to autoradiography this technique decreased the time required for detection, improved resolution, and gave less non-specific background and chemically stable hybridization probes.

Besides that non-isotopic techniques have been developed using DNA probes labeled with amino acetyl fluorene (AAF), mercuration, and sulfonation, which are detected after hybridization by affinity reagents. Recently a very effective system has been described that uses digoxigenin-labeled nucleotides detected by antibodies carrying fluorescent or enzymatic tag. The non-isotopic labeling techniques have also been successfully applied for detection of highly repeated DNA sequences in plant chromosomes. The non-isotopic detection of low- or single-copy genes, however, has not been successful.

Chromosome painting – competitive hybridization using entire chromosome – specific libraries for chromosomes as probes and human genomic DNA as the competitor was one of the first applications of FISH (Fig. 16.1). It provided intense and specific fluorescent staining of human chromosome in metaphase spread and interphase nuclei. A translocation t(9;22)(q34;p11) was first identified in human neoplasia leading to Philadelphia chromosome.

Fluorescence in situ hybridization (FISH) began with the discovery that nucleic acids could be chemically modified to incorporate a hapten such as biotin or digoxigenin, which in turn could be detected with a fluorescently labeled reporter molecule such as avidin or anti-digoxigenin. Since then probe preparation and labeling

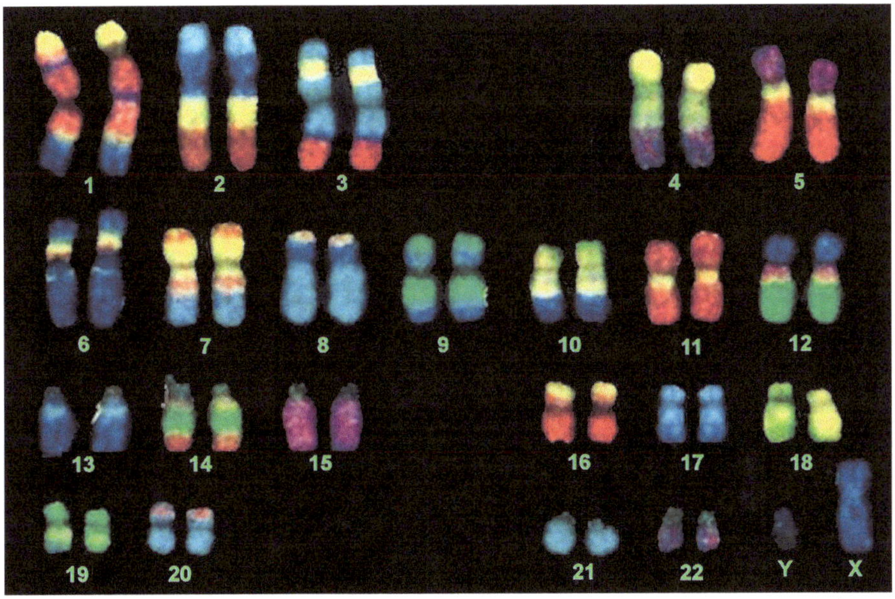

Fig. 16.1 Fluorescent in situ hybridization (FISH) identification of human chromosomes through chromosome painting. DNA probes specific to regions of particular chromosomes are attached to fluorescent markers and hybridized with a chromosome spread. The picture shows a computergenerated "false color" image, in which small variations in fluorescence wavelength among probes are enhanced as distinct primary colors. The combination of probes that hybridize to a particular chromosome produces a unique pattern for each chromosome. This makes it particularly easy to detect segmental deletions and translocations among chromosomes (Taken from https://www.mun.ca/biology/scarr/FISH_chromosome_painting.html)

techniques have been modified and simplified. Now nucleotides can be labeled with fluors directly and incorporated into FISH probes, eliminating the often laborious detection steps.

16.2 Advancement in Fish Techniques

Fluorescence in situ hybridization (FISH) can detect specific sites of specific DNA sequences in metaphase or interphase cells. This technique, initially developed for mammalian chromosome, was first applied to plant chromosomes by Schwarzacher et al. (1989) and Yamamoto and Mukai (1989). FISH has been used to detect 18S.26SrRNA and repeated DNA sequences in plant chromosomes such on *Aegilops*, *Hordeum*, *Oryza*, *Arabidopsis*, *Brassica*, soybean, and barely chromosome.

GISH (genomic in situ hybridization) is a technique in which genomic DNA is used as a probe. In this technique, genomic DNA from one species is used as the labeled probe, while unlabeled DNA from the other species under test is used as the

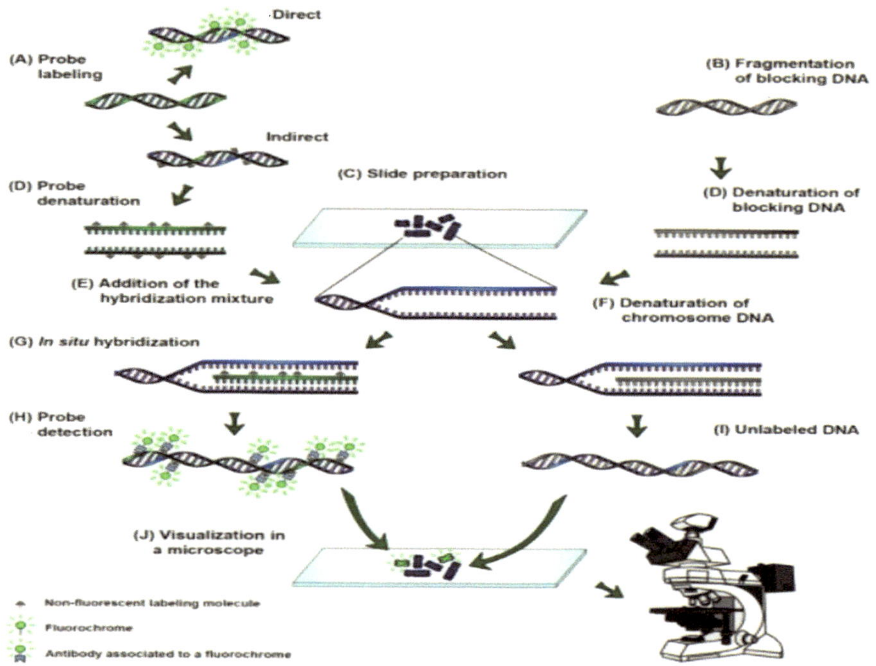

Fig. 16.2 The main steps involved in the genomic in situ hybridization are (**a**) direct or indirect labeling of probe, (**b**) blocking DNA fragmentation, (**c**) preparation of slide, (**d**) denaturation of probe and blocking DNA in a hybridization mixture, (**e**) addition of the probe and the blocking DNA with the hybridization mixture, (**f**) chromosome DNA denaturation, (**g**) hybridization of blocking DNA and probe in the target sequence of the chromosome, (**h**) detection of the probe in the chromosome of one parent, (**i**) chromosome DNA molecule of the second parent related to the unlabeled blocking DNA, and (**j**) visualization of hybridization signals in a fluorescence microscope. Unlabelled chromosomes are visualized with a counterstain (*blue*) (Taken from http://www.slideshare.net/kskuldeep1995/genomic-in-situ-hybridization)

competitor at a much higher concentration (Fig. 16.2). The technique is very useful for cytological identification of foreign chromatin in interspecific hybrids at the molecular level. In plant molecular cytogenetics, GISH has also been used to detect parental genomes in natural allopolyploid species such as *Millium montianum*, *Triticum aestivum*, *Aegilops triuncialis*, and *Nicotiana tabacum*, and also alien segments in translocations.

Now availability of several probe labeling procedures has enabled detection of two or more sequences in the same cell by using fluorochromes of different colors. Reid et al. (1992) were able to visualize seven different DNA probes on human metaphase chromosomes simultaneously by FISH using combinatorial fluorescence and digital imaging microscopy. The multicolor FISH technique has been extensively used in plant molecular cytogenetics. Leitch et al. (1991) demonstrated two highly repeated DNA sequences simultaneously in rye chromosomes. Mukai et al. (1993), using multicolor FISH with total genomic probes and highly repeated sequences, reported simultaneous detection of three genomes of an allohexaploid wheat.

The combination of biotin, digoxigenin, and fluorescein labeling has allowed us to detect multiple probes and to map sequences relative to each other in single cells. Mukai (1995) detected five DNA probes with different colors on a single chromosome.

16.3 Principle Involved in Fish

The basic principle involved is hybridization of nuclear DNA of either interphase cells or of metaphase chromosomes affixed to a microscopic slide, with a nucleic acid probe. The probes are either labeled indirectly with a hapten or directly through incorporation of a fluorophore. The labeled probe and the target DNA are mixed together after denaturation, which allows annealing of complementary DNA sequences. In case the probe had been labeled indirectly, an extra step of enzymatic or immunological detection system will be required for visualization of the non-fluorescent hapten. Finally the signals are evaluated by fluorescence microscopy (Fig. 16.3). The enzymatic detection system involves fluorochrome, which emits colored signals at the hybridization site. The immunological detection system is based on binding of antibodies to specific antigens, which is then demonstrated with a colored histochemical reaction visible by light microscope or fluorochromes with ultraviolet light.

For direct detection, the most frequently used reporter molecules are fluorescein (fluorescein isothiocyanate, FITC), rhodamine, Texas Red, Cy2, Cy3, Cy5, and AMCA. For indirect detection method, the reporter molecules typically used are biotin, digoxigenin, and dinitrophenol.

16.4 Preparation of Probes

One of the most important steps in FISH analysis is the choice of probe. A wide range of probes, extending from whole genomes to small cloned probes (1–10 kb), can be used. There are basically three types of probes, each with a different range of applications, whole chromosome painting probes, repetitive sequence probes, and locus specific probes, which are briefly described below.

1. Chromosome painting refers to the hybridization of fluorescently labeled chromosome-specific composite probe pools to cytological preparations. This enables visualization of individual chromosomes in metaphase or interphase cells and the identification of chromosomal aberrations. The whole chromosome painting probes are complex DNA probes derived from a single type of chromosome that has been PCR amplified and labeled to generate a "paint" which homogeneously highlights the entire chromosome. With this probe, the cytologically visible structural and numerical chromosome rearrangement in metaphase becomes obvious. The chromosomal paint is, however, not helpful in the analysis of interphase cells. Whole chromosome painting is now available for every human chromosome, allowing the simultaneous painting of the entire genetic

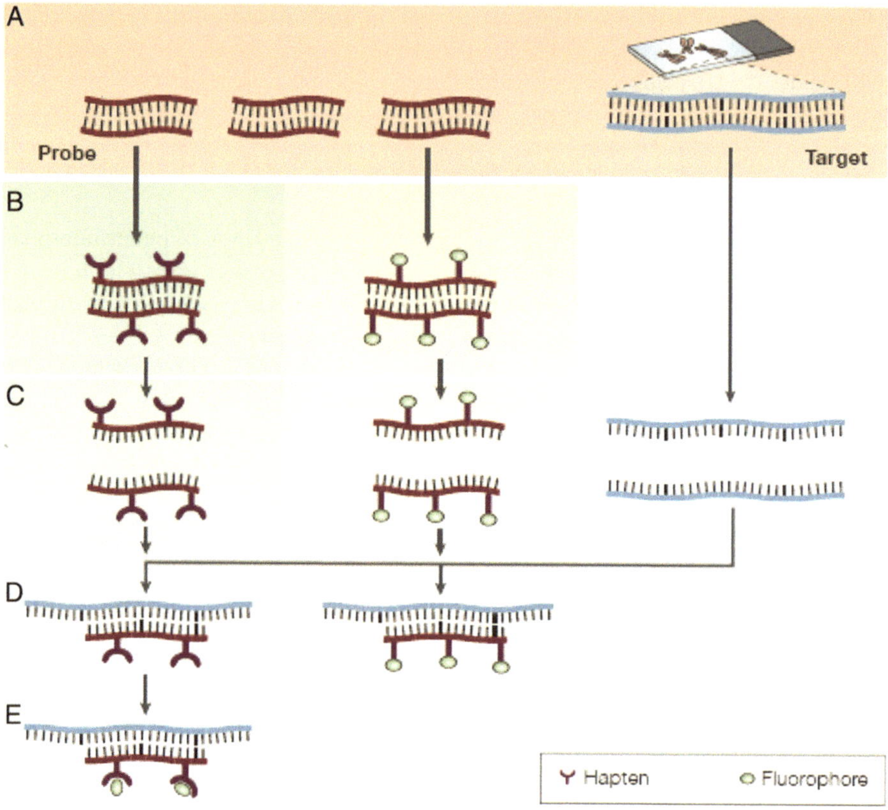

Fig. 16.3 The principles of fluorescence in situ hybridization. (**a**) The basic elements are a DNA probe and a target sequence. (**b**) Before hybridization, the DNA probe is labeled indirectly with a hapten (*left* panel) or directly labeled via the incorporation of a fluorophore (*right* panel). (**c**) The labeled probe and the target DNA are denatured to yield single-stranded DNA. (**d**) They are then combined, which allows the annealing of complementary DNA sequences. (**e**) If the probe has been labeled indirectly, an extra step is required for visualization of the nonfluorescent hapten that uses an enzymatic or immunological detection system. Finally, the signals are evaluated by fluorescence microscopy (Taken from http://biohorizons.oxfordjournals.org/content/early/2010/02/26/biohorizons.hzq009/F1.expansion.html)

complement in 24 colors. This has led to the development of two independent FISH techniques – multicolor FISH (M-FISH) and spectral karyotyping (SKY) – which have important diagnostic and research application values.

2. Repetitive sequence probes hybridize to specific chromosomal regions or structures that contain short sequences, which are present in many thousands of copies. For example, pan-telomeric probes target the tandemly repeated (TTAGGG) sequences present in all human chromosomes ends. Centromeric probes target the α- and β-satellite sequences, flanking the centromeres of human chromosomes. Satellite DNA probes hybridize to multiple copies of the repeat sequences present at the centromeres, resulting in two very bright fluorescent signals in both metaphase and interphase diploid cells. These centromere-specific probes

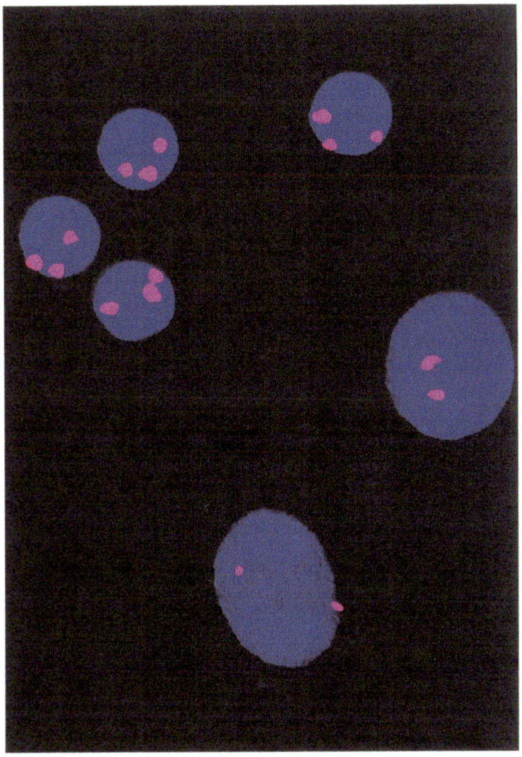

Fig. 16.4 Fluorescence in situ hybridization for trisomy 12. Depicted are the nuclei of NLC (*large ovals*) and CLL cells (*small circles*) examined for trisomy 12 by FISH. The two large NLC nuclei have only two bright fluorescence signal spots, whereas the four CLL cell nuclei each have three bright signal spots, reflecting the presence of trisomy 12 (Taken from http://www.bloodjournal.org/content/96/8/2655?sso-checked = true)

are useful in detection of monosomy, trisomy, and other aneuploidies in leukemias and solid tumors (Fig. 16.4).

3. Locus-specific probes are usually genomic clones, which vary in size depending on the nature of the cloning vector from plasmids (which can carry 1–10 kb) to the larger PAC (P1 bacteriophage-derived artificial chromosome, which can carry 100–300 kb), YAC (yeast artificial chromosome which can carry 150–350 kb), and RAC vectors (which can carry 80 kb to 1 Mb). These probes are particularly useful for detection of translocations, inversions, and deletions in both metaphase and interphase.

16.5 Methodology Involved

In situ hybridization (ISH) involves the following major steps:

16.5.1 Cytological Preparation

Well-spread out and flat preparation ensures best morphology and highest hybridization signals. Most of the ISH studies of plant chromosomes have been made on mitotic root tip preparations. The root tips fixed in ethanol/glacial acetic acid are stained with 1 % acetocarmine and then squashed in 45 % acetic acid on the slide. The slides can be stored in −80 °C freezer for at least 1 year. After thawing the chromosomes are dehydrated on the slide before hybridization.

16.5.2 Probe Labeling

Several methods for labeling DNA probes for nonradioactive in situ hybridization have become available. The most common approach is to label the probe with reporter molecules (haptens). A variety of haptens are available in the market: biotin, digoxigenin, dinitrophenol, fluorescein, rhodamine, AMCA, and coumarin. These haptens can be incorporated as labeled nucleotides by tagging technique of nick translation, random primer labeling, or PCR according to the routine procedures. Detection of hybridized digoxigenin probes is mediated by anti-digoxigenin antibodies conjugated to enzyme or fluorochrome. The labeled DNA may be separated from unincorporated nucleotides using the spin column or ethanol precipitation methods. The random primed labeling method is based on the hybridization of a mixture of all possible hexanucleotides to the DNA to be labeled.

16.5.3 In Situ Hybridization

16.5.3.1 Nonradioactive In Situ Hybridization

For nonradioactive in situ hybridization, the chromosomal DNA is denatured on the slides in 70 % formamide, 2XSSC at 68–70 °C for 2 min. The slides are dehydrated and then air-dried. The hybridization mixture containing DNA probe (20–50 μg/ml) is added to the slide and incubated at 37 °C for 6–12 h. For detection of hybridization sites, the slides are washed in 2XSSC and then PBS. The slide is incubated with 0.6 % streptavidin-horseradish complex at 37 °C for 30 min. After washing, 0.05 % diaminobenzidine-tetrahydrochloride (DAM) and 0.01 % H_2O_2 are placed on the slide and incubated at room temperature in the dark for 5–20 min. Slides are rinsed with PBS and counterstained with 2 % Giemsa for one minute and air-dried. Positive hybridization sites should appear dark brown.

16.5.3.2 Genomic In Situ Hybridization (GISH)

The purified isolated genomic DNA is sheared by passing through an 18-gauge hypodermic needle or by ultrasonication. 1 μg DNA is labeled with biotin-16-dUTP through nick translation and then purified by spin column or through ethanol precipitation. The genomic DNA is denatured on the slide by immersion in 70 % formamide-2XSSC solution at 68–70 °C for 2 min. The slides are rapidly dehydrated and air-dried. Hybridization mixture containing labeled total genomic probe (1 μg/ml) is added to each slide and incubated in moist plastic chamber at 37 °C for 6–12 h. For detection of hybridization sites, the slides washed in 2XSSC and then PBS are incubated with 0.6 % streptavidin-horseradish complex at 37 °C for 30 min.

After washing, 0.05 % diaminobenzidine-tetrahydrochloride (DAM) and 0.01 % hydrogen peroxide are placed on each slide and incubated at room temperature in the dark for 5–20 min. Slides are rinsed with PBS and counterstained with 2 % Giemsa for 1 min and air-dried. Positive hybridization sites should appear dark brown.

16.5.3.3 Fluorescence In Situ Hybridization (FISH)

For FISH, the chromosomal DNA is denatured on the slides in 70 % formamide-2XSSC solution at 68–70 °C for 2 min. The slides are dehydrated and then air-dried. The hybridization mixture containing DNA probe (20–50 μg/ml) is added to the slide and covered with cover slip and incubated in moist plastic chamber at 37 °C for 6–12 h. Slides are washed, dried and then immersed in blocking buffer (1X PBS, 0.1 % Triton-100) for 2 min, and rinsed in PBS for 5 min at room temperature. The semidried slide is treated with 100 μl of 1:100 rabbit antibiotin antibodies and incubated in humidity chamber at 37 °C for 5 min. Slides are washed with PBS and immersed in 100 μl of diluted antibody (FITC-conjugated goat anti-rabbit antibody, 1:100 in dilution buffer) and incubated in the humidity chamber at 37 °C for 30–60 min. After washing, 60 μl of an antifade solution (p-phenylenediamine 10 mg/ml, 90 % glycerol, and propidium iodide 1 μg/ml as a counterstain) is added on each slide. The slide is observed with fluorescence microscopy using B2 or B-2A filter cassette.

16.6 Diversification of Fish Techniques

Ever since widespread recognition of FISH as a physical mapping technique to support massive nucleotide sequencing is involved in the Human Genome Project; it has become a more convenient and popular technique in other areas of biological and medical research including clinical genetics, neuroscience, reproductive medicine, cellular genomics, and chromosome biology.

The diversification of the original FISH protocol into a variety of remarkable procedures developed over the years has come about due to the improvement in sensitivity, specificity, and resolution of the technique (Volpi and Bridger 2008). These improved techniques along with the advancements in fluorescence microscopy and digital imaging have helped in better understanding of the chemical and physical properties of nucleic acids and chromatin.

Some of the techniques listed below, which have been inspired by the glossary of Volpi and Bridger (2008), show the versatility of FISH.

16.6.1 Centromere-FISH (ACM-FISH)

ACM-FISH is a multicolor FISH assay for detection of chromosomal abnormalities in sperm cells. The abbreviation ACM refers to the simultaneous hybridization of DNA probes for the alpha (centromere), classical (1q12) satellite and midi (1p36.3) satellite of chromosome 1 for the specific detection of duplications and deletions of 1pter and 1cen and for the identification of chromosomal breaks within the 1cen-1q12 region in human sperm. The discovery of chromosomal break/damage in the human sperm provided explanation for infertility in oligozoospermic men.

16.6.2 armFISH

armFISH is a 42-color M-FISH variant that allows the detection of chromosomal abnormalities in the p- and q-arms of all 24 human chromosomes, except the p-arm of the Y and acrocentric chromosomes.

16.6.3 Catalyzed Reporter Deposition-FISH (CARD-FISH)

CARD-FISH refers to the fluorescein tyramine signal amplification mediated by horseradish peroxidase (HRP)-labeled oligonucleotide probe (Fig. 16.5). This technique is very useful for detection, identification, and quantification of microorganisms involved in bioleaching processes.

16.6.4 Cellular Compartment Analysis of Temporal (Cat) Activity by Fish (catFISH)

catFISH uses FISH to immediate early rRNA genes and confocal microscopy to identify neuronal population activated at two distinct times. This technique is used to determine the interactions of neuronal populations associated with different behaviors.

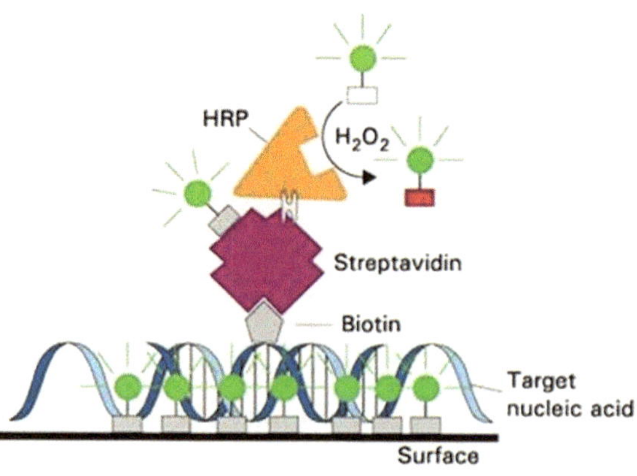

Fig. 16.5 Schematic representation of mRNA in situ hybridization detection using tyramide signal amplification (T5A) in the presence of horseradish peroxidase (HRP) and hydrogen peroxide; tyramide radicals are formed (*red box*) that can covalently react with nearby residues (Taken from http://www.authorstream.com/Presentation/chhabra61-443431-insitu-hybridization/)

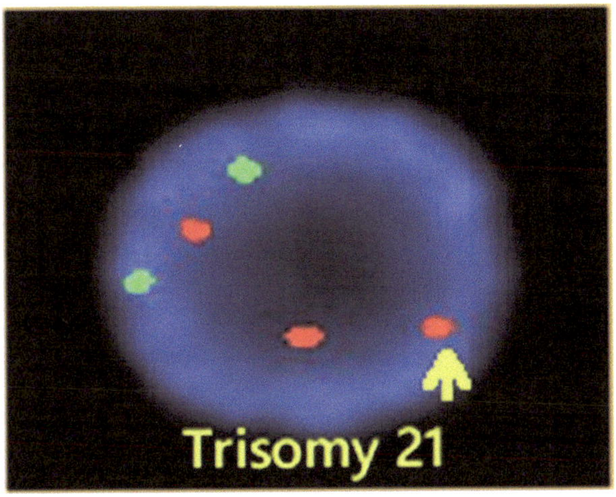

Fig. 16.6 *Above left* Normal FISH with labeled fluorescent probe demonstrating two copies of chromosomes 21 and 13 (*normal*). *Above right* Labeled fluorescent probe demonstrating an additional copy of chromosome 21 (*trisomy 21*) (Taken from http://www.obimages.net/genetic-markersoverview/information/)

16.6.5 Cytochalasin B (CB-FISH)

CB-FISH involves hybridization on binucleated cells in which cytokinesis has been blocked by treatment with cytochalasin B (CB). Figure 16.6 shows increased ratio of mosaic diploid cells in vivo in trisomy 21 cases. Analysis of the chromosomal content of micronuclei can be facilitated by combining the standard CB-FISH protocol with the 24-color SKY technology.

16.6.6 Chromosome Orientation (CO)-FISH

CO-FISH uses single-stranded DNA probes labeled with 5-bromodeoxyuridine during S phase to produce strand-specific hybridization. It allows to determine the relative orientation of two or more DNA sequences along a chromosome. Initially, this technique was designed to determine the orientation of tandem repeats within centromeric regions of chromosomes. This technique has also been useful in assessing chromosomal translocations and inversions.

16.6.7 Combined Binary Ratio (COBRA)-FISH

COBRA-FISH enables recognition of all human chromosome arms on the basis of color and mapping of gene and viral integration site in the context of chromosome arm painting. COBRA-FISH protocol brings together combinatorial labeling which allows different ratios of label to distinguish between probes. This permits the use of fewer fluorochromes to produce up to 48 color combinations for differential painting of human chromosome arms within a specimen.

16.6.8 Chromosome Orientation and Direction (COD)-FISH

This protocol is similar to CO-FISH except for the information about the directional organization of telomeric sequences. It can also stand for concomitant oncoprotein detection-FISH which allows visualization of loci signals for a particular oncogene and also the protein product derived from this gene. Another technique that has been termed COD-FISH is the combined $CaCO_3$ optical detection-FISH, in which FISH is used to detect calcifying microorganisms in open ocean.

16.6.9 Combinatorial Oligonucleotide (COMBO)-FISH

COMBO-FISH is used for specific labeling of genomic sites. It takes advantage of homopurine/homopyrimidine oligonucleotides that form triple helices with intact duplex genomic DNA. This will not require prior denaturation of the target sequence, which is usually a prerequisite for probe binding in the standard FISH protocols. Homopurine or homopyrimidine regions of DNA are usually longer than 14 bp, representing 1–2 % of the human genome, with an average of 150–200 of such stretches in a 250-kb segment of the genome. Accordingly, specific probe sets can be constructed to target genomic regions of interest in that size range.

16.6.10 Comet-FISH

Comet-FISH is a combination of comet assay and FISH analysis. It is used to detect genome region-specific DNA damage. It involves attachment of DNA onto agarose-coated microscope slide prior to in situ hybridization and allows specific sequences to be delineated in the comet head or tail. This will permit the assessment of sensitivity to DNA damage/breakage in the specific genomic region, which has been shown to be associated with the gene density of a chromosome rather than the chromosome size. This technique has been successfully used to determine the sensitivity of telomeres to damage.

16.6.11 Cryo-FISH

Cryo-FISH makes use of ultrathin cryosections (150 nm thick) of sucrose-embedded cells. The spatial interrelationship of chromosome territories and the genome organization in the cell nucleus has been successfully studied with this technique.

16.6.12 Double Fusion FISH (D-FISH)

In this FISH, a secondary color is observed since the adjacent colors overlap. The secondary color will be present or absent in the cases under study (Fig. 16.7). An example is the detection of BCR/ABL translocations, where the secondary color indicates disease. The opposite situation, where the absence of secondary color is pathological, is illustrated by an assay for translocation where only one of the breakpoints is known. Locus-specific probes are made for one side of the breakpoint and the other intact chromosome. In normal cells secondary color is observed, but only the primary colors are observed when the translocation occurs. This technique is called "break-apart FISH" (Fig. 16.7).

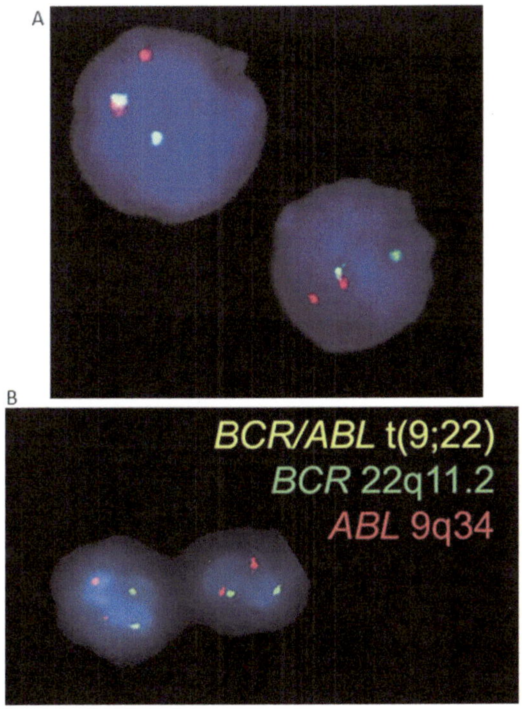

Fig. 16.7 (**a**) Interphase FISH on bone marrow nuclei containing the translocation t(11;19)(q23;p13) using a dual-color break-apart probe. *Green–red* fusion (*yellow*) signals indicate a normal cell. Separate *green* and *red* signals indicate the presence of translocations. (**b**) FISH strategy to detect the t(9;22) uses two differently labeled probes. A normal interphase nucleus (*left*) reveals four separate signals, two for each allele of BCR (*green*) and ABL (*red*). The appearance of a red-green fusion signal (*nucleus to right*) indicates the presence of BCR-ABL and is diagnostic of CML (Taken from http://biohorizons.oxfordjournals.org/content/3/1/85/F6.expansion.html and http://www.actacytol.com/feature/2005/feature062005.php)

16.6.13 DNA Breakage Detection FISH (DBD-FISH)

DBD-FISH has been used to determine DNA fragmentation levels in sperms. Cells are normally stabilized in agarose beads and incubated with the unwinding buffer to form single-stranded DNA in the sample that can be hybridized with the appropriate probes.

16.6.14 e-FISH

e-FISH is a BLAST-based FISH simulation program, which can predict the outcome of hybridization experiments. This program was developed as a bioinformatics tool for selecting appropriate genomic probes for hybridization experiments.

16.6.15 Fiber-FISH

Fiber-FISH is a technique in which DNA fibers or chromatin fibers are released from cell nuclei by salt or solvent extraction and stretched on a microscope slide prior to hybridization. This technique allows high-resolution mapping of chromatin fibers or DNA such as physical ordering of DNA probes, assessment of gaps and overlaps in contigs, and copy number variants.

16.6.16 Flow-FISH

In this technique, the in situ hybridization is combined with flow cytometry for measurement of the telomeric signals from cells in suspension. The PNA-labeled telomere probes are used to visualize and measure the length of telomere repeats. This technique has been used in aging studies.

16.6.17 Fusion-Signal FISH

This technique was initially used for identification of the 9;22 Philadelphia translocation in peripheral blood and bone marrow cells of CML patients to detect minimal residual disease after bone marrow transplantation. *BCR* and *ABL* gene fragments, each flanking one of the two breakpoints, were used as probes for the detection of the BCR/ABL fusion product, hence the name fusion-signal FIS.

16.6.18 Halo-FISH

In halo-FISH the cells are first permeabilized and then extracted with high salt to remove soluble proteins. The chromatin/DNA that is not fixed to an internal structure within cell nucleus is consequentially released, forming a halo around a residual nucleus. FISH can then be performed on these preparations using any type of probe to delineate specific DNA sequences such as α-satellite, telomeres, scaffold attachment regions (SARs), matrix attachment regions (MARs), gene loci, and whole chromosomes.

16.6.19 Harlequin-FISH

Harlequin-FISH is a method for cell cycle-controlled chromosome analysis in human lymphocytes that allows a precise quantification of induced chromosome damage for human biodosimetry. This technique combines FISH painting with differential replication staining of sister chromatids, either with Giemsa and/or fluorescent dyes, after BrdU treatment of lymphocyte cultures. After a few cell divisions, the chromosomes acquire an asymmetrically striped appearance, to which the term harlequin refers.

16.6.20 Immuno-FISH

Immuno-FISH is a combination of standard FISH and indirect or direct immunofluorescence. With this technique, the antigens can be visualized within the sample. Moreover, both DNA and proteins can be analyzed on the same sample. It is often used to investigate co-localization of genomic regions with proteins in the interphase nuclei such as nucleoli or promyelocytic leukemia (PML) bodies.

16.6.21 Locked Nucleic Acids (LNAs)-FISH

The in situ hybridization efficiency is remarkably improved by using locked-nucleic-acid (LNA)-incorporated oligodeoxynucleotide probes (LNA/DNA probes) without compromising specificity. LNA/DNA oligonucleotide heteroduplexes show a structural shift from a B-like helix toward an A-type helix, which has higher thermal stability. LNA/DNA probes are more useful for the detection of mRNA and genes on the chromosomes.

16.6.22 Multiplex (M)-FISH

One of the most fascinating aspects of FISH technology is the ability to identify several regions or genes simultaneously using different colors. The entire chromosome can be painted in a single hybridization by labeling with a different combination of fluorophores. This technique consists of labeling each probe with a unique combination of five spectrally separable fluorochromes in a 1:1 ratio. Originally these probes were used for simultaneous detection of the 24 human chromosomes

(22 autosomes and the X and Y chromosomes), but was subsequently used to analyze specific chromosomal subregions, like centromeres and sub-centromeres. M-FISH and SKY differ only in the method of discriminating differentially labeled probes. SKY uses CCD camera and Fourier transform spectrometry.

16.6.23 Multilocus or ML-FISH

The ML-FISH refers to the simultaneous use of multiple probes in multicolor FISH. This FISH assay was initially designed to screen for multiple microdeletion syndromes in patients with unexplained developmental delay and/or mental retardation.

16.6.24 Premature Chromosome Condensation (PCC)-FISH

PCC-FISH is used for determination of chromosome damage after irradiation. It relies on the use of chromosome-specific painting probes. This technique refers to the effect obtained by virus-mediated cell fusion or phosphatase inhibitors to prematurely condensed chromosomes of cells in G1 and G2 phases. PCC-FISH was initially devised as an assay to estimate/predict the in situ radiation sensitivity of individual human tumors. It has subsequently been used to estimate the effect of whole-body high- or low-dose exposure to human peripheral lymphocytes

16.6.25 Peptide Nucleic Acid (PNA)-FISH

PNAs are synthetic analogues of DNA in which the deoxyribose phosphate backbone is replaced with a noncharged peptide backbone. As a result of this unique structural property, there is no electrostatic repulsion when PNA oligomers hybridize to complementary DNA or RNA sequences. The PNA-DNA and PNA-RNA duplexes become more stable than the natural homo- or heteroduplexes. FISH with PNA probes was first used to measure individual telomere lengths on metaphase chromosomes.

16.6.26 Quantitative-FISH (Q-FISH)

This method has been used mainly for measuring the number of telomere repeats on a particular chromosome, using PNA-conjugated probes (Fig. 16.8). Typically, metaphases are imaged and then analyzed using software TFL-TELO. Q-FISH has become an important tool in studying the role of telomeres in aging and cancer.

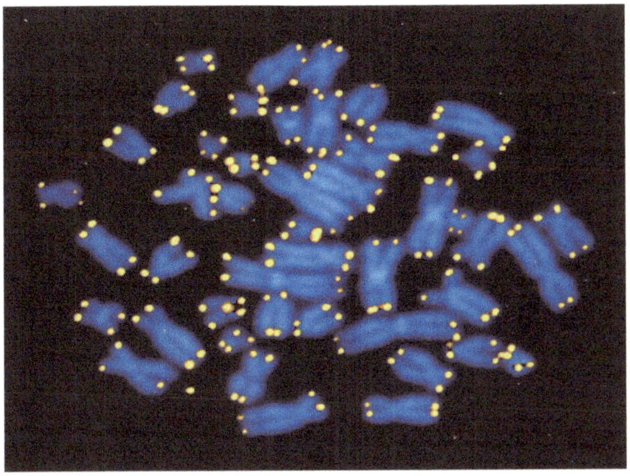

Fig. 16.8 The length of telomere repeats at individual chromosome ends is highly variable. Telomere repeats in a normal human lymphocyte are visualized using quantitative fluorescence in situ hybridization (Q-FISH) using peptide nucleic acid probes. Telomeres are shown in *yellow*, whereas the DNA of chromosomes, counterstained with DAPI, is shown in *blue* (Taken from http://physrev.physiology.org/content/88/2/557)

16.6.27 Quantum Dot (QD)-FISH

Quantum dots are nanometer-sized inorganic fluorophores, characterized by photostability and narrow emission spectra. These have been successfully used for FISH analysis on human metaphase chromosomes, human sperm cells, and bacterial cells. QD-FISH has also been used to detect subcellular mRNA distribution in tissue sections.

16.6.28 Rainbow-FISH

Rainbow-FISH allows simultaneous detection and quantification of up to seven different microbial groups in a microscopic field. This protocol uses specific 16S rRNA-targeted oligonucleotide probes for discrimination of different phylogenetic groups of microbes. As a result, by the combined application of seven DNA probes, each labeled with up to three fluorochromes, seven kinds of microbial strains can be distinguished simultaneously.

16.6.29 Raman-FISH

It is a technique in which FISH is combined with Raman microspectroscopy for ecophysiological investigation of complex microbial communities. The shift in the resonance spectra in Raman microscopy, after anabolic incorporation of ^{13}C isotope, compared with ^{12}C, into microbial cells is the basis of this procedure. This metabolic labeling with stable isotope is combined with in situ hybridization with specific 16S rRNA probe for identification of microbial species. This allows structural and functional interrelated analyses of microbial communities at a single-cell resolution.

16.6.30 Replicative Detargeting FISH (ReD-FISH)

The replication timing of specific sequences can be determined by ReD-FISH. If BrdU is incorporated in the sequence of interest, the newly formed DNA strand will be detargeted, and each oligonucleotide probe will only be able to hybridize to one of the parental strands, and only one chromatid will display a signal. However, if the sequence of interest has not replicated and has not incorporated BrdU, then a FISH analysis will reveal the standard double signal on both chromatids of the metaphase chromosome. ReD-FISH provides qualitative and quantitative information about replication timing, including the relationship between defects in replication timing and defects in chromosome condensation, sister chromatid cohesion, and genome stability.

16.6.31 Reverse-FISH

Reverse-FISH is the process whereby the FISH probe comprises DNA from the material of interest. Reverse-FISH has been useful for characterizing marker chromosomes and chromosome amplifications in cancer.

16.6.32 Recognition of Individual Genes (RING)-FISH

RING-FISH utilizes high concentrations of polynucleotide probes in order to increase the visualization and sensitivity of any part of the genetic material in a bacterial cell, regardless of copy number. It was designated as ring-FISH because of the characteristic halolike, ring-shaped hybridization signal in the cell periphery obtained with this method.

16.6.33 RNA-FISH

RNA-FISH allows simultaneous detection, localization, and quantification of individual mRNA molecules either in the nucleus or cytoplasm at the cellular level in fixed samples. This RNA FISH technology provides a method to achieve allelic-specific expression on a single-cell basis. It has the potential for investigating gene expression profiling in single cells.

16.6.34 Cross Species Color Banding (Rx)-FISH

RxFISH, also known as chromosome bar coding, is based on sequence homologies between human and the apes, such as gibbon (98 %). This technique produces, by cross species hybridization using differentially labeled gibbon chromosome probes, a specific banding pattern on human metaphase chromosomes. If the probes are labeled with a number of fluorochromes, usually three, this allows a colorful and reproducible banding to be observed and analyzed. The color bands make it easier to see intrachromosomal rearrangements, compared to G-banding. However, in combination with G-banding, RxFISH can provide detailed information about the chromosomal breakpoints.

16.6.35 Split-Signal FISH

It is a dual-color FISH assay for detection of frequently occurring chromosome translocations affecting specific genes in hematopoietic malignancies. The assay involves differential labeling of two probes on the flanking regions of the translocation breakpoint. The signals normally co-localize and appear fused, but they split in the translocative event. This technique has been used for the detection of Burkitt translocation in B cell lymphomas and mantle cell lymphomas.

16.6.36 Stellaris RNA FISH (Single-Molecule RNA FISH)

It is a method of detection and quantification of mRNA and other long RNA molecules in a thin layer of tissue samples. The binding of up to 48 fluorescent-labeled oligos to a single molecule of mRNA provides sufficient fluorescence to detect and localize each target mRNA. Figure 16.9 shows RNA FISH for Cre mRNA in genetically identical cells, in which expression is epigenetically controlled.

16.6.37 T-FISH

The three versions of T-FISH – tyramide-FISH, tissue-FISH, and telomere-FISH – are discussed in the order of their arrival in the field.

Tyramide-FISH: Tyramide is a compound that binds to peroxidase and greatly increases the sensitivity in FISH experiments, with the use of only one or two layers of reagents for visualization. The first layer uses a peroxidase-conjugated antihapten antibody or a compound such as streptavidin to bind to the labeled probe (Fig. 16.5). Fluorochromes or haptens, such as biotin, are conjugated to tyramine derivatives. This leads to massive buildup of fluorochromes that make the visualization and detection ultrasensitive. The technology has been used to map gene loci and look for specific transcripts in cell.

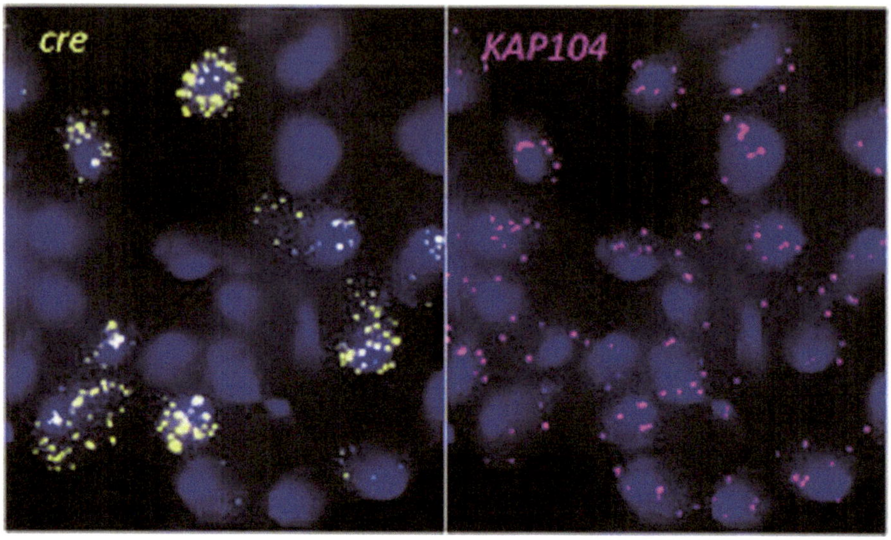

Fig. 16.9 RNA fluorescence in situ hybridization (FISH) for Cre mRNA in genetically identical cells in which expression of Cre is epigenetically regulated. KAP104 is a control genes whose expression is not epigenetically regulated (Taken from https://mcb.berkeley.edu/faculty-andresearch/research-spotlight/rna-fluorescence-x-fish-cre-mrna)

Tissue-FISH: Tissue samples collected from patients or experimental animals are frozen, fixed, or embedded in paraffin wax and used for FISH analysis.

Telomere-FISH: It is FISH using telomeric probes.

16.6.38 3-D FISH

3-D FISH has been developed to analyze spatial positioning and relative organization of chromosomes and sub-chromosomal regions within the cell nuclei. Paraformaldehyde is usually used as cross-linking fixation reagents to preserve nuclear architecture and chromatin organization. Due to cross-linking of proteins, an efficient permeabilization step would be required to allow the probes to penetrate the sample.

16.6.39 Zoo-FISH

Zoo-FISH, also known as cross species chromosome painting, involves hybridizing libraries of DNA sequences of one species to the chromosomes of another species, to identify regions of synteny. The first Zoo-FISH study used human and mouse whole chromosome painting probes on primates, rodents, even-toed ungulates, and whales.

16.6.40 Comparative Genomic Hybridization (CGH)

One of the most significant developments in FISH technology in relation to genome-wide screening was the introduction of comparative genome hybridization (CGH) in 1992. IN CGH, the genomic DNA from the specimen and the control DNA extracted from an individual with a normal karyotype (46,XX or 46,XY) are differentially labeled with green and red fluorochromes, respectively, mixed in equal amounts and co-hybridized to reference human metaphase chromosomes (Fig. 16.10). The relative difference in DNA content between the normal and specimen DNA is represented by a difference in the green/red fluorescence ratios. For example, if the chromosomal material is present in identical copy numbers in both the reference and the specimen genome, the observed fluorescence is a blend of an equal contribution

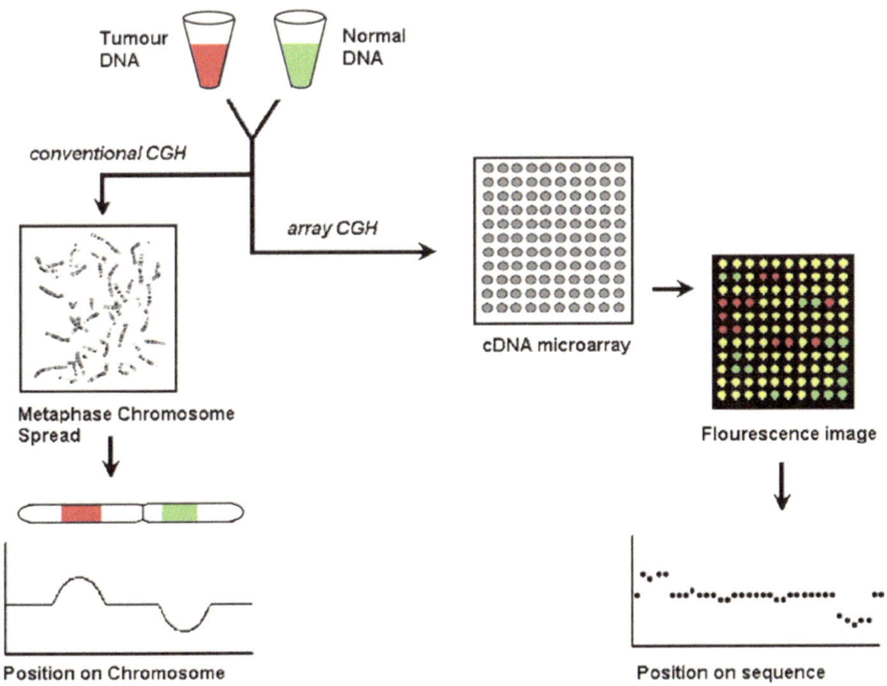

Fig. 16.10 Comparative genomic hybridization. Genomic DNA is isolated from both the tumor sample and the normal reference sample, labeled with different fluorochromes and mixed in the presence of excess Cot-1 DNA to prevent binding of repetitive sequences. In conventional chromosomal CGH, these are hybridized to normal metaphase chromosomes, and the ratio of fluorescence intensities along each chromosome is analyzed. Increased DNA copy number (amplification) in the tumor sample will be detected by increased red fluorescence, whereas decreased copy number in the tumor sample will allow more binding of the normal DNA and increased green fluorescence. On the *right*, a similar hybridization to a cDNA array permits measurement of copy number at a higher resolution. The *red* and *green* spots on the fluorescence image represent increased and decreased copy number changes, respectively (Taken from http://biohorizons.oxfordjournals.org/content/early/2010/02/26/biohorizons.hzq009/F7.expansion.html)

of red and green fluorescence. If chromosomes are lost or chromosomal subregions are deleted in the specimen genome, the resulting color is shifted to red. A gain in the certain chromosome in the specimen, such as amplification of oncogenes, is reflected by a more intense green staining of the respective chromosome in the reference metaphase preparation. The ratios of the test to reference fluorescence along the chromosomes are quantified using digital image analysis.

In array CGH, metaphase chromosomes are replaced as the target by large number of mapped clones that are spotted onto a standard glass slide greatly increasing the resolution of screening for genome copy number gains and losses. In array CGH, the test and the normal reference genomes, which are used as probes, are differentially labeled and co-hybridized to a microarray before being imaged. The fluorescence intensities are calculated for each mapped clone, with the resulting intensity ratio reflecting the DNA copy number difference (Fig. 16.11). Despite some limitations, array CGH has become one of the most widely used cytogenetic techniques in both basic research and molecular diagnosis. This technique has enabled us to understand that tumors of the same type have similar patterns of DNA gains and losses and that the frequency of changes increases with tumor progression.

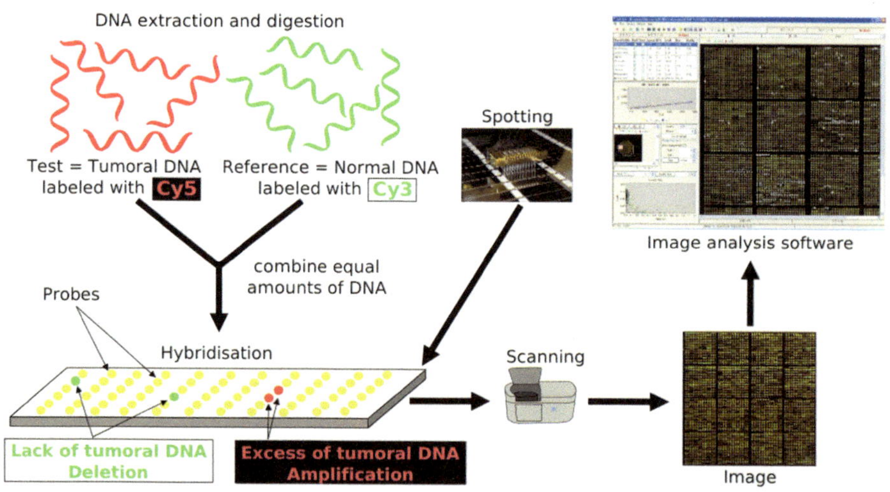

Fig. 16.11 It is a schematic overview of the array CGH technique. DNA from the sample to be tested is labeled with a red fluorophore (Cyanine 5), and a reference DNA sample is labeled with green fluorophore (Cyanine 3). Equal quantities of the two DNA samples are mixed and cohybridized to a DNA microarray of several thousand evenly spaced cloned DNA fragments or oligonucleotides, which have been spotted in triplicate on the array. After hybridization, digital imaging systems are used to capture and quantify the relative fluorescence intensities of each of the hybridized fluorophores. The resulting ratio of the fluorescence intensities is proportional to the ratio of the copy numbers of DNA sequences in the test and reference genomes. If the intensities of the fluorochromes are equal on one probe, this region of the patient's genome is interpreted as having equal quantity of DNA in the test and reference samples; if there is an altered Cy3:Cy5 ratio, this indicates a loss or a gain of the patient DNA at that specific genomic region (Taken from https://en.wikipedia.org/wiki/Comparative_genomic_hybridization#/media/File:Array-CGH_protocol.svg)

16.7 Applications of Fish

FISH has now become an essential tool for gene mapping and characterization of chromosome aberrations. Since the target DNA remains intact, unlike in molecular genetic analysis, information is obtained directly about the positions of probes in relation to chromosome bands or to other hybridized probes. Using differentially labeled probes, chromosome aberrations on particular chromosomes or chromosomal regions can be easily defined. The diseases that have been diagnosed using FISH include Prader-Willi syndrome, Angelman syndrome, 22q13 deletion syndrome, chronic myelogenous leukemia, acute lymphoblastic leukemia, Cri-du-Chat syndrome, velocardiofacial syndrome, and Down syndrome. The analysis of chromosomes 21, X, and Y can identify oligozoospermic individuals at risk.

In medicine, FISH can be used for diagnosis, evaluation of prognosis, and evaluation of remission of a disease such as cancer. FISH can be used to detect diseased cells more easily than standard cytogenetic methods. High-resolution FISH mapping and ordering of probes relative to one another can be performed on released chromatin fibers and is termed fiber-FISH. Fiber-FISH has a wide range of resolution (I kb–I Mb).

One of the major advantages of FISH over conventional molecular biology is the provision of molecular information in the context of cell morphology. Targeting nuclear RNA and the corresponding genes within cells or within a single cell or from a single allele can provide important information about gene expression, processing, and transport of transcripts in normal and mutant cells. The use of RNA FISH for studying the intracellular localization of RNA has increased over understanding of in situ physical characteristics of DNA transcription and transport of RNA transcripts. Similarly FISH can be used to examine many interesting biological questions about nuclear organization. Three-dimensional nuclear DNA FISH can provide high-resolution information about sub-chromosomal domains, gene position, and the relationship of genes and their transcripts in different cells and during different stages of the cell cycle. Accurate analysis of three-dimensional FISH is highly dependent on excellent quality confocal microscopy and image analysis procedures.

FISH technology also allows genome-wide screening of chromosomal gains and losses, which is comparative in in situ hybridization (CGH). It is based on the comparison of genomic DNA from two different genomes and identifies chromosomal gains and losses of one genome relative to the other. CGH is performed in normal chromosome metaphase spreads, which is a distinct advantage for studying tumor samples. The resolution of identifying chromosomal gains and losses on metaphase chromosomes is several Mbs. However, this technique has been modified to increase the resolution to several Kbs by the technique of matrix or array CGH, in which the targets are cloned DNA fragments immobilized on the glass surface. This allows detection of low copy number gains and losses and may be used diagnostically to identify microdeletions or amplifications affecting only one or two genes.

Cancer cytogenetics has benefitted greatly from FISH technology, and hence the clinical laboratories have benefitted from the technique, since it is rapid and can be performed on tissues (fresh frozen or formalin-fixed paraffin-embedded), touch preparations, cytospins, or cell cultures. Since it is usually difficult to get chromosome spread from tumor cells, the use of interphase FISH directly on tumor samples (biopsies, section, and archived paraffin-embedded material) enables the determination of chromosomal aberration without the need for interphase chromosomes preparations. Numerical chromosome aberrations, chromosome deletions, and translocations can all be identified in interphase nuclei providing important diagnostic and or prognostic information.

The advent of spectral dyes and imaging has made FISH more colorful and even more powerful. Using multiple probes simultaneously provides important additional information that can now be obtained for a single sample using multicolor FISH techniques. The techniques allow for both a genome-wide screen of aberrations and a gene or chromosomal regain-specific analyses of specific aberrations in chromosomes and can be adopted for use in the analysis of interphase nucleic. Similarly, genome-wide screen for mRNA expression differences or for genomic aberrations can be performed by microarray FISH, which is based on the comparative hybridization of two samples onto arrays that represent either specific sets of genes or the whole genome. The targets used come as oligonucleotides, cDNA, or genomic arrays.

Glossary

A

Allopolyploid An individual or strain whose chromosomes are composed of more than two genomes, each of which has been derived more or less complete but possibly modified from one of two or more species.

C

Chromosome painting The use of fluorescent-tagged chromosome-specific dispersed repeat DNA sequences to visualize specific chromosomes or chromosome segments by in situ DNA hybridization and fluorescence microscopy.

Confocal microscopy It is an optical imaging technique for increasing optical resolution and contrast of a micrograph by means of adding a spatial pinhole placed at the confocal plane of the lens to eliminate out-of-focus light. It enables the reconstruction of three-dimensional structures from the obtained images.

H

Hapten Haptens are small molecules that elicit an immune response only when attached to a large carrier such as a protein; the carrier may be one that also does not elicit an immune response by itself.

I

Immunodetection The use of antibodies to identify proteins or other chemicals.

N

Neoplasia The presence or formation of new, abnormal growth of tissue.

Nick translations Nick translation is a tagging technique in molecular biology in which DNA polymerase I is used to replace some of the nucleotides of a DNA sequence with their labeled analogues, creating a tagged DNA sequence which can be used as a probe in fluorescent in situ hybridization or blotting techniques. It can also be used for radiolabeling.

References and Additional Reading

Bishop R (2010) Applications of fluorescence in situ hybridization (FISH) in detecting genetic aberrations of medical significance. Biosci Horiz 3:85–95

John H, Birnstiel M, Jones K (1969) RNA-DNA hybrids at the cytological level. Nature (London) 223:582–587

Mukai Y (1995) Multicolor fluorescence in situ hybridization approach for genome analysis and gene mapping in wheat and its relatives. In Proceedings 8th international wheat genetics symposium, Beijing, 45

Mukai Y, Friebe B, Hatchett J, Yamamoto M, Gill BS (1993) Molecular cytogenetic analysis of radiation induced wheat-rye terminal and intercalary chromosomal translocations and the detection of rye chromatin specifying resistance to Hessian fly. Chromosoma 102:88–95

O'Connor C (2008) Fluorescence in situ hybridization (FISH). Nat Educ 1:171

Pardue ML, Gall JG (1969) Molecular hybridization of radioactive RNA to the DNA of cytological preparations. Proc Natl Acad Sci U S A 64:600–604

Schwarzacher T, Leitch AR, Bennett MD, Heslop-Harrison JS (1989) In situ localization of parental genomes in a wide hybrid. Ann Bot 64:315–324

Speicher MR, Carter NP (2005) The new cytogenetics: blurring the boundaries with molecular biology. Nat Rev Genet 6:782–792

Trask BJ (2002) Human cytogenetics, 45 years and counting. Nat Rev Genet 3:769–778

Volpi EV, Bridger JM (2008) FISH glossary: an overview of the fluorescence in situ hybridization technique. BioTechniques 45:385–409

Yamamoto M, Mukai Y (1989) Application of fluorescence in situ hybridization to molecular cytogenetics of wheat. Wheat InfServ 69:30–32